Construction Contract Administration

RALPH W. LIEBING

Registered Architect

Prentice Hall
Upper Saddle River, New Jersey   Columbus, Ohio

Library of Congress Cataloging-in-Publication Data
Liebing, Ralph W.
  Construction contract administration / Ralph W. Liebing
    p. cm.
  Includes index.
  ISBN 0-13-381591-9
  1. Construction contracts—United States—Management. I. Title.
KF902.L5 1998
692'.8—dc21   97-2167
                                                                               CIP

Cover photo: Pictor Uniphoto
Editor: Ed Francis
Production Editor: Stephen C. Robb
Design Coordinator: Julia Zonneveld Van Hook
Text Designer: Pagination
Cover Designer: Russ Maselli
Production Manager: Laura Messerly
Electronic Text Management: Marilyn Wilson Phelps, Matthew Williams, Karen L. Bretz,
  Tracey B. Ward
Illustrations: Christine Marrone
Marketing Manager: Danny Hoyt

This book was set in Dutch 801 and Swiss 721 by Prentice-Hall, Inc. and was printed and bound by Courier/Kendallville, Inc. The cover was printed by Phoenix Color Corp.

 © 1998 by Prentice-Hall, Inc.
Simon & Schuster/A Viacom Company
Upper Saddle River, New Jersey 07458

All rights reserved. No part of this book may be reproduced, in any form or by any means, without permission in writing from the publisher.

Printed in the United States of America

10 9 8 7 6 5 4 3 2 1

ISBN: 0-13-381591-9

Prentice-Hall International (UK) Limited, *London*
Prentice-Hall of Australia Pty. Limited, *Sydney*
Prentice-Hall Canada, Inc., *Toronto*
Prentice-Hall Hispanoamericana, S. A., *Mexico*
Prentice-Hall of India Private Limited, *New Delhi*
Prentice-Hall of Japan, Inc., *Tokyo*
Simon & Schuster Asia Pte. Ltd., *Singapore*
Editora Prentice-Hall do Brasil, Ltda., *Rio de Janeiro*

*To my family, my colleagues,
and my students*

Brief Contents

1. The Insight   1
2. Definitions   11
3. The Concept   45
4. New Concepts   67
5. The Design Professional   99
6. Obligation, Education, and Liability   153
7. Forms for the Process   199
8. The Construction Manager   257
9. The Owner   275
10. The Contractor   363
11. The Subcontractor   317
12. The Building Inspector   325
13. Conclusion   339

APPENDICES

A. Field Observation Services and Procedures   355
B. Sample Regulatory Specifications Section   363
C. Detailed Meeting Formats, Procedures, Agenda, and Planning   369
D. Complete List of AIA Documents   377
E. Standards-Producing Organizations   383

Index   397

Contents

Notice to the Reader xiii
Introduction xv
Preface xvii
Acknowledgments xix
The Context xxi

1 The Insight   1

Context 2
Beginnings 3
Characteristics in Practice 4
   Areas for Emphasis   *5*
   Limitations in Practice   *5*
Final Observations 7
Exercises 8

2 Definitions   11

Serving the Interests of the Project Owner 13
Cooperation in Project Management 13
   On-Site Observation   *14*
Performing Contract Administration 15

    Establishing Responsibility   16
    Language of the Contract   16
    Working with the Client   17
    Legal Considerations   18
    Working with Contract Documents   18
    Effect of Case Law on Contracts   21
    Contracts for Renovation and Alterations   23
    Revising Contract Language   24
        "THE DESIGN PROFESSIONAL'S RESPONSIBILITIES FOR SITE OBSERVATION AND SAFETY DURING CONSTRUCTION," BY HERBERT C. SMITH, II   24
    Types of Services Offered   27
Defining Terms 30
Exercises 43

3   The Concept   45

Contract Administration: A Basic Service 46
Precedence of Contract Documents 51
Benefits of Contract Administration to Professional and Owner 52
Importance of Reviewing Case Law 53
    Summary of Recent Case Law   54
When the Owner Does Not Want Contract Administration Services 56
Code Compliance 58
    The Inspection System   60
Exercises 65

4   New Concepts   67

Defining Responsibilities During Construction 70
    Some Sample Contract Definitions   71
    The Importance of Scheduling   74
    Defining the Professional's Functions   75
Successful Administration of Construction Projects 76
    The Professional's Relationship to the Project   77
Changing Attitudes and Concepts 79
    Scheduling and Productivity   79
The Need for Better Understanding and Cooperation 80
    Consulting with Contractors   81
    The Influence of Project Documents   82
    Abilities and Attitudes of Contractors   82
    Duties of the Contract Administrator   84
The Complicated System of Communication 89
Exercises 96

Contents

5 The Design Professional 99

Contractual Obligations 100
    Contract Options 102
    Differences Among Professional Disciplines 106
        "LEGAL CONCERNS IN CONSTRUCTION FIELD ADMINISTRATION," BY PETER C. HALLS 113
Chief Coordinator for the Entire Project 128
    Responsibilities of the Design Professional 128
    Responsibilities of the Administrator 130
Feedback 139
Project Meetings 140
On-Site Techniques 145
Exercises 151

6 Obligation, Education, and Liability 153

How Contract Text Developed 154
The Influence of Case Law 169
Assigning Liability 171
Alternative Dispute Resolution 174
Abstract of Legal Liability Paper 178
    "THE CRUMBLING TOWER OF ARCHITECTURAL IMMUNITY: EVOLUTION AND EXPANSION OF THE LIABILITY TO THIRD PARTIES," BY JEFFREY L. NISCHWITZ 181
    Limiting Professional Liability 192
The Professional's Role in Dealing with Problems of Liability 192
Exercises 198

7 Forms for the Process 199

Importance of Documentation 200
Exercises 228

8 The Construction Manager 257

Role of the Construction Manager 258
Development of the Construction Manager System 259
    Agency vs. GMP Construction Management 263
Contractual Obligations 264
    Authority of the Construction Manager 265
Avoiding Impediments of Work Progress 273
Exercises 274

9 The Owner 275

Education of the Owner 276
   The Owner as Project Leader   277
   Development of the Education Process   277
Protection of the Owner's Interests 286
   Policing the Contract   287
   Decision Making   288
   Cost Control   289
Provisions of AIA Document A-201 290
     "An Owner's Response to the Use of the A-201 (1987) General Conditions: Major Problem Areas—Owner Recommendations," by Mary J. McElroy   291
Exercises 302

10 The Contractor 303

Contractual Obligations—The Many Things Included 304
Resolving Information from Others 310
Ensuring Good Work 312
Coordination and Responsibility for Subcontractors 313
Tempo of the Job 315
Exercises 316

11 The Subcontractor 317

Contractual Obligations 318
Responsibility to Every Member of the Design/Build Team 318
Protection of Subcontractor's Rights, Interests, and Business 319
Exercises 224

12 The Building Inspector 325

Background 326
Parameters of the Job 329
The Need to Understand Various Construction Systems 331
The Authoritative Bystander 333
Exercises 337

13 Conclusion 339

Building for Safety 340
Fire Safety 341

Inspection and Code Compliance 346
Tasks and Exercises 351

Appendices

A. Field Observation Services and Procedures   355

B. Sample Regulatory Specifications Section   363

C. Detailed Meeting Formats, Procedures, Agenda, and Planning   369

D. Complete List of AIA Documents   377

E. Standards-Producing Organizations   383

Index   397

Notice to the Reader

The author and publisher make no representation or warranties of any kind, including but not limited to the warranties of fitness for particular purpose or merchantability, nor are any such representations implied with respect to the material set forth herein, and do not take responsibility with respect to such material. Neither the author nor the publisher shall be liable for any special, consequential, or exemplary damages resulting in whole or in part from the readers' use of or reliance upon this material.

The author and publisher do not warrant or guarantee any of the products described herein or perform any independent analysis in connection with any of the product information contained herein. The author and publisher do not assume and expressly disclaim any obligations to obtain and include information other than that provided to it by the manufacturers.

The reader is expressly warned to consider and adopt all safety precautions that might be indicated by the activities described herein and to avoid all potential hazards. By following the instructions contained herein, the reader willingly assumes all risks in connection with such information.

Introduction

There is no one system, scheme, plan, or configuration for construction contract administration that has been established for use on every project; there are no standards, formats, or "designs." Just as projects themselves and the participants change, so does the contract administration program. Often this is client-driven, to meet specific requirements or procedures. It can be said that certain aspects change very little because they form inherent parts of every administrative program, but each project has its own set of requirements and procedures.

Construction Contract Administration has been written as a general overview of the current methods and concepts of performing contract administration as a basic professional service. This book is published as an introductory instructional aid to all those interested in or engaged in contract administration. Even though there is no specific target readership other than the college classroom, we feel there is value to many others. The material herein, while heavily dependent on architectural sources, can be used by other professions, on any project, and in varying configurations.

Although this book is directed toward potential professionals and professionals in practice, it is not conceived as a "how-to" book. One must bring knowledge of construction and architecture/engineering to this text. We portray things as they are or as they should be done. There is no attempt to develop skills in actual construction work, but only in methodology, which should be used to ensure the work is appropriate as to function, style, size, appearance, stability, and quality. It is a guide to ensuring that the contract obligations are met and the contract documents faithfully executed.

In addition, nothing herein is presented or intended as, nor shall it be construed as, legal advice. The inclusion of legal (case law) citations, legal papers, and other legal information in this book is not intended as, and should not be construed or used as, legal advice. The legal material herein is for general illustrative purposes and as a method of showing legal directions and decisions in some professional matters. However, each set of specific circumstances should be discussed with legal counsel to ascertain local legal parameters and liability. The author and the publisher make no warranties or representations and assume no responsibility for any act or consequence resulting from the use or application of the data contained in this book. The book is meant to reflect some of the conditions that exist today, and in no way excludes other contractual configurations or lines of responsibility. It is impossible to portray every conceivable situation. This is general background material, not directed or mandatory technical advice.

This book uses established, standard construction vocabulary, which is replete with words of masculine gender. This reflects a traditional use of the words. As far as practical, the use of gender-specific words has been avoided throughout this book, or both genders have been used. Since the English language does not include neuter personal pronouns, where masculine pronouns are used, it is the author's intention that they apply equally to women. Also, female readers should know that many construction associations now cater especially to the interests of women in this industry. In many cases, inquiries can be made locally to chapters of national/regional organizations.

Preface

This effort seems to have had some predestination, in many respects, as well as several rebirths.

Stephen A. Kliment, FAIA, former editor of *Architectural Record*, gave the initial impetus for this book. While he was architecture editor at John Wiley & Sons, we approached him about a renewal of some sort for our previous book, *Systematic Construction Inspection*. That book "lived" a normal life, for a professional book, but had been included in more bibliographies since it went out of print. Mr. Kliment felt that this effort could be the basis for a book on construction contract administration. We are grateful to him for his insight, help, and continued support in developing this project.

For various reasons, the original situation with Wiley went by the boards. Later we did find a new lease on life for this effort through Ed Francis at Prentice Hall. Mr. Francis and his reviewers provided new insight and suggested that, in lieu of a professional "how-to" book, we prepare an introductory, college-level text. We responded.

My dear friend and colleague Russ Groves (see Acknowledgments), who wrote the first chapter, and I believe that contract administration is greatly misunderstood by professionals as well as by their clients. Very limited time in the schools of architecture is given to practice instruction. We felt a college text was most appropriate, so long as it was gauged to concentrated, short-time instruction. We believe future professionals will benefit from this information, and may well find it to their advantage in their practice. What will the practice of the future look like? The ever-increasing amount of computer software will, no doubt, bring fantastic changes to the practice and augment the skills of the professionals. However, it is fair to assume that field operations will still need to ensure that the requirements of the project documents are met and the project is built as the client desires. No one can sit in the office and forecast success without being on site and seeing the actual work and its relationship to the design and contract documents. Some professionals (and their staffers) will still have to "get muddy" and have their suits ruined by overspray from concrete pumping. Without definite decision-making capacity, self-generated, it does not appear that any computer or other machine will replace the contract administrator: the human who can see, ascertain, evaluate, understand, inform, and resolve.

Professionals may still take issue with this concept. Some will still seek to reduce or eliminate construction site liability by not engaging in contract administration. Other third parties (construction managers, general contractors, etc.) may thrive, but future clients and their projects will not. For the very slight reduction in fee, the client who deletes contract administration by the professional will have projects reduced in some aspects and to some degree for the lack of the professional's on-site insight and ability. The projects will not be brought on-line as conceived and intended! Proper, insightful interpretation and meticulous execution will suffer from the lack of the professional's incisive knowledge of the project. The owner may receive less than dollar value and never really be aware of it. Any problems that may arise will easily be attributed to the professional, who may well not be responsible, and could have resolved the problems quickly, decisively, and with no impact on the project as a whole, while keeping the aesthetics and concept intact.

Without an ongoing presence on the project site and direct input (or any input) to the project as it is constructed, the professional is at a loss to prevent manipulation and diminution of the contract intent, the design concept itself, and proper execution of the contract obligations, both legal and architectural. For this reason, it is difficult to explain how the design professional can "lose" contract administration when it is a basic included service in the standard design service contracts. Unless the client is very well informed (and wary) and adamant, or the professional is less than enthusiastic, it seems that contract administration should come to the project "naturally"—perhaps even unchallenged—as part of the normal contractual process.

If the client should challenge the total fee and seek reduction, total elimination of contract administration is not, by far, a well-founded answer. The professional sincerely desires the design to be built intact. The owner wants better "value." But, in many cases, neither wants the very process that can ensure both!

I sincerely hope this effort can bring new understanding and new emphasis to contract administration among new professionals. In addition, I hope this book will interest established professionals and their clientele. Why not use this book as an educational guide with your client?

Ralph W. Liebing
Cincinnati, OH

Acknowledgments

The author wishes to recognize the valuable information provided by the three major papers included in this text, and the kind cooperation of the authors.

- **Peter C. Halls** is a partner in the firm of Faegre & Benson in Minneapolis, MN. His practice focuses on the construction industry. Mr. Halls is involved in contract drafting, preventive legal advice, bid protests, claims handling, arbitration, and litigation. He has been involved in cases around the country and regularly represents architects, engineers, owners, general contractors, subcontractors, and sureties. Mr. Halls, who received his law degree, magna cum laude, from the University of Minnesota, frequently speaks and writes for groups in the construction industry.
- **Jeffrey L. Nischwitz** is a partner with DiLeone, Nischwitz, Pembridge & Chriszt in Cleveland, OH. His practice focuses on general corporate and commercial litigation, including ERISA, securities, real estate, construction, and bankruptcy litigation. He graduated with high honors from Ohio State University College of Law. He was awarded the order of the Coif and was editor of the *Ohio State Law Journal*. He is a member of the bar of the State of Ohio, numerous federal courts, and the U.S. Supreme Court. He is also a member of the American, Ohio, and Cleveland bar associations. His article in this book was previously published in the 1984 *Yearbook of Construction Articles* and, in part, in the 1985 *Legal Handbook for Architects, Engineers, and Contractors*.
- **Mary J. McElroy** is currently a sole practitioner in California where she specializes in architectural and construction law. An attorney since 1977, she has represented owners, architects, contractors, and subcontractors both in private practice in Chicago and Los Angeles, and as Construction General Counsel for Homart Development Co., the former nationwide real estate developer of regional shopping malls, office buildings, and other large commercial projects. Ms. McElroy has drafted and negotiated hundreds of architectural and engineering services agree-

ments as well as construction contracts for all forms of project delivery. Ms. McElroy is the former Chair of the American Bar Association's Forum on the Construction Industry, the largest construction law organization in the country with over 5,500 members. Prior to that time, she served on the Steering Committee and as Chair of the Contract Documents Division of the Forum, and on the Forum's Governing Committee. She is also a past Chair of the Chicago Bar Association's Land Development and Construction Committee, a member of the construction law sections of the Los Angeles and California Bar Associations, and a founding member of the American College of Construction Lawyers where she has served as vice-chair of the Contract Documents Committee as well as a member of the Board of Governors. Ms. McElroy holds a Bachelor of Architecture degree from Tulane University and a Juris Doctor, magna cum laude, from DePaul University in Chicago. She is the author of "Construction Damages and Remedies," *Construction Law* (Matthew Bender, 1986, 1990), "Responsibilities and Liabilities of Architects and Engineers for Construction Failures," *Construction Litigation* (Ill. Inst. for CLE, 1984; Supplement Co-Author 1987, 1989, 1993), and *Construction Industry Forms, Second Edition* (John Wiley & Sons), as well as prior cumulative supplements for the first edition. Ms. McElroy has spoken at numerous professional educational programs and conferences including meetings of the Forum on the Construction Industry, the American Bar Association's Real Estate Section, the Illinois Institute of Continuing Legal Education, the American Arbitration Association, the National Association of Corporate Real Estate Executives, the Chicago Chapter of the American Institute of Architects, the School of Engineering of the University of Wisconsin, the Construction Law Institute of Illinois Institute of Technology's Chicago-Kent College of Law, and the Los Angeles Bar Association. Ms. McElroy has served the American Arbitration Association as a member of the Chicago Regional Advisory Committee and as a member of the AAA's panel of construction arbitrators in Chicago and in Los Angeles.

John Russell Groves, Jr., AIA, is a long-time and very good friend. He is both a registered architect and an attorney. He is associate dean and teaches in the University of Kentucky School of Architecture and in the School of Law. He travels internationally as a consultant, and he acted as the building code official for the Commonwealth of Kentucky immediately after the Beverly Hills Supper Club fire disaster (he reorganized the state's code system in its entirety and the personnel as well). He has practiced privately, written extensively, served on several AIA committees, serves as a Brigadier General in the Kentucky National Guard, and acted as the Commonwealth's Adjutant General. He is a tremendously talented and versatile man. His patient, sincere, insightful, enthusiastic, and cheerful support for this project and the author is much appreciated.

Sincere gratitude is also offered to the following people who generously reviewed, discussed, and commented on the material regarding construction management in this book: **L. H. Waller, Claude Gibson,** and **Robert Renshaw, Jr.,** of Alliance Corporation, Glasgow/Bowling Green, KY.

Two other people deserve a tremendous vote of thanks, both for their insight and for their contribution—**Sharon and Dean French**. This husband-and-wife team worked with the author on one project and were an inspiration. Dean's 40 years of major construction experience and Sharon's office expertise, skill, and management provided a smooth, proper, finely honed operation—nothing amiss, everything well executed. Special thanks to Sharon for her cheerful, expert, and speedy transcription of the final manuscript. A masterful job, so very well done!

The Context

Contract administration and construction management are different!

This concept is fundamental to understanding the context of this book. As a basic text, it emphasizes contract administration: a system of activity that brings a project to completion with all participants in harmony. The job of the contract administrator is to see that the project is completed in virtual compliance with the contract documents and that each participant is satisfied with the progress, the performance, the product, and the profit derived.

With several very divergent attitudes among the project participants (owner, contractor, subcontractor, supplier, manufacturer, and so on) there is a need for someone to keep the focus on the contract and to maintain a working harmony toward satisfactory completion. This function falls to the contract administrator.

The administrator is *not* a party to the Owner-Contractor agreement, but through a separate contract with the owner is a full participant in the process. Certain duties, functions, and responsibilities are given to the contract administrator by the owner. Usually the responsibilities are different from those given to the construction manager. Nonetheless, the owner expects the administrator to fully engage in the process on behalf of the owner and to fulfill *all* duties assigned. Basic to this is the idea that no construction project has ever been constructed precisely as it was designed. Through the years, courts have decided that construction is more art than science and hence is imperfect at best. This "imperfection" demands that there be some form of control to assess and resolve deviations in the work. These deviations—nominal (of no consequence), minor (some contractual adjustments are required), or severe (failure to meet major design levels)—occur continually. Many are easily resolved, but others lead to the excesses of arbitration and even litigation. While unable to prevent the latter, the contract administrator can deal with the other deviations and may be able to minimize their number and severity.

It is odd that although a group of individuals has gathered to create a new entity (the project), there is often an air of suspicion, distrust, wariness, and rivalry/competitiveness that can undercut or even destroy the very concepts so necessary to success—harmony, understanding, cooperation, the common good. This is one of the primary issues that have been addressed in the newer concepts and contract arrangements (discussed later) in the construction industry.

Since the topic discussed in this book crosses several occupational lines, it is necessary to point out that much of this discussion is directly usable by any of the major project participants.

Even from a cursory perspective, a distinction can be drawn among architect, engineer, and construction manager. The difficulty is to delineate the distinction without even slightly demeaning their individual contributions.

This, certainly, is not the forum to discuss, much less resolve the unfortunate dispute among architects, engineers, and construction personnel. Suffice it to say that (1) things are different from the past and changing constantly, and (2) the three major project participants need to act in behalf of one thing—a satisfactory (to the client) project, in every respect. Although the activities, procedures, directions, vision, and goals of the participants may differ, for varying reasons, in the end (hopefully) they aim for the same conclusion. Variation does not mean invalidation.

This book, then, attempts to describe the work of all three project participants during the construction phase. Surely the term *professional* applies. The dictionary describes a professional as "one having great skill or experience in a particular field or activity; engaged in a specific activity as a source of livelihood; one who has an assured competence in a particular field or occupation." We ask your understanding in this, since we wish to be just and fully descriptive of the work of each; certainly we do not wish to denigrate the work of any of the three.

We have made every effort to use the terms *design professional* or *professional*. Usually design professional refers to architects and engineers, since they do the actual design work; however, in some instances, the construction manager can (and does) perform as discussed. We intend professional to refer to any of the three; our use of the term *administrator* is also intended to refer to all three.

In the past, when things were simpler, the distinction among jobs still was made. It was easier to simply say that

- The architect, after proper programming to ascertain the client's needs and desires, created the overall scheme and then designed, detailed, and documented the building.
- The engineer designed the structural and mechanical systems and fitted them into the building.
- The construction manager helped with cost data and advised about construction materials and methods, bidding, scheduling, and field coordination.

Of course, these roles are intermixed. Some engineers "design" whole structures. Some architects do structural and building systems design work. Construction managers are sometimes general contractors and are fully able to produce projects with minimal documentation.

It is rather difficult, though, to use one name or designation to include the architect, engineer, and construction manager. Much of this text is directed toward the design professionals (architects and engineers) and their legal, ethical, and contractual obligations. There are, however, many aspects of contract administration that these professionals share with the construction manager. Still, there are vast differences.

They all come together (but retain their individual perspectives) during construction and completion of projects. Being relatively new, the field of construction management does not have the legal precedents of the professions. Additionally, the managers are used on different bases. Often, because of their unique status via their contract and the variety of their services, creating the project team is more difficult (there is no one scenario). Varied services frequently prove disruptive.

The Context

One other caution: This book is not intended to be a primer, for owners and others, on contractual matters, i.e., contract, content, intention, configuration, appropriateness, and so on. The text on contracts is included here only for general understanding of all the scenarios that one *could* encounter in administration today.

It is our sincere desire only to instruct and inform—not to be judgmental. The construction industry today has many, many relationships and configurations. Certainly most of them work; some may fail in concept, but not because of who is involved.

Integrity is synonymous with professionalism. There is simply no room for chicanery, deviousness, overzealous competition, divisiveness, or jealousy. The primary mission of the design/construction *team* is to be just that—a team—which produces a project that satisfies the client by appearance (image), function (of both building and the space within), maintenance (wearability), and budget (low first cost, low continuing costs). Of course, a decent level of profit in the pocket of each professional also is a good just reward, along with pride in work well done.

We fully intend that all of the professions be given their proper due for the contribution they make to each project. Types and depths of services may vary, but it is only the combination of all of these talents, abilities, skills, and backgrounds that produces quality projects.

It is our concept that, in contract administration in particular, there is no caste system, no classes of citizenry; all of the participants are necessary, and their individual contributions are requisite to a successful project. If nowhere else, certainly in the administration of constructing the project only a proper and concerted effort is acceptable. The players, their titles, and their methods may vary, but the foremost goal, a fully successful project, never changes.

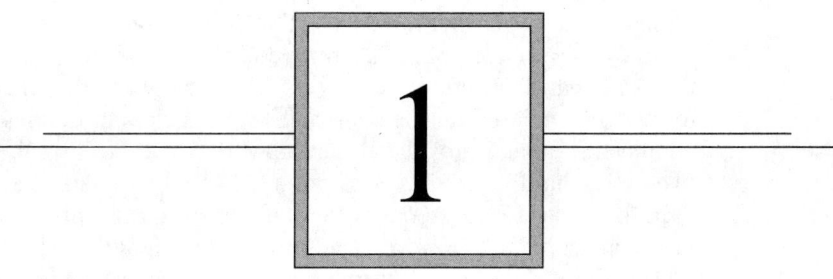

The Insight

Vision looks inward, and becomes duty.
Vision looks outward, and becomes aspiration.
　　　　　—Stephen S. Wise, American civic leader

Design professionals know construction contract administration as the last phase of normal basic services. Standard form contract documents such as those published by the American Institute of Architects (AIA) assign the term *contract administration* to services rendered during construction. This chapter defines construction contract administration and provides the framework for the technical, legal, and administrative topics that follow.[1]

In the United States today, several universities offer programs entitled "Construction Management." However, the subject here should not be confused with the content of those programs. University curricula in construction management generally portray a separate discipline and discuss scheduling, estimating, managing, and other areas of knowledge related to overseeing or supervising the construction process, whether by the owner, contractor, subcontractor, construction manager, or designer/builder. In this text, contract administration is that service (or collection of services) provided to owners by design professionals during construction. The goal of those services is to maximize the probability that construction will proceed in accordance with the contract documents and with a minimum of disruption.

Mistakes, weather, misunderstandings, poor coordination, and honest disagreements are but a few of the problems that can beset the construction process. When these conditions arise, a quick and knowledgeable response is required. This text attempts to describe the range of responses that have been shown to be helpful. Of equal importance, however, is the preparation that can anticipate and eliminate the developments that result in disruption. Proactive measures by the design professional's team members help reduce these unwanted occurrences (see Figure 1.1).

CONTEXT

Contract administration is included in the contract documents as a distinct phase of the services provided by the design professionals. The AIA documents assign 20 percent of the work effort (and fee) of design professionals to this phase. The actual time spent may be more or less.

However, for planning and managing time and resources, 20 percent is only a guide. But a grasp of contract administration requires that the relationship between contract administration and earlier phases of work by the design professionals be understood.

Unintended conditions may become apparent during the construction phase as a result of errors or omissions in the documents or mistakes by the contractor(s). However, the message at the heart of this book is that contract administration need not be (and normally is not) adversarial. The key to that nonadversarial relationship and clear understanding of the benefits contract administration can provide is increased communication and opportunities for problem solving.

Contract administration is interrelated with each aspect of professional services. Where contract administration is understood and appreciated, it is integrated into the overall process, from initial client contact; as a marketing tool; an avenue for mutual discussion of contractual terms involving the owner, contractors, and design professionals; a context for material selection and evaluation; and a source for estimating cost data. It can also be a means for improving construction details, conditions, and systems placed in the contract documents, and a method by which design professionals acquire knowledge of changing technologies. Thus, contract administration is much more than a detached phase of professional services.

It usually lacks the glamour and prestige associated with the design process. It is seldom discussed in critical assessments of works of architecture, unless one consid-

---

1. Prior to starting a study of this text, it is suggested that the reader refer to and read "The Context," which precedes this chapter.

Chapter 1 The Insight   **3**

There's Honor in the Field

It is no secret that the years have seen steady erosion of the architect's authority in a most critical phase of the construction project—when all the design creativity and technical prowess is translated on site into the physical building.

The same years have seen the watering down of words, from "inspection" and "supervision" to weak words such as "observe," "advise," "consult." Look through a typical owner-architect agreement, and the number of times you see "shall not" after "architect" boggles the mind.

The history of backing off has gone hand in hand with the litigiousness of the owner and of third parties. Not so many years ago architects were typically paying out 4 percent of net billings for e&o insurance premiums. The insurance companies demanded cautious contracts.

The consequences anyone could predict. Where angels feared to tread, others rushed in: entrepreneurial individuals willing to trade extra risk for extra fees—construction managers, engineers, value engineers, design/build entrepreneurs.

But the climate has eased; the courts have established a more stable relationship between owner and architect.

That's exactly the time for architects to recoup and reach anew for a more active role in the field. It will take time. As architect John Webb, a devoted laborer in this vineyard, said to me: "This authority was a long time losing; it will take a long time to get back." How to start?

1. Don't let your grasp exceed your reach. Assess your firm's skills. (Call . . . the AIA to try out its self-assessment program.)
2. Keep on the lookout (or advertise) for men and women who like the construction side of architecture, who glory in the sight of a shop drawing, product data, and samples, and love the on-site dirt and bustle as the building goes up. Encourage those staffers; have them understudy your top CA people.
3. Attend one of dozens of workshops on the subject. Read books, such as Jay Bannister's . . . *Building Construction Inspection, A Guide for Architects* (Wiley, 1991) or Culbertson et al., *Contract Administration Manual for the Design Professions* (McGraw-Hill, 1983).
4. Schools should offer compulsory courses in building construction.
5. Consider, in this age of increasing niche services, setting up an architectural practice specializing in administering construction.
6. As the AIA goes into its new cycle of document revision, become involved, and press for development of documents that reflect a greater construction phase role for the architect.
7. Make sure the fee reflects the risk.

Someone must do the job—why not the architect?

STEPHEN A. KLIMENT

FIGURE 1.1 Recapturing the architect's site role has its risks, but the rewards are worth the effort. An editorial by Stephen A. Kliment.
Used with permission of *Architectural Record* magazine from the December, 1990 issue. © The McGraw-Hill Companies, Inc., 1990.

ers case law and legal opinions as critical assessments—then the discussions are replete. Sections that follow will discuss selected legal issues and concepts.

BEGINNINGS

The history of architectural practice makes reference to the architect (and engineer) as master builders for whom design and construction were a single process. It is possible that during the evolution of building design and construction, and the emergence of architecture as a separate profession, the description of master builder may have been apropos. Some see this as a "golden age" when architects were able to control all aspects of the work and were at their zenith of prestige.

However, a closer look at architectural practice in the United States in the last one hundred years indicates that as the profession of architecture has become more established and dependent upon formal education as a means of preparation, the status of master builder has declined. Indeed, it is arguable that architects in recent history were never master builders. History may give to certain architects the aura of master builder, but realities of practice and the division of construction responsibilities argue otherwise. Even Frank Lloyd Wright did not view, in person, many of the sites for houses he designed, much less see the finished works.

From these historical conditions of practice, it could be concluded that one condition not applicable to architects in the discharge of contract administration services is mastery of construction techniques and procedures on par with the construction trades.

With few exceptions, design professionals do not have the breadth of skills to actually construct buildings. The education and experience necessary to manage the growing complexity of architectural process alone underscore this condition. The fact that in modern times education in architecture has favored art over construction also adds to this division. Recent information shows little (if any) change from studies in the late '70s and early '80s in which architecture graduates rated the three (of fourteen) educational subject areas receiving "too little" emphasis in the schools as:

- materials/specifications (65.1%)
- professional practice/management (62.7%)
- construction methods (58.9%)

Thus, contract administration may be seen as an integral and interrelated phase of architectural and engineering design services. Accepting this premise, a design professional may provide contract administration, as stated in the contract documents, without the fear that only those highly experienced, knowledgeable, or skilled in construction can be qualified to do so.

CHARACTERISTICS IN PRACTICE

After the completion of construction documents, knowledge of the status of work in progress is important to the design professional. Having overseen the preparation of the documents, the design professional will be in a key position to solve problems that may arise during construction. This is the opportunity to see the work executed as intended.

Briefly, contract administration calls upon the design professionals to:

1. Know the standards of quality that are to be achieved in the construction process.
2. Know the abilities of the design team to describe and depict conditions that can be realistically constructed.
3. Continuously improve the contract administration services through on-site experience for staff at every opportunity.
4. Integrate lessons learned during contract administration into subsequent design and document production phases.
5. Resist making statements to potential clients about the firm's experience in contract administration, which may later be belied by an inability to perform competently during the construction phase.

To meet these goals, a process of continuing education can pay big dividends. As a plan of contract administration education is implemented, general terminology should give way to specific tasks and demonstrations of competence should be required of those designated as contract administration representatives.

Areas for Emphasis

Architects increasingly need to be familiar with provisions of the law that may affect their practice. Along with a working knowledge of building codes and other construction regulations, and a general understanding of registration laws, law as it pertains to practice in the design professions is logically a part of professional study or continuing education. However, only a small fraction of the efforts in those areas cover the legal environment in practice. At least one area of the law deserves mention here: the legal standard of care for design professionals (other legal discussions are located throughout this book).

The standard of care is a legal doctrine that refers to the description of professional conduct and competence which the law is likely to require in practice. The primary source for determining these professional behavior standards is the collection of legal decisions called case law. Through a survey and compilation of the findings, with emphasis on decisions that seem to be common under roughly similar circumstances, it is possible to make assumptions before litigation as to the probable outcome of a claim.

When design professionals encounter situations that involve risk, uncertainty, or conflicting inclinations, it is likely that case law has, in part at least, addressed those matters. Knowing in general terms the legal precedent for an issue at hand is a good start for a design professional. If conditions warrant, further detailed research or investigation can take place prior to any decision or action. The likelihood may then be that the potential for claims or litigation will be reduced.

Continuing education programs on this subject are offered frequently and can reveal areas of practice needing revision or reorientation. In addition, selecting an attorney to routinely assist the firm is good practice when a confrontation between the design professional and other parties may be in the making.

Limitations in Practice

Possibly one of the greatest shortcomings of design professionals is the pursuit of work beyond their ability or capacity to perform. Under the aegis of "get the work, first, last, and always," the capability of producing a successful project may simply be overwhelmed by the demands imposed. Further, contract administration is one of the major areas where deficiencies may reside under these conditions.

The ability to observe work for faults, make accurate estimates of the extent of completion of the work in order to verify the contractor's request for payment, evaluate and process change orders, estimate costs for substitutions or changes, apply a firm but fair and knowledgeable presence on the job site, anticipate problems, solve problems in a firm but sensible manner and act as the owner's agent to resolve disputes fairly, even when the owner is involved, requires a seasoned outlook. Granted, it may be possible to hire individuals for such purposes. However, when a design professional becomes merely a broker for such services without the ability to oversee and generally monitor the flow of events, problems are more likely.

If becoming proficient in the tasks associated with contract administration is the goal, how can it be accomplished? Experience is a good teacher under many circumstances. However, the realities of practice also require education and exposure on the subject, which may not always be "on the site" (desirable as that may be).

For many firms providing architectural or engineering services, normally one partner or major shareholder will be involved in contract administration, either as a supervisor of others or in actual performance (or both). Learning the details of contract administration under these circumstances may be a give-and-take proposition. In a young firm, the principals may have established themselves as designers, but may need assistance in the construction phase. In more established firms, the challenge may be to train additional staff to provide contract administration in the context of growth or personnel turnover. In any situation, however, someone will have to take charge of both contract administration and the process of educating the staff who will be involved, whether in the processing of documents or in construction observation.

There is a general consensus that where design professionals have performed contract administration, especially full-time, the resulting experience is invaluable. As summarized by J. Russell Groves, Jr., AIA, assistant dean, School of Architecture, University of Kentucky,

> . . . contract administration is not likely to diminish as part of architectural services. Mixing prudence, if not jurisprudence, with practicality on the job site and in the office during the construction phase can serve to bring the project to a rewarding and satisfying completion.

While demands of all but the most basic forms of construction normally require an experienced contract administration representative, the opportunity to expose staff members to the process on the job site, even if only for short periods, should be taken, especially if a full-time representative is present who can act as guide and mentor. Stephen A. Kliment, FAIA, editor of *Architectural Record,* agrees. In the June, 1990 editorial, Kliment says,

> It's time . . . to cease backing out of responsibility for construction-contract administration and, by training and selecting skilled site people, build back the reputation of the architect as a key player on the building team.[2]

When design professionals provide contract administration services, they tend to impose conditions or requirements on the contractor beyond the contractor's contractual obligations. They may find that they cannot resist the desire to tell the contractor and trades how to do work. Others will simply observe and record, and attempt to solve problems as they arise in accordance with the contract documents.

The latter approach is more in keeping with the normal contract requirements. This approach normally includes observing the work for deficiencies that can be reasonably detected, but does not require continuous testing or checking built conditions to assure conformity to the contract documents in every respect.

The field component of contract administration normally includes a pattern of spot checks, which can vary depending on the status of the work and the design professional's judgment as to where emphasis should be placed. Checklists can be helpful, but they should conform to the particular needs of the project and the extent of observation to be provided under the terms of the contract. Full-time contract administration will mean a continuous presence on the job site. Even so, the work will not be observed continuously. On-site record keeping, job meetings, and coordination with the design professionals are likely to be administrative tasks performed by the full-time contract administration representative in addition to observing the work.

Where contract administration is not full-time, the on-site portion will largely involve solving problems brought to the design team's attention by the contractor(s)

---

2. Used with permission of *Architectural Record* magazine from the June, 1990 issue. © The McGraw-Hill Companies, Inc., 1990.

and trades. Searching for deficiencies beyond those observable during routine site visits by the design professionals will usually be secondary. At the same time, office aspects associated with contract administration involving reviewing the contractor's request for payment, processing change orders, evaluating material samples, reviewing shop drawings, dealing with the normal flow of general telephone and correspondence communications, and coordinating with field activities.

FINAL OBSERVATIONS

If a tendency to go beyond the normal contractual requirements of contract administration exists, it is likely to be in the form of directions from the contract administration representative to the contractor, or trades, as to how to do the work. Contract documents, not the off-handed preferences of the design professional, must indicate the intended finished conditions of the work. Conditions in the contract documents are those to which the contractor and trades are legally bound. Likewise, confirmation from the contract administration representative to the contractor and trades as to methods and means of construction to be employed normally is not appropriate. If guidance in this category is appropriate, it should be only to the extent described on the contract documents.

Where the contract administration representative dictates methods and means of construction beyond the stated requirements of the contract documents, the representative may be legally performing "supervision." This connotes a much broader scope of control over the work than contract administration requires. "Supervision" carries with it a major increase in risk for the design firm and potential litigation associated with deficiencies in the work and management, which, absent "supervision" by the contract administration representative, would remain legally the sole responsibility of the contractor. Issues such as scheduling of the work and safety on the job site are matters that normally and properly are assigned to, and accepted by, the general contractor and trades.

A final point is that few, if any, clients are inclined to let their projects be used as the testing ground for services (e.g., contract administration) not previously attempted by the design firm. Likewise, if such a condition were to develop, the design professional could face adversity on other grounds. Opportunities for future or related work with the same client or using that client as a reference may be unlikely if the experience becomes acrimonious. Also, claims and litigation may result. According to Gerald G. Weisbach, FAIA, architect/attorney in San Francisco, CA,

> With this breakdown in the construction process, buildings are either not being built on time, or have to be rebuilt to get the project correct; delay damages result to cover the additional costs of construction; and often the architect becomes the target in this conflict.

A frank and complete discussion with each potential client regarding the overall capabilities and resources of the firm may result in short-term losses, but long-term gains. The ability to provide contract administration should be a part of those discussions.

As a brief abstract supporting this statement, refer to pages 340–346. Here one supremely important fact comes out: The design professions have but one opportunity to bring a building project on-line in a condition that meets all of the parameters imposed at the outset. As seen in the quote in Figure 1.2, there is no tolerance or opportunity for testing, evaluating, and rebuilding a building. The synthesis of every project must be set forth, established, and depicted *before* any construction starts. (See Figure 1.3.) To successfully reach the correct (and legally required) finale, the professionals have but one program available to them. That hallmark program is **construction contract administration.**

> Think about this: Is a delay in a project a better signal to an owner, if it produces a project in a cheaper, more cost-effective manner?
>
> In a building project, unlike a manufactured product, there is no scenario for trial and error—for undoing bad design. But if there is a direct effort to be "sure" from the outset, the project will benefit (as will the owner and all other parties).
>
> For example, an error in design may cost $10 to fix in that stage, but may cost $100 to fix in the production phase (due to complexity of interrelationships where multiple things must be changed). Such an error left undetected or unresolved could cause a $10,000 expenditure for changes made in bidding. Allowed to "process through" and caught and resolved in the construction phase, the error could easily cost perhaps $100,000.
>
> This is the rule of tens!
>
> —paraphrased from
> Continuous Process Improvement
> by George D. Robson
> The Free Press, 1991

FIGURE 1.2 Not to do something right the first time costs 10 times as much to find and fix each time it escapes to a subsequent stage of handling.
Paraphrased from George D. Robson, *Continuous Process Improvement* (The Free Press, 1991).

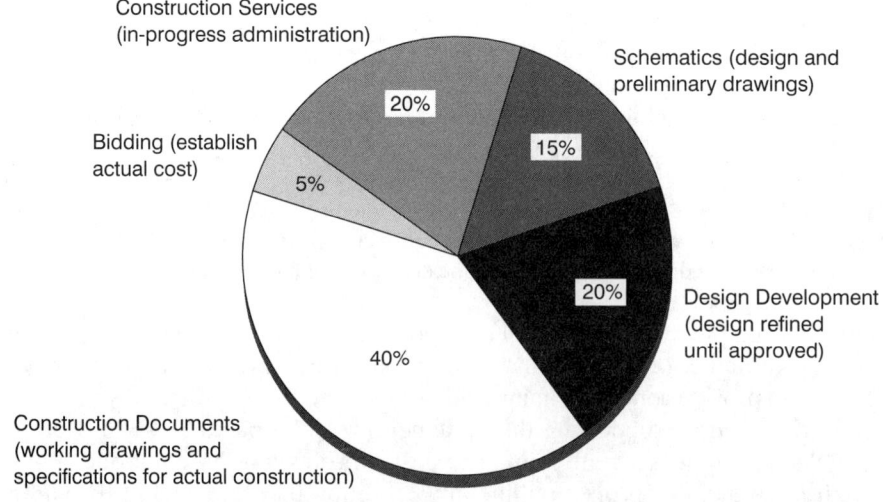

FIGURE 1.3 No person or profession can decry, excuse, rationalize, or justify the creation of an adverse situation, when the providing of the full range of services by the person or profession would prevent that situation from occurring. This is why the American Institute of Architects strongly advocates that clients be provided with a complete and comprehensive array of professional services, those that take the project full circle, from inception to completion. This figure shows the various project phases, set by contract, the general area of work involved, and the percentage of fee usually allocated.

EXERCISES

The following tasks are based on the concepts and information in the chapter. All of the material or information required to complete the tasks may not be contained in the chapter text. Independent research is required. Consult your instructor for guidance.

1. What is the primary goal of construction contract administration?
2. Research the concept of the master builder and give a brief presentation to the class.
3. What is the "standard of care," and how would it affect you as a project representative?
4. Discuss how proficiency can be achieved in contract administration. List tasks, courses, readings, or discussion that would be helpful.
5. List some of the limits that tend to "focus" the work of the contract administrator.
6. Develop a scenario or curriculum appropriate for addressing the concept of contract administration in an engineering or architectural program.

2

Definitions

Why do most contracts fail to *specifically* state *how* the design professional is to perform?

—Professor Justin Sweet, College of Law, University of California—Berkeley

> One of the nobler professional goals of every architect should be: that every project is completed, closed out, and delivered with the same verve, intensity, anticipation, interest, effort, attention to detail, dedication, enthusiasm, skill, poise, and concern for the project that was present and employed when the project first came into the office; all to the client's complete satisfaction and pleasure.

FIGURE 2.1 One professional goal given too little attention.

Short of performing all the functions and work her/himself, the design professional has but one sure method for ensuring that the project will be completed properly, as required by the approved contract documents, the client, and the design professional who conceived the design concept and delineated and documented the construction required. This method is called construction contract administration.

Construction contract administration is a basic, follow-up service intended to be performed by the design professional(s). It is an exciting, challenging, dynamic enterprise, not just a recording of current and past happenings. Those persons directly involved in this process usually gain high levels of self-esteem and satisfaction from the proper completion of the projects. This is due, usually in no small part, to their direct, experienced, personal, and poignant efforts. As Robert Frost said,

> My object in living is to unite my vocation and my avocation . . . for there, greater happiness dwells.

The ultimate goal of construction contract administration is the delivery of construction projects to their owners, which are complete, on time, on or under budget, and fully faithful to the contractual obligations (see Figure 2.1). This is the terminal aspect of the professional services contract between design professional and owner—a project ready for occupancy, use, and operation. In the final analysis, it is how the legal provisions of the contract/agreement and the associated documents (prepared by the professional) are met that is the true measure of the project. In simple terms, the owner's requirements have been ascertained, programmed, resolved, designed, depicted/described (by the professionals), and set into a legal framework (contract documents), which binds the owner with several other parties. Project success comes with full accord and execution within the contract obligations. J. Patrick Rand, FAIA, associate dean/research at the School of Architecture, North Carolina State University, says,

> Concept and implementation must be seamlessly related if the built work is to be architecture, and not merely a building.

The achievement of this ultimate goal is the culmination of a process that was quite exciting to various parties over a period of time. For the professional this was a sequence of sine-wave-like experiences. Of course, the securing of the service contract initially brought the new project into the office. This was an exciting time in that it provided yet another opportunity for the firm to exercise its collective talent and expertise in pursuit of another quality, and hopefully, fully successful and profitably produced building; any design or other awards along the way are the "icing on the cake."

The documentation of the design concept took place over a period of months and, by its very nature, involved a lot of detail and careful analysis and planning of every aspect of the construction. To those committed to production work, this was a period of exciting but hard work, as well as stress and strain. The goal was to finish on time and in a fully appropriate manner (good documents, at a profit). Completion of the production phase was a bittersweet experience. Renewed excitement came with bidding and the start of actual construction, bringing concept to reality. In this process, the contract administrator had the opportunity to exercise his/her skill and felt the rush of excitement in aiding the actual process of building the project. Not all professionals participate in this aspect of the work, and hence will not experience the adrenaline flow that an on-site representative feels while bringing the project on-line.

SERVING THE INTERESTS OF THE PROJECT OWNER

Since the owner is the primary resource for the project (funding, needs, desires, parameters, intentions and expectations) it behooves all participants to ensure that the project runs as smoothly as possible; thus the owner's best interest is served. It is wise to remember that acquisition of a built facility is a major expenditure and investment for any owner, whether an individual, corporation, or public agency. The project is expected to satisfy the objectives within the constraints as set out by the owner, *and* relevant regulations. Yet this is not a quest for perfection! Reality and the owner's expectations must be resolved, with an understanding that so many factors resist perfection, the best one can expect is the most faithful fulfillment of the contract obligations. Contract administration exists to guide and facilitate this fulfillment.

Obviously, every owner is concerned with the project from inception to finale. This entails a cycle longer by far than just the construction itself. It occurs across the entire life of the project, from the identification of the need to build, throughout its useful life until the point where the project is no longer of value to the owner; i.e., a period of time starting well before actual construction and lasting until well after initial construction. Hence, all the processes and techniques of planning, design, and construction of the project must be programmed from the viewpoint of the owner. Quality of work and performance is crucial to the success of the project, in the owner's eyes. Therefore, focusing on the viewpoint of the owner eliminates the competitive nature of the services provided by the various participants. Understanding this and the entire process of project conception and construction allows all participants to respond properly and effectively to the owner's needs.

COOPERATION IN PROJECT MANAGEMENT

Improvement of project management through innovative and effective procedures and techniques is not academic, but is necessary for individual project construction and, in fact, aids the construction industry as a whole. The first management incentive is to replace confrontation and adversarial relationships with a spirit of joint endeavor and accomplishment. The biggest hurdle in this establishment of effective construction management is the continuation of the historic divisions between planners, designers, and constructors. From the outset, then, coordination and cooperation are necessities, not mere "good ideas." With the fragmentation of project management (again a necessity), good communication, cooperation, and coordination become prerequisites for success. Each of the numerous participants must acknowledge, understand, and respect not only their own particular perspective but the different perspectives of the other participants.

Unfortunately, much of the writing in this area of professional practice is aimed at small, isolated, specific areas and special problems in the process. Although specialty knowledge is essential to each project, it must be considered within the context of "project" as an entity unto itself; i.e., being adapted to meet both the legal and the tangible expectations of the owner. Usually, the professional will complete the design and documentation and will then proceed with on-site observation during

Chapter 2 Definitions

construction, in keeping with the contractual obligations contained in the Owner-Architect (Engineer) service contract.

On-Site Observation

Although on-site observation is part of the comprehensive basic services package provided under the standard contract, this service remains the subject of much discussion, reevaluation, and reconsideration. Two polarized positions currently represent the status of the service as assessed by professionals. Larger, more established firms seem to understand the value of this service and want to continue providing it. Others, for varying reasons, seek to eliminate it in an effort to reduce professional involvement and liability and lower costs to the owner. There is no doubt that divesting oneself of this service directly reduces the liability level by eliminating the need to address the problems and nuances of the construction functions. There seems to be a basic philosophical difference involved. One hopes there also is a valid evaluation of the trade-off between lower liability and complete, comprehensive service to the client.

Over the last few decades, owners desiring to reduce costs determined that reducing professional fees would aid that effort. Thus, professionals in some situations have become increasingly isolated from the construction process, both by the owner's choice and, at times, by their own choice. To some degree the increased use of value engineering has contributed to this change. Here the owner and contractor can combine and choose to absorb the cost of changing the design and construction without the involvement and added cost of the design professional. The result is that the finished project is quite different from the first design, and perhaps is either unsatisfactory to the owner or meets a revised version of the owner's program. Slightly higher initial professional fees undoubtedly would produce more responsive projects for the owner, by allowing for adequate time for more thorough programming, design, and documentation.

Additionally, the threat of litigation and the lack of knowledge by architects and engineers regarding new construction techniques has caused many professionals to remove themselves from contract administration. The professional, at this juncture, is no longer in control of the project. This causes the professional to become an antagonist of the general contractor.

Since each phase of contract work entails some liability the cessation of one phase eliminates only a portion of the threat. There remains a scenario in which the decisions made during construction (without consultation with the professional) can be made to reflect adversely on the professional; hence, a liability threat, perhaps reduced, still exists. The staff of the firm, however, is not tied up with the construction sequence and can move on to other, more profitable work. Hence, the firms offer "plans and specs" professional services, whereby they fulfill their entire contract when they turn completed drawings and specifications over to the client. No more professional work follows; the contract is payable in full at this point.

The professionals in this situation, while relieving their liability, are also abdicating their ability to monitor, control, and administer construction. This prevents them from ensuring that the project is constructed in full accord with the contract documents. Here, the client is left to her/his own devices to ensure such compliance, a process that the client may be ill-prepared to pursue. In this event, a number of issues could be raised, several involving the design professional and the services provided. Obviously, if the client is dissatisfied and not fully aware that contract administration was available but deleted, there could be adverse reactions. This situation puts the professional in jeopardy, even though the liability exposure is not present; other legal aspects could be brought into play.

With these two different philosophies of professional practice, there is a need for analysis and discussion. To be fully capable of understanding the program and resolv-

FIGURE 2.2 A list of the major functions of contract administration.

> Contract administration should:
> - Address contract compliance
> - Provide for faithful construction of the design concept
> - Maximize exchange of information
> - Monitor costs, schedule, changes
> - Facilitate continual progress of work
> - Coordinate/cooperate at all levels
> - Be solution oriented

ing its use, one must know its parameters and impact: what contract administration is intended to do and what cessation means, legally and professionally. (See Figure 2.2.)

PERFORMING CONTRACT ADMINISTRATION

Most private and many large corporate and governmental organizations opt to have contract administration done by those most qualified for such work. Although they may have expert staffs that could play a continuing part in the construction, the design professional is deemed to be the party most likely to understand the project. No one, save the owner, has been involved in the project (now ready for construction) longer than the design professional. No one, including the owner has better insight and deeper understanding of the details of the project's program and construction than the design professional. Therefore, that person or firm is best prepared to ensure correct delivery of the project as programmed, conceived, designed, approved, documented, and bid.

Contract administration as a reduced form of total project management is the art of directing and coordinating human and material resources through the course of the construction project by using modern management techniques and objectives of scope, cost, time, quality, and participation satisfaction. Field personnel working for the professional are usually activated at the implementation (construction) of a project and do not have extensive input in the design process itself. Hence, a project manager who is experienced and well educated in fundamental construction and contract administration principles is far more helpful than one trained only in a specific area.

The concept seems fundamental, straightforward, and relatively simple. Not so the continuing professional discussion and interest and the process itself, in many instances. It is a rigorous, involved, often stressful process, certainly requiring full cooperation, continued work, and dedicated personnel. Reasonable parties and good understanding enhance the process, but hard work is still required, both physically and administratively.

Construction contract administration utilizes the design professional by having that person engage in certain activities during the execution and implementation (construction) of the contract documents and the project's design. Contract administration further emphasizes complete and faithful compliance with the Owner-Contractor Agreement and the ancillary documents. Contract administration is not "metaphysical certitude" (i.e., ultimate perfection, as defined by one noted television news analyst). But neither is construction a perfect science! Rather, contract administration is a professional service based on the design professional's duty to the client to exercise professional skill to the end that the project will be constructed as programmed, planned, designed and approved.

In fact, as professional design services have evolved, the design professions have seen fit to retain this basic service, standard to each project. Contract administration is the primary means available to the professional, and to the client, to assure that the project will be constructed as programmed, designed, detailed, bid upon, and expected (contractually). Any reduction of the full contract administration service as provided in standard documents of the American Institute of Architects (AIA) and the documents committee (EDJDC) of the National Society of Professional Engineers (NSPE), or even a seemingly limited modification, may adversely affect the manner and scope of this phase of services. Most owners simply do not have the expertise to provide administration service comparable to that which is normally available from the professional who prepared the documents. Even those owners with technical (construction-oriented) staff members cannot fully assume the role of the design professional in terms of observing and understanding the rationale for the work in progress, and as a result determining compliance with the contract documents.

Neither does a contractor normally have a relationship to the planning and programming of the project. Therefore, he/she is not likely to have the full understanding of the project necessary to address the impact of the proposed changes. A client will be ill-served when a project is completed that is not in accordance with the design professional's intent and overall expectations, as expressed in the contract documents. Participants in the project, other than the design professional, have different perspectives that may not be fully or knowledgeably challenged unless objective contract administration is provided by the professional on behalf of the owner.

Establishing Responsibility

Fundamental to the process of contract administration is the establishment and acceptance of responsibility. This is an active process, not a "paper exercise." It is necessary, early on, to set forth the responsibilities of each project participant (along with some limits). This is a crucial process, and one that will serve well all participants in the long run.

This process is achieved through contractual provisions and through specifications, memorandums, letters of agreement, and so on, listing specifics of the responsibilities. Of course, there is a deeper legal aspect to all of this. It certainly should not be a process whereby one or more of the team members are "hung out to dry." This is the "glue" that will hold the entire project team together as it progresses through the stages of construction.

The basis of all this is the acceptance, by each team member, of his/her full and proper share of responsibilities. Ideally, the team would be self-enforcing as to this aspect, but at times, contracts do get in the way of some members "leaning" on others. Overall, though, each member should be clear as to what is expected, when, and in what form.

Language of the Contract

No set of contract documents is perfect, so a system of interpretation and clarification of the intended meaning of the documents is often useful, especially to avoid a legal claim. When the design professional is not involved during construction, the project may take a new and unintended direction, which may be contrary to the interests of the owner. Departing from the requirements of the contract documents is contrary to the Owner-Contractor Agreement. When considering the relative responsibilities and project involvement of the parties, the design professional is normally best suited to clarify ambiguous conditions arising from the documents, in accordance with the project's concept and intent and the owner's desires and expectations.

Chapter 2   Definitions

Working with the Client

Construction contract administration by the design professional is based in the perceptions and impressions of the client/owner. For it is here that the basic criteria are created, rightly or wrongly. The client naturally expects that the professional will (1) ensure the client receives everything being paid for and to which the client is entitled; (2) watch over all of the work, seeing that all contractors do their work properly, in a timely fashion, and in keeping with the client's desires and needs; and (3) be responsible if the contractors do not perform as required or anticipated by the client/owner. Of course, the design professional has a different view of the responsibilities involved. The situation needs discussion, explanation, education, understanding, and mutually accepted resolution!

This portion of the professional's contract administration work holds a greater risk of liability than the other phases. With the client's expectations and the introduction of third parties, the professionals find that most of their problems occur in this phase of the work. Most of the current standard documents for both architects and engineers have provisions that give the execution (the actual construction work) of the design concept solely to the contractor. Therefore, if the client has other ideas or priorities in mind for the professional, these will have to be fully resolved and incorporated into the design services contract at the beginning. The situation simply cannot languish, or it will become a wedge between client and professional. This cannot be tolerated, especially since the professional, by contract, is to be an "agent" of the owner, acting in that person's behalf. Any divergence of opinion or impressions will strain this relationship.

There is only one proper resolution to this situation: in-depth conversation. This entails full listing and exposure of all issues pertinent to the project, meaningful explanation of the work and the options that may apply, definition where appropriate, full understanding of the issues and their interrelationships, and careful and specific listing of all obligations, nuances, and conditions of the relationship between professional and client *in the contract document* itself. This could entail modification of a standard document, which requires even more care, prudence, and meticulous rewriting.

Defining Services

The issues of ambiguities, differing opinions/perspectives, and their correct resolution arise before job site problems are encountered. The professional should become aware of and resolve *all* inappropriate, inapplicable, open-ended, or ambiguous language in the agreement being proposed to the client. Exactly what services will be provided and descriptions of each service phase should be included. Equally important is enumeration of what is not included in the services. Both the inclusions and the exclusions are important to the contract administrator, who must vary his/her work to match the variant requirements. Administering several projects at once can involve meeting several different sets of parameters and requisites. Just as many offices use a system of standardized construction details, a firm should develop its own glossary of service terms, with a detailed description of all services, phrases, and words used in the contract documents (including the contract itself and the General Conditions). This should include varied definitions that can be used interchangeably as clients may choose. Such variations would allow quick resolution of contract language and eliminate the need to negotiate each term for each project. Nor would the professional always have to research and investigate each term and the ramifications of each client's desires.

Using Standardized Contracts

The use of standardized contract document forms is advisable wherever possible. These forms have been extensively researched, are well developed, openly negotiated, made available on the highest of professional terms, and are time proven. However, they should not be conceived as a panacea to be used blindly, without analysis. They should be modified to suit the professional, the client, and the particular project. As we will see later, the contract documents are the "measuring stick" against which a court will view the performance of the parties. This concept requires a system of variables that will allow the professional to consummate any contractual arrangement in the shortest period of time, to the full satisfaction of the client, and in the professional's best interest. Simply, the professional should be prepared to adapt his/her procedures to the client and the project; this demands definition, discussion, determination, and resolution early on.

Legal Considerations

Another important situation must be resolved. While there is a legal obligation on the part of the professional to see that the project meets the contractual parameters, these are set forth as private considerations. There is, in addition, a public consideration—the various codes and other regulations in public law that affect construction. The professional is responsible for the project's compliance with all applicable laws, ordinances, codes, regulations, rules and procedures. These should all be reviewed and solutions resolved within the production of the design and contract documents. This responsibility, as seen in the case law in Chapter 3, is not transferable, and not something to be ignored or avoided.

After the professional produces a set of documents that will, if fully and properly executed, create a fully compliant project, how does he/she ensure that the complying construction will be built or installed? Where the professional has no function or input into the construction process, he/she has no way to forestall decisions that might downgrade, disrupt, or destroy the compliant nature of the project or some portion of it. Yet the heavy legal responsibility of the public law lies with the professional. If the project does become noncompliant, a long and involved legal process could ensue. The professional may be required to expend large amounts of time and money simply to have his/her correct relationship and responsibility affirmed. The professional's work lies in the hands of others, over whom the professional has no authority or control, much less a legal remedy (short of litigation).

This scenario shows the necessity for construction contract administration by the professional. There is also crucial value in proper execution, expenditure of the available funding, aesthetic control, maintenance of function and form in the project. None of this may cause as much legal trouble as lawfully constructing the project and meeting public law obligations. Although there may be numerous ways to meet code requirements, there is a need to ensure that whatever mode of compliance is chosen, that it be installed or a proper and viable substitute used. Random substitution without rechecking the code/regulation requirements *and* the owner's requirements can be damaging to the project and the progress of construction. Yet decisions made by those other than the design professional often are not made thoroughly or correctly. Further, remember that part of the owner's basic obligation to the general public is to produce a project in keeping with public law.

Working with Contract Documents

A design contract that manifests itself in numerous large sheets of drawings and a set of project manuals, each inches thick, obviously has a good deal to say. These are the documents used to bring the concept to realization. All of this information—drawings, instructions, forms, specifications, details, etc.—needs administration. There is no automatic or effortless manner in which these documents can be turned over to the contractors and have the result turn out exactly as the designer envisions

Chapter 2   Definitions

(and the client desires). Within this array of material, no matter how well detailed and meticulously done, there are discrepancies, glitches, gaps, needs for interpretation, conflicts, redundancies, and flat-out errors that must be resolved. Without the dynamics of contract administration, this process will not occur; problems will prevail, not disappear.

The "voice" with which the documents speak is seated in many things. Certainly the array of information on the drawings and in the specifications carries very extensive and important data required on the site during construction. However, despite the size and scope of the documents, a large accumulation of information is not contained within the contract documents. If this information were included, the documents would become so massive and so complex that havoc and confusion would occur on the site.

Trade associations, industry standard writing organizations, and even governing authorities promulgate documents that deal with small portions of the materials, systems, and assemblies that could be incorporated into the design and construction of the project. (See the listing of such organizations in Appendix E.) These documents are referred to as Reference Standards. They set the standard for the particular items involved: the way they are made, how they should be treated, installed, handled, and so on.

This is one of the key issues of expertise that the design professional brings to every project. The professional has access to a proliferation of data that may have a direct bearing on the construction of any project, but is not needed on every project. Much of what is done on a project is defined in various types of documents produced by an array of sources. The documents explain the development, testing and production of the items, their proper uses, and their installation. This information provides a base for evaluating what the product should be and should do. In essence, these documents are a form of quality control, which can be relied on and referred to if necessary to resolve issues involving the device, material, or system.

The documentation of projects has developed over the years in a way that incorporates many of these standards, but by referring to them by name, title, or number only, not by reproducing the entire text of the standard in the specifications. One simple reference to a document can thus make mandatory (as a contract requirement) the contents of a document several hundred pages long. Obviously, had all such standards been fully copied into the specifications, a literal truckload of documents would arrive at every project.

The design professional, primarily through the specifications writer, will be aware of these documents and how they are to be utilized. Part of his/her professional expertise is to know which documents apply, how, and when. This knowledge is lost if the professional is not a continuing presence on the construction site. No one else has the intrinsic, detailed depth of knowledge.

Since contractual obligations do not change with the size of the project, it is obvious that contract administration, as incorporated in the standard contract form, is a necessary activity. Of course, many projects are contracted on a "plans and specs" basis, but these usually will shortchange the owner in the long run. Either owners delude themselves into believing they can handle the administration of the contract, or they have utmost confidence in their contractors. Under such "administration," performed by a different player with a different perspective from that of the professional, the element of impartiality is lost, since no party in the project has the contractual bent of the professional. The professional, of course, is interested in seeing the project built in complete accord with the concept and the approved documents. Although this provides a profit to the professional, it is not a "contracting" profit. As the owner's agent, the professional has an inherent obligation to ensure compliance and true value in the project. No other participant carries this specific viewpoint.

When the professional broaches the subject of contract administration with the client, there is an opportunity for explanation, definition, and discussion of all this.

The definition lies in two areas: definition of the entire process, and detailed definition of the various relationships, obligations, parts, tasks, and nomenclature used in the process. The latter discussion should carry definitions far beyond those commonly found in dictionaries. Terms should be translated into the professional context as used in the construction industry. Essentially, simple words used in the documents must be fully defined, explained as to precise meaning and implication, and proper understanding achieved as to what, *exactly,* the professional will provide in the way of service and what the client should expect. Without this interaction ambiguities will survive and could easily cause unnecessary disruptions, arguments, misunderstandings, anger, and even litigation.

Incorporating precise definitions in the contract documents should not be summarily dismissed as unneeded. Much more often than not this information leads to quick, equitable, and favorable claims resolution. (The legal books listed at the end of Chapter 6 are invaluable resources for all contractual parties.) Using imprecise words and meanings can lead to time-consuming and often costly legal consequences for all involved.

A single word often can mean the approval of a claim and substantial added cost. For example, one owner was to demolish a building on the construction site "to grade." The work was completed but left the foundation, slab, and stairs. The contractor, by specification, was to provide "clearing and grubbing" and removal of resulting debris. The contractor submitted a claim for the removal of the foundation, slab, and steps, noting that these were not taken "to grade" as indicated. The appeals board agreed that "to grade" meant to ground level. Similarly, another project specification defined "rock." When a claim was submitted for removal of "rock," the owner balked, said that the material was "unclassified," and disallowed the claim. The review panel approved the claim based on the definition of the word "rock" in the specifications. A proper definition in the contract documents should always be supported.

This type of interaction with the client needs to take place on every project, since the results could differ from project to project, even with the same professional involved. Defining the terms of the job is particular to the client, the associated project, and the circumstances of the project, not to the status or operating procedure of the professional; it is a service to the client! The professional should not rely on the resolution of this situation simply by using standardized contract forms and other such documents; these forms contain many of the very words and phrases that require detailed explanation and definition. The entire process starts with a determination by the professional as to the service level to be offered in regard to the liability and risk the professional is willing to assume.

The reluctance of professionals to engage in contract administration is seated not so much in the extended effort required, as in the risky, complex aspect of the work. Simply, extreme care must be taken in every part of the work to ensure that the contract is met, but in a manner that does not inappropriately expand the overview of the professional or the level of liability. Here money, time, and satisfaction are intermixed to varying degrees; claims, disputes, and litigation are the downside issues. This is unfortunate. Currently there are many efforts toward wider use of less costly and less time-consuming procedures such as arbitration, which may relieve the penchant for filing suit.

In any event, it is essential that the professional have a thorough working knowledge of the legal exposure, the types of situations that have been litigated (and therefore are avoidable), the complete interface of all legal obligations imposed by the contract(s), and how each party is affected and is expected to react. Professional firms need to ensure that all their personnel understand this process, since an unintended comment, a slight misstatement, or incomplete or slightly erroneous information can become the crux of a large problem. The reading of case law (legal precedents) reveals the meager basis of some suits, which never should have been pursued.

Effect of Case Law on Contracts

Case law is continually changing. Any "package" of case law of fairly recent date, however, gives some incisive direction to design professionals in regard to on-site activities, site services, and obligations. In reviewing the following cases, consider how the details of each case varied, leading to different decisions and delineation of the various responsibilities, as the courts saw them:

Kleb v. Wendling, 67 Ill. App. 3D 1016, 385 N.E.2d 346 (1979)

First National Bank of Akron v. Cann, 669 F.2d 414 (6th Cir. 1982)

Shepard v. City of Palatka, 414 So.2d 1077 (Fla. App. 1981)

South Burlington School District v. Calcagni-Frazier-Zajchowski, Architects, Inc., 138 Vt. 33, 410 A.2d 1359 (1980)

Hunt v. Ellison & Tanner, Inc., 739 S.W.2d 933 (Tex. Civ. App. 1987)

Moundsville Independent School Dist. No. 621 v. Buetow & Assoc., Inc., 253 N.W.2d 836 (Minn. 1977)

Council of Co-owners, Atlantis Condominium, Inc. v. Whiting-Turner Construction Co., 308 Md. 18, 517 A.2d 336 (1986)

Sweeney Co. of Maryland v. Engineers-Constructors, Inc., 823 F.2d 805 (4th Cir. 1987)

Krieger v. J. E. Greiner Co., 282 Md. 50, 382 A.2d 1069 (1978)

Al Johnson Construction Co. v. United States, 854 F.2d. 467 (Fed. Cir. 1988)

Chiaverini v. Vail, 61 R.I. 117, 200 A. 462 (1938)

Sheetz, Aiken & Aiken, Inc. v. Spann, Hall, Ritchie, Inc., 512 So.2d 99 (Ala. 1987)

C. P. Robbins & Associates v. Stevens, 53 Ala. App. 432, 301 So.2d 196 (1974)

Mayor v. City Council of Columbus v. Clark-Dietz & Associates-Engineers, Inc., 550 F.Supp. 610 (N.D.Miss. 1982)

U.S. Home Corp. v. George W. Kennedy Construction Co., 610 F.Supp. 759, D.C.III (1985)

Watson, Watson, Rutland/Architects, Inc. v. Montgomery County Board of Education, et al., 559 So.2d 168 Ala. (1990)[1]

Other cases are discussed elsewhere in this book (and at length in Sweet's *Legal Aspects of Architecture, Engineering and the Construction Process,* 5th ed.; see listing in Chapter 6). Some insight comes from a quick review of some cases.

In *Kleb,* the court held the design professional to a "supervision" standard of contract administration despite the use of the less than comprehensive term "administration [in the documents]; must prevent gross carelessness or imperfect conditions." The decision seems to imply that continuous monitoring is required. This is supported by the older case of *Pancoast v. Russell,* 148 Cal. App. 2d 909, 307 P.2d 719 (1957), where the court found "no real value in supervision, unless the same be directed toward securing a workmanlike adherence to specifications and adequate performance on the part of the contractor."

The court in *First National* found that the design professional's duty (in Article 3.4.3 of the AIA form) was "to inspect and monitor in a manner that would uncover the defective conditions, and failure to do so perform constitutes a material breach of the contract documents." By issuing a certificate of payment, the professional represented to the owner that "to the best of his knowledge, information and belief the quality of the work was consistent with the plans."

---

1. This case cites a number of related cases that provide more insight in similar matters.

In *Shepard* the court found that the contract (a standard AIA form and text) created no duty on the design professional and clearly protected the professional when he did not discover misuse of drywall and other contractor omissions; inspections were made in accord with the contract. This point of view is supported by the *South Burlington* decision where the court found that absent expert testimony regarding the community's professional standard, and where the standard AIA contract provisions were in effect, "the design professional had no duty to supervise the project." In fact, the Owner-Contractor contract "made the contractor solely under duty to supervise and direct the job." Also, the contractor had hired a special superintendent to supervise the project.

More recently, *Hunt,* provides a summary and clarification of the issues in the other cases. Basically, the court found that the professional carries on-site responsibilities, by contract, for viewing, familiarizing, determining, informing, and endeavoring to guard; however, none of this requires "construction work" by the professional, but merely "required that he provide information." Further, this court decided that the exculpatory clauses noting that the professional would not be held responsible for failures by the contractor to perform properly were meant to show that the professional is not "the insurer or guarantor of the general contractor's obligation." In essence, the court said that the professional who performs within his/her contractual obligations cannot be held responsible, but at the same time, no automatic relief from responsibility is provided by merely adding exculpatory clauses to the contract.

Simply, *Hunt* tells the professional to see to it that the contract text is correct and accurate and contains only those provisions, words, terms, and phrases that he/she (the professional) is prepared to accept and perform. This goes directly to the issue of full definition, so the owner has the same understanding of the services to be provided as the professional. Also, the *Hunt* case shows that simply writing in clauses claiming exclusion from responsibility does not make that happen.

Four general principles arise from these court decisions:

1. There is a breach of contract if the professional does not perform as promised as to site visits, their frequency and their intensity.
2. If the breach is the substantial cause of harm to the client, compensation for the losses must be paid.
3. If the professional exceeds the contract requirements (more visits, more added intensity, more work done, etc.) and the owner comes to rely on this to his/her detriment, then the professional must compensate the client.
4. Exculpatory clauses provide no protection if the professional observes contractor failure to comply with the contract and/or building and safety laws and takes no action to ensure that corrections are made to provide compliance in the finished project (and third parties on or about the site and adjoining property suffer no harm).

Along with the inherent liability that each professional carries is the implied threat that there could be further implication and liability for the contractors' mistakes. An array of legal cases have left this situation in a most contradictory manner. Hence, the best direction for the professionals is to modify the standard documents carefully to define the requirements for many aspects of their work, not the least of which is construction observation. Legal experts specializing in construction law strongly suggest this to all professionals. For example, where AIA contract form B141 is used for outlining professional services, the following changes are examples of items to be modified to suit:

- List a specific number of visits to the site, any time frame or cycle involved, and/or the critical work items necessitating observation.

- Outline the limits regarding what portion of the contractor's work will be reviewed.
- Specifically exclude work that cannot be observed, such as that performed when the architect is not present.
- Narrow or limit the duty to observe work that is repetitive and that occurs whether the architect is present on-site or not.
- Have owners retain separate consultants (professionals) for asbestos and other hazardous materials work.
- Strictly limit responsibility for shop drawings to that which the fee warrants.
- Ensure payment in the event of delay or suspension of the project.
- Include interest for late payment of professional fees and legal fees for any collection work necessary. Ensure timely payments.
- Provide for mediation and if necessary arbitration as the primary methods of dispute resolution.

The last provision, when accepted by both parties, is a real indication that they both want the contract to work. This is most important, and if it can be solved early (at the time of the contract execution) it goes a long way in aiding the progress of the project and establishing confidence in the professional. All contracts need to be fair; it is risky to place too heavy a burden on other parties. A fair contract makes the entire process workable and manageable for both parties, and places risk where there is the best ability to control harm and to bear the risk.

Contracts for Renovation and Alterations

Renovation and alteration/addition work generally is similar to new projects, but such work presents new aspects of the contract that require further modification. For example, how is information regarding the present facility to be gathered, and by whom? Usually the owner will have at least minimal documentation, and this should be made available (through explicit contract terms) and then enhanced to provide a proper basis for the new work design. The contract should state that the owner provides all this work, utilizing those resources the owner deems appropriate. Of course, the professional can be retained to provide these services through added contract language, but it is best left to the owner, who can control costs and personnel used.

Next comes a consideration of existing features that might affect the new work. Since it is best to avoid surprises (and their inherent cost), some process or system should provide for added investigation and verification of the property. Again, this can best be done by the owner with guidance and direction from the professional. The contract text should reflect in detail what is to happen, how, when, and by whom. This can entail anything from checking locations and dimensions to destructive testing, as the circumstances require. The owner is well advised to seek this information as an important adjunct to any basic information available.

There should be no illusion, however, that every nuance, condition, or situation will be uncovered. Existing buildings simply contain too many such conditions, which may vary from the documents or which have changed, naturally or by design, with no documentation. The more time and money spent in this initial investigative work, the less time and money will be needed to rectify the conditions later. This must be undertaken as a joint effort by owner and professional, with proper contract provisions and agreement about the work.

Once the investigative work is complete, the next step is evaluation of the conditions. What can be saved, reused (in whole or in part), renovated, or augmented? What must go? This service requires yet more definitions to describe the activities of each party involved. It may seem overly "persnickety" to include all this in the contract, but better to explain the interrelationships than to rely on assumption.

Revising Contract Language

The contract administrator, while not a participant in contract making, must be aware of all contracts and how they function. AIA document B162 (used in conjunction with document B161) lists various services that the architect can provide. "Basic" (included) services should be described as well as those that are "added"; knowledge of what is *not* included also is extremely helpful. Since the contract administrator may be required to work through all of the contract work, knowledge of the contract provisions is requisite to proper administration.

Best advice and the case law cited above show that the wisest path for the professional is to include only those contract provisions that one is willing and prepared to assume. The standard AIA agreement provides that contract administration is a basic service (see Section 2.6.2) and is part of that agreement, *unless* modified or deleted. Here the professional has to decide whether or not this service will be provided, and in what format.

If the professional deems that some configuration other than that in the AIA form is advisable (and in the professional's best interest), or that deletion is proper, the standard contract form and all associated documents must be thoroughly reviewed and similarly revised to include the correct provisions. If this is not done, the contract will be flawed and could be held, by the courts, against the professional. This is particularly true where the professional is the drafter of the contract. This is an extremely important issue, since it becomes an integral part of the responsibilities assigned to the professional by the contract, and part of the duty thus owed to the client/owner. Further, it defines the course for the contract administrator. The accompanying reading discusses how some courts have interpreted such contract provisions.

The Design Professional's Responsibilities for Site Observation and Safety During Construction

Herbert G. Smith, II

The administration of a construction contract is the one phase of a design professional's performance where the nature and scope of his services are most often misunderstood and likely to generate litigation.

The design professional's role during construction is complex, requiring him not only to observe construction, but often to mediate and adjudicate "disputes" between the owner and the contractor in an effort to ensure that the project is constructed in accordance with the contract documents.

Sometimes there can be less incentive for design professionals to attend to the details of their performance during construction with the same precision they do while engaged in performing the earlier phases of their employment. This risk arises when, under a typical owner/design professional payment schedule, the design professional has been paid 80 percent of his fee by the time construction begins.[1] Thus the owner's expectations must reasonably conform not only to the risks and responsibilities actually assumed by the design professional, but also to the compensation paid for his construction phase performance.

This article examines only two aspects of a design professional's many duties and responsibilities during construction of a project: specifically, observation of the progress of construction and any accompanying responsibility for jobsite safety.

Observation of Construction

Article 2.6 in AIA Standard Form B141 (1987) sets forth the obligations of the architect for site observation during construction of a project.[2] Article 2.6 provides that the architect shall "advise and consult with the owner" during construction and, "at intervals appropriate to the stage of construction," the architect shall *observe* the work in an attempt to "guard the owner against defects and deficiencies in the Work." The architect, however, is not required to make exhaustive or continuous on-site inspections to check the quality or quantity of the work. The distinction between observation and inspection has become critical to the courts when considering whether to impose liability on the design professional for faulty construction.

In a recent case decided in Alabama, the contract language dictating the architect's duty for observation of construction was derived from a clause in AIA Standard Form B141.[3] The clause at issue required the architect to "make periodic visits to the site" and "familiarize himself generally with the progress and quality of the Work." Although the architect did visit the site to observe the progress of construction, he failed to discover a defect in the construction of the roof. After the building was completed, the roof developed extensive leaks. The owner sued the architect for breach of contract based on the architect's failure to discover the contractor's deviation from the plans and specifications.

Reversing a jury verdict in favor of the owner, the Supreme Court of Alabama ruled that the contract only required the architect to observe the construction. The court reasoned that an architect who discovers a defect has a duty to inform the contractor and owner of that defect. Where, as in this case, however, the architect failed to discover the defective construction, the court heard evidence on the issue of whether the defect should have been obvious to an architect whose contract required only a general duty to observe construction.

The architect presented testimony establishing that the obligation imposed by the contract between the owner and architect did not require the architect to discover the contractor's mistake under the specific facts of the case, i.e., the defective construction was not obvious. Due in part to the owner's failure to come forward with any evidence to the contrary, the court determined that the owner had failed to prove the architect breached the agreement.

The implication of the Alabama case was that if the architect had undertaken specific obligations for inspection rather than a simple duty to observe, the result would likely have been an affirmation of the jury verdict.

Jobsite Safety

Until 1961, both the standard AIA owner/architect agreement and the General Conditions gave the design professional the obligation to "supervise and direct" the construction work. Until 1963, the AIA documents also gave the design professional the power to stop construction work if he was dissatisfied with the contractor's performance. When considering this broad contractual authority, courts in many jurisdictions held architects liable for injuries that occurred on a construction project because the architect was provided contractual authority to stop work if he observed construction that did not conform to the contract documents.[4] The courts reasoned that the authority to stop work carried with it the duty to provide a safe workplace and a safely constructed structure for all participants on the construction project as well as to third parties. Largely as a result of these court decisions, AIA Standard Form B141 was revised.

The revised documents now grant the design professional authority only to "reject" nonconforming work;[5] the right to stop work is given to the owner under the AIA General Conditions. Moreover, Article 2.6.6 provides that the architect is not responsible for safety precautions. Article 2.6.6 is supplemented by Article 4.2.3 in the General Conditions that provide similar exculpatory language placing jobsite safety squarely on the shoulders of the contractor.

Where contract language varies from Standard Form B141 to provide the design professional with authority to do more than observe or reject work, courts are more likely to impose liability on the design professional for jobsite safety. For example, a Florida court recently held that an engineer was liable to an injured construction worker because of the engineer's contractual power to stop work in the event the engineer observed hazardous defects or construction methods.[6] After considering expert testimony, the Florida appellate court held that the obligation or authority of the engineer to *stop* work carried with it a duty to ensure that every step of construction was completed in a safe manner.

Thus, unless the design professional takes on duties beyond those prescribed by Article 2.6, courts generally hold that design professionals are not responsible for jobsite safety. The majority rule is that a design professional would only be liable for negligent supervision and jobsite safety if the design professional had the contractual duty or authority to stop work or actually undertook that responsibility despite his limited authority.[7]

Avoiding Common Pitfalls

The most effective method of assuring that the design professional does not become entangled in litigation because of an alleged failure of the design professional to inspect the contractor's means and methods of construction is to address the issue of site inspection responsibility prior to entering the owner/design professional contract. Any number of claims arise because of miscommunication between the owner and design professional during the drafting process.

Misunderstanding may result from overly aggressive and optimistic "marketing" by the design professional and the potentially unrealistic expectations of the owner generated by such marketing. The design professional should be careful to temper the desire to market his services by recognizing the fact that many owners may not understand the scope of services that the design professional can or will deliver. Often it is the initial heightened expectations of the owner that give rise to litigation.

As for jobsite safety, after the design professional's contract has been executed, but before construction begins, the architect should call a preconstruction conference to explain to the contractor the chain of command and discuss difficult areas of construction. In concert with these and other topics, the design professional should also discuss his role with respect to site visits to ensure that the contractor understands that the design professional's authority and obligation to reject nonconforming work does not carry with it a duty to ensure a safe workplace or safe construction.

Having carefully explained his role to all participants on the project and prepared his contract accordingly, the design professional should be in a position to administer construction of the project without assuming the risk of unanticipated liability.

Herbert G. Smith, II, is an associate in the McLean, Virginia, law office of Venable, Baetjer, and Howard, where he specializes in architect-engineer malpractice. Used with permission of *The Construction Specifier* magazine, April 1991 issue.

1. The "disincentive" is deminimus for most design professionals because of their professional pride, careful budgeting practices, and insurance provisions that mandate requirements for design professionals during construction.

2. Article 2.6 also addresses certification of applications for payment, change orders, and the review and approval of shop drawings. These responsibilities are equally litigation-plagued, but their discussion is beyond the scope of this article.

3. *Watson, Watson, Rutland/Architects, Inc., v. Montgomery County Board of Education,* 559 So. 2d 168 (Ala. 1990).
4. See, e.g., *Erhart v. Hummonds,* 334 S.W. 2d 869 (Ark. 1960); *Miller v. DeWitt,* 226 N.E. 2d 630 (Ill. 1967).
5. It is interesting to note that shortly after the L'Ambiance Plaza collapse, Senator Weicker introduced S. 2518, the short title of which was "Construction Safety and Health Improvement Act of 1988." Although the bill failed to pass, of significance was the fact that the bill burdened project architects and engineers with jobsite construction safety.
6. *Moore v. PRC Engineering, Inc.,* 565 So. 2d 817 (Fla. App. 1990).
7. For a survey of cases, see *Krieger v. J.E. Greiner Co., Inc.,* 382 A. 2d 1069 (Md. 1978).

There is nothing in this determination, however, that forces or requires a given action by the professional. There is full freedom to protect one's rights and self-interest, while still reacting to the client's needs or requirements. There is a risk that certain clients will refuse to enter a contract if construction contract administration is included, but the professional should be prepared to let the competition have that project, or propose some other contract arrangement suitable to both owner and professional.

In making the decision that determines the configuration of the services (and sets the definition of the process) the professional should assess the following choices:

- "normal" contract administration (per the basic service provisions of the standard AIA documents; to "observe")
- "modified" contract administration (appropriate changes to the standard forms; to "supervise," "monitor," or "inspect" as the professional and client agree)
- full-time, on-site representation (changes in standard forms that convert contract administration from a basic service to an "additional" service with commensurate increase in fee; define functions and extra payments)
- no contract administration, and the impact of such a decision on the professional, the client, and the project (explanation of strict adherence to these provisions and no use of sporadic or partial service)

The professional should also assess the effect of the agency relationship that will be created. Although agency is not often a major issue in contract administration, it should be given attention by architects and engineers before the fact. In practice, the actual performance of contract administration frequently is not undertaken by a principal of the firm. Therefore, the persons acting as contract administrators carry the authority of the firm, consistent with the doctrine of agency. It is one thing for a drafter to be an agent, subject to the daily (almost continuous) supervision of a principal. It is quite another for the field representative, who may not be a registered design professional, much less a principal. The field representative may often have to make important decisions quickly and without benefit of consultation with a principal. While the architect's or engineer's agency powers are limited, the actions taken under that relationship are extremely important, crucial to the success of the practice, and vital to each of the projects served. Everyone involved must understand the status, function, authority (and limits), and impact of the field representative on the project. Further, this relationship should be fully and properly delineated in the agreements and in the ancillary documents that deal with the field representative.

Types of Services Offered

Normal Contract Administration

Normal construction contract administration is an expression of what is to be expected under the terms currently written into the standardized contract documents, without deletion or modification. It still requires a good deal of explanation and definition, but, in general, provides for site visits by the professional or some proper representative. This is expressed as visits to the site to see that "general compliance of the work with the contract documents" is in effect. Further, there would be some measure of the amount of work completed, for which payment should be made. These two functions or parameters should be made known to the owner and to the contractors. There should be no function, operation, or attempt to interfere with the contractors' work or methods. Only the functions enumerated in the contract form should be exercised.

Through the years, courts have taken three different approaches toward the functions and duties of the professional. Older cases appear to hold more toward a duty of supervision (in the court's terms) over the work, to prevent or ascertain fraud or negligence by the contractors. More recent court decisions hold more toward basing the functions of the professional on the provisions in the contract under which the services are offered and performed. This reinforces the need for full disclosure and definition of those functions and their limits. Other courts, though, looking primarily at the nature of the construction impropriety and the opportunity for its recognition by the architect or engineer during site visits, make other findings.

There have been ongoing efforts to restrict the liability of the design professional through contract language and provisions. These efforts have been successful, for the most part, where the professional meets the provisions. One remaining issue is ensuring that the owner understands what is written, what needs to be provided, and how it all affects the project. The words "full definition" come to mind. Courts have been moving toward a tighter rein on liability issues regarding the professional in field operations. Thus, it is necessary to give the courts the correct format and text on which to base their decisions. It is crucial to the successful conclusion of a lawsuit that the professional act totally within the bounds of the contract; to exercise all authority granted the contract for normal contract administration.

Modified Contract Administration

Modified contract administration is simply some form of operation between normal and full-time administration. Usually this will arise from the client's desire to have something more than the basic service. For the most part, it is aimed at tighter control of the project by the professional. The client will usually feel this is necessary if the budget is tight, the project is complex, or if for some reason untried or suspect contractors are involved. (This is particularly true for public sector projects on which any and all contractors can bid and the lowest and "best" bid must be accepted.) The request for modification of contract administration should always be a "red flag" to the professional, not to indicate a problem project but to show the need for extra effort and added risk. Of course, as noted previously, the modifications are basically negotiated to suit both client and professional.

Full-Time Contract Administration

Full-time construction contract administration affects the professional's liability. The primary responsibility of the professional is to the client/owner. That concept is examined here with reference to all forms of contract administration, including the effect on liability caused by the professional providing a full-time field representative for contract administration (see Figure 2.3).

Professional On-Site Responsibility and Authority.   Normally, the professional wishes to exercise on-site authority. The professional takes pride in his/her work and wants the project completed in full accord with the contract documents, to reflect favorably upon the professional's standing, talents, and abilities. An unsophisticated

Chapter 2 Definitions

```
SUPPLEMENTARY CONDITIONS                          DOC 00800/4

                           ARTICLE 4
                    ADMINISTRATION OF THE CONTRACT

  4.2  ARCHITECT'S ADMINISTRATION OF THE CONTRACT

     DELETE   Subparagraph 4.2.10, and substitute the following:

  4.2.10  A Project Representative will be employed at the site
  by the Architect. The Project Representative's functions and
  limits are as set forth in AIA Document B-352, entitled
  Duties, Responsibilities, and Limitations of Authority of the
  Project Representative.
```

FIGURE 2.3   Example of a change in the general conditions to the contract, contained in a specification's supplementary conditions. This adapts a standard form to specific project circumstances.

owner wants the professional on the site for obvious reasons, and the contractor prefers dealing with the professional, rather than the owner.

To some extent, the professional will act as an arbiter or buffer, as well as an interpreter of the plans and specifications. These latter functions (which are contract obligations) will be carried out more promptly and satisfactorily with full-time representation.

The professional's on-site role encompasses three areas: owner, contractor, and third parties. The owner expects the professional to represent him/her on the site and to prevent errors or breaches. More affirmatively, the owner expects the professional to see that the contractor complies with the contract documents.

The problem is that the professional's compensation, under normal circumstances, is not fashioned to cover full-time supervision. Normally, the site visits are required only for specific, limited purposes. Hence the visits are periodic, to see that progress is made in accord with the contract documents and schedule. If more comprehensive contract overview is required, the contract must be expanded to reflect the full-time professional presence. In any event, the professional's responsibility does not extend to everything the contractor does. This would impose an impossible burden, a duplication of effort, and in essence make the professional a "turnkey" contractor.

It is obvious that even a full-time representative cannot be in all places at all times. The professional cannot, then, be held responsible for literally "everything" that occurs on the site. Reason must prevail. The owner's ultimate recourse against contractor errors is litigation.

It is necessary to precisely describe and carry out the professional's on-site authority, since this ultimately establishes the liability between the contracting parties, i.e., it determines who is liable for poor job execution and injuries by third parties. There has been lively discussion within the design professions on how much on-site responsibility the professionals are to assume, and how to express it to limit "make good" and third-party liability claims.

One school of thought maintains that words such as "supervise" and "inspect" entail more than the professional intends to cover or is hired to cover. This being so, the argument goes, those words should be deleted in favor of "observe," so as to more accurately reflect what actually takes place at the project. When the professional observes an error, he/she has the authority (via proper channels) to pursue other contract remedies. If the professional does not detect an obvious error, he/she can be held liable. But should the owner or the professional wish to go beyond this rather passive role, the more inclusive words "supervise" and "inspect" can be used. The professional should understand, however, that this exposes him/her to increased liability.

Another school of thought holds that the professional should first consider the policy question (i.e., what is the best and perhaps proper relationship acceptable to the professional between the level of liability, the final quality of the project, and the client/owner involved?) and then decide how to handle the situation. For example, the increase in liability may be deemed worthwhile if the project is prestigious, if the owner is a repeat client, or if the owner holds good promise for future work.

Accordingly, a considerable number of professionals prefer to retain the more inclusive authority implicit in the words "supervise" and "inspect." Moreover, it has been suggested that the cases to date do not indicate monetary losses commensurate with the loss of or change in authority caused by the change in terminology.

There is even a parallel trend to expand services to improve opportunity for the professional. This is typified by some professional firms offering construction management, design/build, or turnkey services. Of course, the risks change with each of these programs. It is possible, under a design/build or turnkey arrangement, that the professional's average standard of care increases to that of a contract commitment, which may include warranties of suitability.

Other trends to increase professional functions depart from the conventional triangular pattern. To that extent, an aura of professional concern associated with the master-builder concept is inapposite. Assessment of professional liability in such an assignment is more appropriately measured with reference to the particular contract provision.

Regardless of format, the professional's on-site duties are normally due and owing the owner. When the professional omits the performance of these duties, he/she may be liable to the owner. However, in the event of omission, there is a correlative concept of whether the owner is liable to the contractor, or for the same reason, whether the prime contractor is liable to the subcontractor.

Clarification of professional construction contract administration responsibilities is the primary way to lessen possible litigation. This involves everything from the basic definition of the words, terms, and phrases denoting the responsibilities to more major issues of control (or no control) over contractors' methods and means of construction, safety and safety law violations and related areas. Progress has been made to confine the liability of the professional, but further discussion in this area is not now left to the individual professional.

It is crucial that the contract administrator be fully conversant with the lexicon and jargon of the construction industry (see Glossary and Dictionary listings at the end of Chapter 6). The administrator must be fluent in these terms since frequently they are different for design and for actual construction work. It is essential that the administrator be able to communicate in the terms used by all the other participants. Quite often the same item or work process will be known by a different term depending on who is referring to it. The exact meaning may require precise wording, because the terms do not always encompass the same range of attributes. The administrator must be careful to use the correct terms properly to refer to the work involved.

DEFINING TERMS

It seems prudent to give some definitions of the material in the remainder of this book. Having attempted a context and definition of contract administration in the first chapter, here we attempt to place additional terms in context.

> **CONSTRUCTION CONTRACT:** An agreement between two parties with a set of standardized provisions, often with modifications, in legal form, which purports or by reference is considered to predict, delineate, direct, prevent, resolve, and provide coverage

Chapter 2   Definitions 31

for all situations, relationships, circumstances, disputes, obligations, interfaces, production, results, and financial matters that will occur when dealing with innumerable relatively unknown and ever-changing conditions, under highly restricted financial, time, and results configurations, and which rely on numerous other agreements and persons for a successful and satisfactory completion.

This definition may be facetious, but is intended to make the point that much is potentially contained in the contract. This gives rise, many times, to unattainable expectations. Even with these considerations, well researched and written as they may be, a possibility for litigation still exists. Therefore, while no contract will be all inclusive, the terms and conditions should be as clearly expressed as possible. In addition, the contract should include an intermediate process for informal dispute resolution short of litigation. In any case, a common understanding of terminology is important.

ADMINISTRATION: The process of directing, controlling, overseeing, arranging, or guiding the completion of a task.

The administration of construction contracts is the final phase of architectural basic services. When provided it has the potential of contributing directly to the success of any project. The importance of contract administration can be first acknowledged by insistence (through a strong negotiating effort, if necessary, by the architect) that it be retained as part of the basic services of the design professional. This will indicate that the parties to the owner-professional contract agree to complete and comprehensive services, which will, more than likely, best serve the owner in the process of constructing a satisfactory project. However, often the owner and/or the professional opt to eliminate or limit contract administration, feeling their financial interests are better served.

The commentary associated with AIA document A201 (General Conditions) notes specifically, at Section 4.2, that "the word ADMINISTRATION is not intended to imply that the architect either supervises or directs the construction effort." This is supported by Section 3.3.1 and its corollary in Section 4.2.3 of the A201 document, where the contractor is named as the party solely responsible for "construction means, methods." In addition, the architect's responsibilities are not without limit; see the complete text of Section 4.2.

Also, the definitions for contract administration found in the following documents should be used: American Institute of Architects (AIA) *Handbook of Professional Practice,* Chapter 9; and National Society of Professional Engineers (NSPE) *Letter Agreement Between Owner and Engineer for Professional Services with General Provisions Attached.*

As a summary of the general language, Article 1.5 of the AIA document B141 is shown in part here:

**1.5 CONSTRUCTION PHASE—ADMINISTRATION OF
THE CONSTRUCTION CONTRACT**

1.5.1   The Construction Phase will commence with the award of the Contract for Construction and, together with the Architect's obligation to provide Basic Services under this Agreement, will terminate when final payment to the Contractor is due, or in the absence of a final Certificate for Payment or of such due date, sixty days after the Date of Substantial Completion of the Work, whichever occurs first.

1.5.2   Unless otherwise provided in this Agreement and incorporated in the Contract Documents, the Architect shall provide administration of the Contract for Construction, as set forth below and in the edition of the AIA Document A201, General Conditions of the Contract for Construction, current as of the date of this Agreement.

1.5.3   The Architect shall be a representative of the Owner during the Construction Phase, and shall advise and consult with the Owner. Instructions to the Contractor shall be forwarded through the Architect. The Architect shall have the authority to act on behalf of the Owner only to the extent provided in the Contract Documents, unless otherwise modified by written instrument in accordance with Subparagraph 1.5.16.

1.5.4 The Architect shall visit the site at intervals appropriate to the stage of construction or as otherwise agreed by the Architect, in writing, to become generally familiar with the progress and quality of the Work and to determine, in general, if the Work is proceeding in accordance with the Contract Documents. However, the Architect shall not be required to make exhaustive or continuous on-site inspections to check the quality or quantity of the Work. On the basis of such on-site observations as an architect, the Architect shall keep the Owner informed of the progress and quality of the work, and shall endeavor to guard the Owner against defects and deficiencies in the Work of the Contractor.

1.5.5 The Architect shall not have control or charge of and shall not be responsible for, construction means, methods, techniques, sequences/procedures, or for safety precautions and programs, in connection with the Work, for acts or omissions of the Contractor, Subcontractors or any other persons performing any of the Work, or for failure of any of them to carry out the Work in accordance with the Contract Documents.

CONTRACT ADMINISTRATOR: The on-site full- or part-time person who is the representative of the design architect or engineer, with duties as defined in AIA document B352 (architects) or EJCDC document 1910-1-A (engineers), who observes work in progress, acts as liaison between the field and the offices, and generally facilitates problem solving and continuous operations. Some of these duties can be shared with or handed over to the construction manager, as provided in the owner's contracts.

This definition shows the direct correlation between the parties who can be involved with contract administration. Their goals are the same, although their titles and methods may differ, as required by their contracts. It is possible to have any one of these parties, two, or even all three as the project demands. Currently, contracts vary widely to meet clients' desires.

SUCCESSFUL PROJECT: A project in which the owner has received the results expected and is satisfied with the services provided, and no claims or litigation remain unresolved.

In essence, this is saying that everything worked well, contractual obligations were met, budget and timing were acceptable, and any problems that did arise were satisfactorily resolved. A successful project also can lead to the owner hiring the design professionals (and perhaps the entire team) in the future, with positive referrals to other potential clients.

PROFESSIONALISM (in the context of contract administration): The simultaneous application of ethical and professional judgment due, in part, to the requirements to work fairly, consistently, and firmly with many participants, each having potentially different perspectives.

In the words of R. E. Onstad,

> Professionalism is a personal attribute that one acquires.
>   It cannot be inherited.
>   It cannot be bequeathed.
> Only they, having made the acquisition, who put to use
>   that knowledge,
>   that skill,
>   and with all of their ability,
>   and complete dedication of purpose,
> Can truly be called a Professional.

Chapter 2   Definitions 33

> **OBSERVATION:**   The fundamental basis of most of the elements of contract administration, in ensuring that all activities are as required. Primary on-site activity of the representatives of the design professionals, and in part, can be performed by the construction manager (where assigned by contract).

The essence of contract administration is for the design professional to seek conformity with the contract documents. Since contract administration itself has been expanded and made part of the construction manager's duties, in varying degrees as the owner desires, the involvement of the design professional also varies.

Realistically, this requires visits to the job site (mandatory under some building codes), which involve reviewing the work, discussing it with the contractors, establishing the status of the work, and being familiar with the work in progress.

This word, *observation,* has been chosen to present a legally sound direction, although one that does cause added liability to fall on the professional. This makes up the contractual obligation of the professional on the site, and is the term used in the standard AIA forms.

Although the term *observation* has been accepted since its uncertain introduction (when the wording was changed from *inspect*), there is still some discussion over it. The search continues for an even better word—both properly descriptive and accurate. Of course, any such term would have to be sound enough to pass legal scrutiny and to receive the endorsement of all interested parties in the construction industry. The following composite definitions of words similar to *observe* are for comparison purposes only:

> **OBSERVE:**   Watch attentively; take notice; watch or be present without participating actively; also, to notice with care; to be on the watch respecting.
>
> **INSPECT:**   Derived from the Latin *inspicere,* to look into; has broader meaning than looking—examine carefully, critically for flaws; investigate and test officially; close/careful scrutiny; prying examination; act of overseeing, official superintendence.
>
> **SUPERVISE:**   To direct and inspect performance of workers; oversee; superintend; to oversee for direction; to inspect with authority. In ordinary use, has substantially the same meaning as to have or exercise the charge and oversight of.
>
> **SUPERINTEND:**   To have charge of; exercise supervision over; manage.
>
> **MONITOR:**   To keep watch over; supervise; scrutinize or check systematically to collect certain specified categories of data.

The most pervasive operative word in construction contract administration is *supervise*. Much has been made, both functionally and legally, of this word and the activities it describes. Documents have been assiduously worded to avoid this word and its impact. The word has severe legal implications in construction, as evidenced by case law and the varied court decisions, which reflect the smallest change in circumstance or function. The professions strongly discourage its use.

One federal court has been somewhat helpful and has provided some insight into the word and its use. In *U.S. Home Corp. v. George W. Kennedy Construction Co.* (610 F. Supp., 759 [D.C.III] 1985), the court ruled that where the owner (and presumably the owner's agent, the professional) "oversees" or "supervises" the contractors' performance (and/or the means and methods of the contractor), the owner shares the responsibility for defective work. However, where the owner just "inspects"[2] the

---

2. This case was decided just prior to the full-blown discussion and implementation of the word *observe* in the 1987 editions of the AIA contract forms, to replace the word *inspect,* which has negative overtones and implications itself.

work for compliance with the contract documents, the responsibility for the sufficiency of the work remains solely with the contractor.

Numerous other court cases have decided slightly different aspects of similar issues. One that is both recent and offers an array of other case citations is *Watson, Watson, Rutland/Architects, Inc. v. Montgomery County Board of Education,* et al., 559 So.2d 168, Ala. (1990).

It is regrettable, though, that all the participants in the construction industry and the courts (who rule on what is presented to them) can't seem to come to some widely accepted agreement, not only on the word *supervise* but on proper wording for all of the documents involved in a construction project. Everything is somehow always open to interpretation; perhaps that is just an inherent human trait. Despite long hours, high intelligence, supposed openness, and rigorous discussions, we still "suffer" from flawed or disputable documents, varied viewpoints and opinions, wide-ranging interpretations, and last but not least, variant court decisions. Often situations are resolved on the most minuscule shred of evidence or change in circumstance; on a tenuous, marginal, or "wobbly" rationale. Nonetheless, legal precedents are created and are relied upon in subsequent cases.

What does this have to do with the individual contract administrator? A personal, professional observation and opinion:

Always thoroughly review and know the contractual nuances and obligations you are to work under. Know (and understand) completely and specifically what is to be done by you; what your rights, duties, obligations, and authority are (and their limits); and always function totally within all these bounds—vigorously, professionally, with your best effort, every contract day!

In simple terms, the function (no matter what it may be called) of the professional in contract administration is to make a series of site visits, both at crucial points in construction and at other times deemed appropriate, to look over the work completed or in progress, in order to establish (1) a professional opinion that said work will meet the contract requirements in the completed project, and/or (2) that said work encompasses the materials and labor required in keeping with the current request for payment received from the contractor. In this effort, the professional acts only within authority given her/him specifically; deals in no matters, such as safety, that are not so authorized; and gives no approval per se, issues no directions, directs no work, suggests or orders no construction methods or procedures. Any and all comment deemed necessary, whether positive or negative, is directed by the contract administrator to the design professional or the contractor's superintendent (if the matter is relatively minor). The entire sequence is to determine *compliance.* How that compliance is achieved is a matter solely in the hands of the contractors, who have a set of contractual obligations in the form of the drawings, specifications, and agreement.

This direction is the true essence of contract administration in fundamental terms. Even here, there are still ambiguous words and phrases that need explanation specific to the project. The administrator simply must know and be aware of the legal entanglements that have been incorporated into the contract by the client, the attorneys, and the principals of his/her firm. These change from contract to contract as the client may require, or as deemed necessary by the firm. Without intimate knowledge of these provisions and how they must be handled, the administrator could easily become immersed in a legal quagmire, both personally and professionally; some actions could be deemed personal actions at odds with and decried by company policy and/or the professional standard, leaving the administrator alone in any defense.

Despite adequate design and well-executed contract documents, the construction process will not take care of itself. In this, though, the professional

- Does not guarantee the contractors' work
- Does not relieve the contractors' responsibilities under the contract
- Must endeavor to protect the owner against defective and deficient work
- Must ensure compliance with the terms and provisions of the contract documents
- Must understand that quality control and assurance, and elimination of defects, cannot occur after completion of the work—it must be ongoing during progress of the work
- Must understand that undiscovered defects leading to damage or injury, even years hence, may still be attributable to construction personnel, including the professional

In essence, the professional and all participating members of the professional staff must understand the contractual provisions for contract administration and its correct delivery. The more preplanning, the better and the more professional the system itself, its delivery, and its impact on the project as a whole. This is necessary for professional success in contract administration, no matter the size or complexity of the project, or the other participants.

Although current document language moves the professional away from the "master builder" mode of operation into what some call a more passive role, this does not imply a lack of activity on the job site. The role has been purposely redefined to avoid court-imposed entanglements and to reduce, to its proper context, the professional's exposure to liability. Of course, different projects and clients will produce varied requirements, so the professional is left to ascertain the best demeanor for his/her business and should see that this is properly reflected in all contract documents. Then the professional and all staffers must operate accordingly on the site. Obviously, the professional is not on the site in a totally passive, nonparticipatory role; but neither is the professional present to manage, direct, or supervise. The professional's function is to (1) collect data, (2) scrutinize for flaws, and (3) watch. With this, it would seem that the word *observe*, which is used to define the function of the professional, should be extrapolated into a more specific meaning.

Observation may be the legally prudent and acceptable term, but the professional should not allow this to be carried on in too casual a manner, for from this activity comes the assurance that the contractual obligations of the professional and others are being met. Should observation become too perfunctory, the various obligations could "slide." Prevention of this can be tied to expanding the professional definition of *observe* to include aspects of viewing (i.e., to evaluate the work for compliance or percentage of completeness). This more specific direction comes from several sources, but its compilation serves to clearly depict what the professional or the on-site representative should be doing during the periodic visits to the site to "observe" the project.

To further the level of mutual understanding among all participants (see Figure 2.4), it might be well for the following (or an appropriate adaptation) to be included in the contract documents:

FIGURE 2.4 Participants and users in construction contract administration

- Owner (Tenants)
- Design professionals (Consultants)
- Construction managers
- Governmental agencies (Building inspectors, etc.)
- Contractors (Superintendents/project managers; subcontractors)
- Lending institutions/agencies
- Product manufacturers (Franchisees/license holders)

The parties to the contract(s) shall agree that for the purposes of clear understanding of the interrelationships of the various actions to be taken under the contracts that the following concepts shall have the following meanings:

Inspection (by the Contractor) is that process by which, through use of his/her own forces, the Contractor assures that his/her responsibility under the contract is being fulfilled in order to prevent rejection, backcharge, or termination of the contract by the Owner.

Observation (by the Design Professional) or *inspection* (by the Owner) is that process by which the Owner is permitted to inquire, for his/her own information, into the adequacy of the system being employed by the Contractor. Any inspection by the Owner shall be without contractual significance, unless such is specifically set forth in writing to the Contractor.

Acceptance (by the Owner) is that act by which the Owner, in reliance upon the representatives of the Contractor, that he has completely and satisfactorily completed the work in accordance with the strict requirements of the contract, agrees that the payment set forth in the contract is payable in full to the Contractor. Nothing in the contract shall diminish any rights surviving to the Owner after final inspection and acceptance.

Figure 2.5 lists key words related to the functions of each party that might be used in contracts.

Since there are other devices and instruments used throughout the construction process, which contribute to the contractual process (but are not necessarily listed in the contract documents), their terms also are best specifically defined. This will further mutual understanding, even beyond those who have direct contractual ties to the project, and with subcontractors, material suppliers, fabricators, and manufacturers. Figure 2.6 shows a sample of the stamp used by many professional firms in the review of shop drawings. The notations on the submittal review stamp should be defined in the contract documents, but are generally accepted as having the following meanings:

1. "Reviewed"—*Final unrestricted release:* The work covered by the submittal may proceed, provided it complies with the requirements of the contract documents; acceptance of the work depends on that compliance.

2. "Reviewed as Noted"—*Final but restricted release:* The work covered by the submittal may proceed, provided it complies with both the architect's notations or corrections on the submittal and with the requirements of the contract documents; acceptance of the work depends on that compliance.

Architect/engineer
Function: ADMINISTER
   Key words: represent, observe, certify, recommend, verify, respond, interpret, review, advise, arbitrate, modify, liaison, final inspection

Construction manager
Function: MANAGE
   Key words: analyze, calculate, formulate, monitor, expedite, aid, coordinate, convene, facilitate, adjust, implement, advise

Contractor
Function: SUPERVISE
   Key words: direct, order, instruct, do, oversee, inspect, comply, locate, purchase, furnish, provide, install, make safe, schedule

Common to all parties
Function: COMPLETE PROJECT, PROPERLY
   Key words: as noted above, plus: communicate, cooperate, coordinate, team concept, directed/mutual effort

---

FIGURE 2.5 Synopsis of functional aspects of construction contract administration.

Chapter 2 Definitions 37

```
┌─────────────────────────────────────────────────────────────────┐
│                   THE SUBMITTAL REVIEW STAMP                    │
│       REVIEWED _____      REVIEWED AS NOTED/CORRECTED _____│
│            REVISE AS NOTED AND RESUBMIT _____               │
│            REJECTED/RESUBMIT AS SPECIFIED _____             │
│                FURNISH (   ) CORRECTED COPIES                   │
│- - - - - - - - - - - - - - - - - - - - - - - - - - - - - - - - │
│           Notations do not authorize changes to contract sum.   │
│                            * * * * *                            │
│                                                                 │
│           Submittal was reviewed for general conformance with   │
│        the project's design concept and information given in docu-│
│               ments. The contractor is solely responsible for, │
│         and this review does not include: confirming and cor-   │
│              relating quantities, dimensions, and fit; selecting│
│             fabrication processes and techniques of construction;│
│                coordinating work with that of other trades; and │
│           performing all work in a safe and satisfactory manner.│
│           Corrections or comments made on this submittal during │
│                   this review do not relieve the contractor    │
│             from compliance with the requirements of the contract│
│           documents or with the responsibilities listed above.  │
│                                                                 │
│      Date _____      By _____ │
│                       XXXX, Architects/Engineers                │
└─────────────────────────────────────────────────────────────────┘
```

FIGURE 2.6 Sample submittal review stamp.

If the design professional desires, when the above notations are indicated, permission may be given for the start of fabrication. Permission can be included on this stamp or given separately; on the stamp is preferred. In the case of the following notations, it should be indicated that fabrication shall *not* start:

3. "Revise as Noted and Resubmit"—*Returned for resubmittal:* Do not proceed with the work covered by the submittal, including purchasing, fabrication, delivery, or other activity. Revise the submittal or prepare new submittal in accordance with the architect's notations stating the reasons for returning the submittal; resubmit the submittal without delay. Repeat, if necessary, to obtain a different action marking. Do not permit submittals with this marking to be used at the project site or elsewhere where work is in progress.

4. "Rejected/resubmit as Specified"—*Returned for resubmittal:* The work covered by the submittal does not conform to the design concept or meet the contract document requirements. Prepare a new submittal in accordance with the contract documents requirements. Do not permit submittals with this marking to be used on the project site or elsewhere where work is in progress.

5. "Furnish Corrected Copies"—*Supply corrected copies:* Where submittals are marked "Approved as Noted" work covered by the submittal may proceed as noted above. However, corrected copies are required for the architect's and owner's files. These are not re-reviewed.

Another check-off can be used, if desired, as follows:

"No Action"—*Other action:* Where the submittal is returned, marked with the architect's explanation, for special processing or other contractor activity, or is primarily for information or record purposes, the submittal will be marked as above.

The shop drawing stamp is but one such issue in which definitions are a tremendous help. It is indicative of the depth to which definition should be taken. This is not to say that the number of specific definitions should be overwhelming or that new, creative, job-specific definitions are required. Simply, all participants in a project must be on the same track for their own understanding and for the good of the project itself. Quibbling over words and interpretations is a time-consuming and fruitless exercise at best, and usually produces nothing more than animosity.

In addition to definitions in the basic contract text, there is a further need for definitions in the project manual. Here there can be refinements and modifications to standard document language, as well as additional administrative and technical definitions. No one seeks to make the project manual a forbidding document, but it should give helpful, clear direction for the personnel.

AIA document A201, General Conditions of the Contract, is a prime example of material included (even if by reference) in the project manual, where added definition is extremely helpful and productive. This document contains a number of definitions throughout its text in pertinent locations: Sections 1.1, 2.1, 3.1, 3.5.1, 3.12.1, 3.12.2, 3.12.3, 4.1.1, 4.3.1, 5.1, 6.1.2, 7.2.1, 7.3.1, 7.3.6, 8.1, 9.1, and 9.8.1, for example. In several of these cases the definition refers only to the context of the base document and is not a general definition, which could be quite helpful if properly expanded and applied elsewhere.

Without any regard for construction jargon (which could be produced as a dictionary of its own), the following is a partial list of other words that can be considered for use in the project manual, usually in the Division 1 sections. Some repeat or modify A201 definitions, others are new:

bidding documents	indicated	allowance
addendum/addenda	typical	directed/requested
bid	contract documents	approve
alternate	project	project site
unit price	testing lab	concealed
design consultant	construction manager	exposed
design professional	installer	owner furnished
owner	architect	engineer
contractor	subcontractor	contract
work	contract amount	contract time
substantial completion	calendar day(s)	change order
modification	shop drawing(s)	written order

This list, while incomplete, indicates the type and detail of definition that is required for true understanding of the project requirements. Some may consider this approach redundant and unnecessary, but more and more General Conditions documents are including some, if not all, of these words. Even though the construction industry is well established, there are still disputes and even litigation over simple words—definitions, meanings, and intentions. One should *never* think that all participants understand a word in the same way.

An excellent case in point is a major suit filed over a two-building complex. The highrise buildings, while slightly different in layout, were built utilizing the same construction systems and materials. However, the painting specifications consistently referred to "building" or "the building"; the specs were compiled on the old "cut and paste" method. The lawsuit, finally resolved in favor of the contractor for several hundreds of thousands of dollars, was based on the premise that these specifications entailed work in but one of the two buildings (because of the use of the singular noun) and did not address both buildings.

Often documents simply will not be able to contain every word, or project manuals will become dictionaries. If a word has but one specific meaning or can be accepted as the dictionary directs, without variation, then, of course, we need not define it further. However, there must be an effort to note, distinctly, where word meanings are varied—where they simply do not mean, in the case at hand, what they may mean elsewhere. Part of this is a product of the regulatory system, where building code or zoning context words often have slightly different meanings from those in the standard dictionary. These differences *must* be noted to prevent needless costly and time-consuming resolution efforts during the progress of the project. Every participant on the project must be "on the same page, at the same paragraph, and on the same word (definition)" for the project to work well.

Figure 2.7 presents excerpts from specifications sections. They show the depth of definition utilized for technical parameters. Such data need to be formulated in a manner appropriate to the project, but these illustrations can be used as guidelines for format and content.

The need for explicit definition is not confined to technical matters, nor to the details of the project and its construction. It extends to the entire concept, administration, and contractual network of the project. Participants need to understand each other and the parameters under which they will (or should) operate. The need for definition extends to the issue of whether or not contract administration will remain as a part of the basic professional services package. Definition is needed even where the service is dismissed, so the proper legal interfaces will survive.

The contractual requirements for architectural services may, of course, be written to totally exclude the professional's involvement during construction. However, at least one source (see Am. Jur. 2d 622, Architects, Section 1) has included "superintend" as being within the practice of architecture. In terms of architectural services, contract administration is a basic service, one that is normally provided. The contractual parties must take specific action to exclude this service; therefore, it can be considered an option by the owner. This holds true even in states where architectural services are required by law for certain threshold buildings. The practice of architecture, in registration law terms, is fulfilled by completing the contract documents and placing the required professional seal and signature on the documents. Most public agencies and clients will opt for contract administration, however. Owners/developers, owners with their own "inspection" staff, speculators, and those simply wishing to reduce the architecture/engineering fee may do otherwise.

Some architecture/engineering firms are very comfortable providing documents only. Legal history makes the point that an architect is not liable to the owner for failure to detect defective on-site installations when he/she is not retained to design the area at fault and the professional's field operations are deleted or limited. Likewise for an architect excluded from contract administration, upon the occurrence of a site mishap. There should be no illusion, though, that this creates a risk-free environment for the professional. Most likely, the professional will be included in any lawsuit, merely to allow the court to assess the work and responsibilities of all parties and their final liabilities. This scenario is explored elsewhere in this book, but suffice it to say here that even with proper definition (as necessary as it is) the quagmire of legal hazards remain, even where services are limited or redirected.

Some firms treat contract administration as necessary but not particularly desirable; many others believe contract administration to be a necessary and indispensable part of the fulfillment of a design. In this latter group, providing contract administration is, philosophically at least, also a way of detecting job site problems of a litigious nature, which can often be solved if the professional is present on a regular basis. Most firms seek to retain contract administration as part of their service package and a serious part of their practice.

Some firms have attempted to become "design only," hoping, in part, to avoid construction-related liabilities. When no contract administration is provided, the architect-engineer still faces the specter of liability premised on negligent design, as shown on the drawings or in the specifications. The matter of separating design

PART 6—SUPPLEMENTARY CONDITIONS

ARTICLE 1—GENERAL

The following supplementary conditions modify, change, delete, or add to the General Conditions. Where any part of the General Conditions is modified or voided by these articles, the unaltered portions of that part remain in effect.

ARTICLE 2—DEFINITIONS

Supplement General Conditions Article 1 as follows:

Architect: Wherever in these Contract Documents reference is made to the Architect, it shall be understood to mean _____ or their duly authorized representatives.

Construction Manager: Person, company or corporation named in the Contract Documents.

Furnish: The term "furnish" is used to mean "supply and deliver to the project site, ready for unloading, unpacking, assembly, installation, and similar operations."

Install: The term "install" means generally, to set in position, connect, test, and adjust for use; includes project site operations such as "unloading, unpacking, assembly, erection, placing, anchoring, applying, working-to-dimension, finishing, curing, protecting, cleaning, and similar, allied operations."

Provide: The term "provide" as used herein means to furnish, install, and pay for.

Installer: An "Installer" is an entity engaged by the contractor, as an employee, subcontractor, or sub-subcontractor, for performance of a particular construction activity, including installation, erection, application, and similar operations. Installers are required to be experienced in the operations they are engaged to perform.

Owner Furnished–Contractor Installed (OFCI): Equipment or components of a system that are purchased by the Owner and furnished to the contractor who shall receive, store, protect, install, connect, and test each item, unless otherwise indicated.

Rock: Hard material, as found in nature, that cannot be dislodged from its bed without blasting or drilling, or continuous use of a ripper or other special equipment. All other material, including paving and paving foundations, is considered to be "earth."

Finish Grade: Required elevations indicated to be achieved during final grading operations. Should finish grades shown by spot elevations conflict with those shown by contours, spot elevations shall govern.

Subgrade: Undisturbed earth or compacted soil layer immediately below granular subbase, drainage fill, or topsoil materials.

Required subgrades shall be true planes, parallel to finished grades, of depths indicated or specified.

Structure: Buildings, foundations, slabs, tanks, curbs, or other man-made stationary features occurring above or below ground surface.

DEFINITIONS

Rough Carpentry is defined as carpentry work incidental to other systems, concealed in place, and used for support or attachment of other construction. Finish Carpentry is defined as carpentry work exposed to view, built or formed and installed in the field, and requiring a field-applied finish.

DEFINITIONS

Gypsum Board Construction Terminology: Refer to ASTM C 11 and GA 505 for definitions of terms for gypsum board construction not otherwise defined in this section or other referenced standards.

FIGURE 2.7 Examples of typical definitions that help to clarify project terms. Suggested for use with others in specifications.

Deterioration of Insulating Glass: Failure of the hermetic seal under normal use due to causes other than glass breakage and improper practices for maintaining and cleaning insulating glass. Evidence of failure is the obstruction of vision by dust, moisture, or film on the interior surfaces of glass. Improper practices for maintaining and cleaning glass do not comply with the manufacturer's directions.

DEFINITIONS

"Paint" as used herein means all coating systems materials, including primers, emulsions, enamels, sealers, fillers, and other applied materials whether used as prime, intermediate, or finish coats.

APPENDIX B
DEFINITIONS

The following definitions related to specifications are presented to clarify terminology used in this Guide:

Bidding documents: The advertisement for bids, the instruction to bidders, the bid form, other sample bidding and contract forms, and the proposed contract documents. The bidding documents also include addenda issued prior to receipt of bids.

Conditions of the contract: Those portions of the contract documents that define the rights, responsibilities, and relationships of the contracting parties and others involved in construction. This portion is also known as the general, supplementary, and special conditions.

Construction contract: The agreement or contract between the owner and contractor for construction of a project, or portions of a project, in accordance with contract documents.

Contract documents: The owner-contractor agreement, the Conditions of the Contract, drawings, specifications, and all addenda issued prior to, and all modifications issued after, execution of the contract. Addenda may be incorporated into the conformed contract documents and therefore may not exist after contract award. Modifications after award are often in the form of change orders or extra work orders.

Drawings: Graphic and pictorial documents showing the design, location and dimensions of the elements of a project. Drawings often contain material lists, material designations, and limits of construction.

Instructions to bidders: Instructions contained in the bidding documents for preparing and submitting bids and obtaining additional information for a construction project or designated portion of a project.

Quality assurance: All planned and systematic actions necessary to provide adequate confidence that a structure, system or component will perform satisfactorily and conform with project requirements.

Scope of work: The document developed in conjunction with specifications which covers the type and location of work, what is to be provided by other than the contractor, and the listing of attached specifications.

Shop drawings: Drawings, diagrams, schedules and other data serving as extensions of the design that are required for manufacture, fabrication and erection of the components of construction and which are prepared by the contractor, subcontractor, or supplier.

Specification engineer: The person responsible for preparation and administration of specifications. The duties and objectives of the specification engineer at various experience levels are described in Appendix F.

Specifications: A part of the contract documents consisting of written requirements for materials, equipment, construction systems, workmanship, and quality assurance. The word "Specifications" in this Guide is taken to be the scope of work and technical specifications, excluding Conditions of the Contract and other nontechnical documents.

Standard: A prescribed set of rules, conditions, or requirements concerned with: the definition of terms; classification of components; delineation of procedures; specification of dimensions, materials, performance, design, or operation; descriptions of fit and measurement of size; or measurement of quality and quantity in describing materials, products, systems, services, or practices.

errors from mistakes in construction fabrication, means or methods, is compounded, however, if the architect/engineer has had no presence on the job site. If the shop drawing and submittals review process has also been eliminated from professional services, the potential for construction incongruity with the document is increased. One publication has advised architect/engineers as follows:

> Rather than abdicate your duties, you should attempt to provide higher quality and more frequent construction review. If you divest yourself of this traditional role, you may find you have succeeded in reducing your public image to that of a mere technician, and in increasing your courtroom visits.

Although this is a view expressed by an insurance company, it probably is more true than not.

There has also been an inordinate increase in the number of third-party (personal injury) claims against professionals. However, two contrasting court decisions leave the issue in limbo, as far as a definitive direction is concerned. This in itself is a concern.

In *Day v. National U.S. Radiator Corp.,* 241 Ga 288, 128 So. 2d, 660 (1961), the court's decision would require a trial judge to examine the contracts between the professional and owner to determine whether the professional had complied with the contract requirements and to determine the professional's function(s). As to the latter, if the examination reveals the function is to protect the owner, the matter of the professional's conduct will not be submitted to the finder of fact (in determining if the conduct contributed to personal injury of a third party).

In *Erhart v. Hummonds,* 232 Ark 133, 334 S.W. 2d 869 (1960), the court held (after lengthy discussion), " . . . the contract further provides: the architect shall have general supervision and direction of the work. He has the authority to stop the work whenever such stoppage may be necessary to insuring the proper execution of the contract." It was a question for the jury as to whether the professional was negligent in failing to stop work until shoring on the east wall was safe for the workers. Since this important issue is still the source of discussion among courts, legal authors, and legal scholars, it focuses on the professional's need to take extreme care in devising contract provisions. There is also a current increased threat to professionals over this issue, in that the Occupational Safety and Health Administration is attempting to indicate that professionals are responsible for job site safety programs, requirements, and precautions. This attempt would severely expand the liability of professionals and is being opposed by professional societies.

Strict liability has not been a major factor in resolving cases of alleged negligence by architects and engineers. Justin Sweet discusses this issue, in part, in *Legal Aspects of Architecture, Engineering, and the Construction Process:*

> If strict liability is based on defective design, it is often difficult to determine whether services are "defective." . . . Even worse, such a standard, if applied to design, especially if the concept is expressed in terms of fitness for a particular purpose will probably become strict, [as] in [the] sense of [being an] absolute. All failure will be attributed to design. The likelihood that the contractor's conduct and other factors may have played a role in project failure, very likely will be ignored.

From this, it is quite easy to see that if a function of contract administration exists, it will be made part of the overall consideration, separate from the contractor's role. A portion of fitness/suitability may be attributed to the contract administration effort, especially if the requirements of the contract documents are not fulfilled. This was well summarized by code consultant Hans W. (Bill) Meier, FCSI, as follows:

> The design professional's stock-in-trade is his knowledge, his judgment, his ability to advise an owner on which material to use and where to use it. . . . An extremely high

Chapter 2 Definitions

degree of professional responsibility is needed to carry a major project through from concept to final acceptance.

The design professional is hired by the owner precisely because of the need for specialized knowledge in making these design decisions. . . .

But . . . if the designer has not clearly specified his choice (or fails to enforce a requirement that he has specified), the actual design decision is made somewhere way down the ladder. . . . if the design professional does not strictly make and enforce the design decisions they will be made and enforced for him and by someone less qualified.

Now if you don't like to make the decisions, you have no business being a design professional. . . . You isolate the problem and arrive at a solution. You decide what the solution should be and you advise your client. . . .

Somewhere within the framework of the contract documents is every tool you need to state and to enforce your design decisions, and if it's not there it's because you didn't put it there.

In the evolving construction scene, the professional must be on guard to ensure that the client/owner does not conclude that construction management is competitive with, or a replacement for, the basic construction contract administration provided by the professional. The professional may have to show that the two programs are complementary and necessary, and that neither can totally compensate for the loss of the other. Some owners tend to place more credence in the construction management work, without realizing the loss to themselves when the professional is no longer involved in the project. Further, some owners miss the fact that each program can perform some functions in better ways than the other. Overall planning, aesthetics, working relationships, programming, and so forth are traditionally best handled by the professional. Cost analysis, scheduling, assessment of construction means, availability of workers and materials, and the like fall easily within the purview of the manager. The owner is the beneficiary when both programs blend and work concurrently. Unquestionably, though, the professional's expertise and insight are indispensable during construction.

EXERCISES

The following tasks are based on the concepts and information in the chapter. All of the material or information required to complete the tasks may not be contained in the chapter text. Independent research is required. Consult your instructor for guidance.

1. Conduct a mini-debate with one side defending the use of the word *observation* and the other challenging with alternative words.
2. The professional intends to use the AIA A201 General Conditions form. However, several changes in the form are required due to the unique nature of the project. What actions are required of the professional?
3. Explain all aspects of *furnish, install,* and *provide,* exactly, regarding a large mixer to be located in the kitchen of a new high school, as per specifications.
4. Explain the contractual purpose, advantages, and disadvantages of shop drawings.
5. What responsibilities do the architect and the contractor have in shop drawing processing?
6. What information do shop drawings show that is *not* normally shown on contract drawings?
7. In what terms does the designer "inspect" work?
8. Define and discuss: level, plumb, straight, even, smooth.

9. List five terms or words you feel require definition or better definition in project specifications.
10. What is the "ultimate goal" of contract administration?

3

The Concept

Each individual has his own kind of living assigned to him, as a sort of sentry post so that he may not heedlessly wander throughout life.

—John Calvin, Swiss theologian

> Design, in architecture, is not self-fulfilling; its expression and intent must be interpreted and converted into terms and documents that are usable by the constructors.
>
> Both the design professionals and constructors then implement the contract documents.
>
> It is this implementation that brings the design to reality, and allows it to be observed and used as architecture!

FIGURE 3.1 How a design is translated into a building.

Many professionals are missing a very important point about contract administration.

It is a prodigious sequence from design concept and documentation to built project. Between these extremes lies not a "magical process" that provides everything necessary for creating reality from a paper exercise, but, rather, a treacherous road, fraught with all sorts of technical and legal entanglements. There is no delicate way to express the process of realizing one's concept of a real-life, usable, occupiable structure. Figure 3.1 focuses on how design is converted into architecture.

CONTRACT ADMINISTRATION: A BASIC SERVICE

Everyone dealing with the project looks toward the start of construction. It is a major milestone and a distinct note of progress. However, it also produces dust, mud, frustration, delivery failures, unforeseen conflicts, misunderstandings, manufacturing glitches, questions about basic concepts/designs, forced changes, added costs. . . . These are all problems of trying to bring a massive amount of isolated, relatively small parts together cohesively and properly to produce the desired result. Often the owner will be thoroughly overwhelmed by the process, having expected a smooth flow of work, no errors, and complete coordination. This scenario needs a knowledgeable moderator who understands the entire process and can deal with it. Figure 3.2 outlines how this can be accomplished.

To meet the need, and unless specifically deleted from the standard AIA Owner-Architect contract (B141) by mutual consent, contract administration is an integral part of the services due the owner. Its cost is included in the basic fee contained in the contract.

Thus, contract administration is not an afterthought or a "paste-on" extra effort, which is an additional cost item. It is, however, a point on which the professional may have to educate the client, so there is an understanding that the professional is not "padding" the contract or seeking added fees. Rather, there is a comprehensive array of services being provided under the basic contract. These services have been formulated to provide at least minimal protection to every aspect of the owner's interests.

The rudimentary concept in the owner-architect agreement is that the architect will act in areas not familiar to the owner, or in which the owner has limited (if any) expertise. There is a direct "agency" effect in this. In concept, the architect takes the place of the owner as various aspects of the project unfold. Basically, this is a "stand-in" scenario, not a replacement; the owner maintains, without reduction of any sort, his/her full status and authority within the total project complex. The owner and the architect will confer and will "act as one," with the architect issuing any orders or other documentation.

There is nothing in the contractual provision for contract administration that can be deleted without directly downgrading the owner's protection, as Figure 3.3 shows. Therefore, contract administration should never be left as an option in the owner's mind. If it is deleted, there will be no direct replacement identical to this service, no matter how any new agreement may be drawn; no one else serves in exactly the same capacity as the original design professional. The level of expertise, the intimacy

I. Prebid
   A. Independent review of plans, specs, and contract documents
      1. Possibly by another designer (architect/engineer)
      2. By staff who will perform construction quality control
   B. Competent cost estimates
   C. Meeting with potential bidders
      1. Review special concerns
         a. Possible soil problems
         b. Safety
         c. Etc.
      2. Review planned quality assurance program
         a. Lines of responsibility
         b. Lines of communication
         c. Methods for establishing change orders, clarifications, etc.
         d. Shop drawings
      3. Record meeting
         a. Possibly issue written minutes or provide meeting agenda to all attendees
II. Preaward
   A. Review bids and resolve any questions or inconsistencies
   B. Conduct meeting with apparent successful bidder
      1. Rediscuss items in prebid meeting
      2. Resolve any questions or concerns about documents, safety, quality assurance, etc.
      3. Record meeting and issue minutes
III. Preconstruction
   A. Get early submittals
      1. Bonds
      2. Insurance
      3. Early shop drawings
      4. Building permits
      5. Schedule
      6. Etc.
   B. Hold meeting with field personnel
      1. Rediscuss major items
         a. Quality Assurance Program (see item C below)
         b. Safety and security
         c. Special concerns (trash removal, etc.)
      2. Hold at site, if possible
      3. May be combined with preaward meeting on small projects
      4. Resolve any questions or concerns
      5. Record meeting and issue minutes
   C. Quality Assurance Program
      1. Staffing
         a. Lines of responsibility and communication
         b. Laboratories or other agencies
         c. Provide project diagrams
      2. Inspection procedures
         a. Who inspects what item
         b. Procedures for nonconformity or rejected items

FIGURE 3.2 The quality assurance program. Review this program to understand what precedes the construction administration phase and how contract administration is a part of this continual assessment program.

            c. Procedures for reporting
            d. Procedures for follow-up
            e. Applicable checklists, ASTM standards, etc.
        3. Submittals (shop drawings, samples, etc.)
            a. Who reviews various items
            b. Timing requirements for submission and review
            c. Responsibility for follow-up
        4. General administration
            a. Filing and log systems
            b. Documentation of paper flow
            c. Procedures for processing—clarification requests, field orders, change orders, pay requests, etc.
        5. Progress meeting
            a. Frequency
            b. Schedule update
            c. Attendance requirements
            d. Responsibility for issuing minutes and performing follow-up
    D. Issue notice to proceed when everything is satisfactory
IV. Construction
    A. Establish one-person responsibility
    B. Follow established lines of communication and responsibility
    C. Conscientiously perform inspections, issue needed paperwork, etc.
    D. Get submittals, samples, mock-ups, and qualification testing early
        1. Provide good review in a timely manner
        2. Reject when defective or incomplete
    E. Process invoices, change orders, etc., quickly
    F. Document work—maintain good files
        1. Photographs
        2. Daily inspections
        3. Test results
        4. Telephone conversations
        5. Etc.
    G. Hold periodic meetings
        1. Record
        2. Issue minutes
        3. Resolve questions or concerns
        4. Note dates for needed responses
V. Closeout
    A. Get a thorough inspection for punch list and date for completion
    B. Get and thoroughly review documents and closeout items
        1. Keys
        2. Manuals
        3. Warranties/guarantees
        4. Lien releases
        5. Bond releases
        6. Final test reports/manufacturers' certifications
        7. Record drawings

FIGURE 3.2   *Continued*

with the details of the project, and the designed interplay of project requirements and construction lie solely with that person. Deleting contract administration by the professional can lead to the loss of valuable information. Things will languish; problems will be unresolved or will have to be "covered" by others, and more than likely at some added expense. The project will suffer, and more importantly, the interests of the owner will be compromised.

The larger implication of the contract administration program is the fact that it keeps the professional "in the loop" (of communication and knowledge) throughout the project. The professional has legal responsibility to ensure the compliance of the actual construction with both the private sector (the Agreement with the owner) and the public sector (the codes and regulations). Although this responsibility is personal in nature (it is imposed on each registered professional who is responsible for a project), it really carries over to all personnel of the design firm involved on the project. Through contract administration the professional is able to function in a reciprocal fashion, receiving information from the project (including shop drawings, requests for changes, etc.) as well as responding, as necessary, for added information or interpretations. This is an essential part of the legal mandate and responsibility, and certainly goes a good way toward ensuring the propriety of the entire project.

The professional's responsibility cannot be delegated, nor can it be cast aside by contract language (for example, declaring involvement only for "plans and specs" and no

---

By contract, the design professional, when acting as contract administrator, is required to function in a number of different roles. Following is a brief listing of the functions under each role:

ROLE	FUNCTIONS
Owner's representative and agent	Explicit: Perform as representative to extent defined in contract documents; if agent, separate written authorization is required denoting extent of authority, including expenditure of funds
	Implied: No indication of an agency relationship; contractor relies on architect/engineer directives as an agent to the owner; owner directs prime professional to perform change order in presence of owner where owner does not object
Evaluator	Assess: Bid submissions; list of proposed subcontractors; quality of work (reject noncompliant); shop drawings/samples (review/approve); tests/inspections ("observe such test"); schedule of values; request for payment/supporting data
Modifier	Issues: Contract modifications; addenda; change orders; field orders (minor change orders)
"Inspector"	Punch list determinations; determines date of substantial completion; certifies final payment
Judge	Assess performance and responsibility of separate primes " . . . determines which contractor . . . "
Interpreter	Address clarification of contract requirements and proper execution of work
Observer	Monitor progress of work ("periodic visits, observes, general compliance")
Certifier	Specific dollar value (progress payments); specific work (substantial completion); specific period
Arbitrator	Acts in court-recognized role in dispute between owner and contractor; cloak of immunity, incidental occurrence

FIGURE 3.3 Legal and contractual functions of the administrator.

construction-related activities), because courts look to the complete array of responsibilities and related features when assessing an errant project. The following is an excerpt from a job description for a design professional of record (either architect or engineer) who is responsible for a project (A/EOR refers to architect/engineer of record):

CONSTRUCTION RESPONSIBILITIES

Since the A/EOR is responsible to ensure that the project is designed in full conformance with all applicable building and other codes and regulations, it is necessary that this person be an integral part of the construction process.

The A/EOR has an obligation to recommend that the proper permits be applied for and received for all work of the project where such permits are required. . . .

The A/EOR has the responsibility, through the professional registration status, to review fabrication drawings and other supplier submittals prepared by others, to check and ensure that they are in general conformance with the design requirements. The onsite personnel shall provide the necessary copies of all required submittals to the A/EOR, in a timely and correct fashion, subsequent to checking the submittals themselves to meet their responsibilities.

The A/EOR is required to make periodic observations of the construction, to observe whether installation and construction activities are being executed in conformance with the construction documents. It is not intended, however, that the A/EOR act as the inspector for construction. To aid the A/EOR in all work during construction, the A/EOR shall receive a copy of the daily construction report. . . .

The A/EOR also has the responsibility of providing construction support in accordance with any specific contractual requirements. No change in the work is authorized until the A/EOR issues an appropriate and fully executed field instruction or change order. Minor discrepancies in the work and/or documents shall be discussed with the A/EOR before any remedial work is done.

Construction site services by professionals are controlled and required by statute or administrative regulations in several states. Foremost among these are Florida, Hawaii, and California; the number of such state regulations is increasing. The states want to ensure the protection of "public health, safety and welfare" and are requiring site presence and observation by the design professionals; efforts to maintain the influence, input, concern and expertise of the professionals in the entire construction process.

Several concerns have been addressed in these efforts. First, the use of the word *supervision* has been given a more specific definition in some cases, which is closely aligned with that used in standard professional contract forms. Also, this work is limited to work completed (checking for proper compliance in the finished project) and is not related to small portions of work in progress. In addition, there is a need for some form of sequence in this effort. If the prime design professional is not under contract (to the owner) for on-site services, there must be a statement filed as to who will provide such services; this does not preclude the prime professional from seeking an additional contract for such services. Some states utilize these provisions only for structures larger than a threshold project. In other instances, the regulations call for a number of mandatory site visits by the design professional (even if a contract administrator is on-site) to verify compliance with the contract documents and the applicable codes and regulations.

While all this may have been influenced by the professionals and their organizations, to some degree, it does show a concern for the construction process, its interface with the public, and the liability exposure of the professionals. No one relishes regulation, but these efforts appear to be rational and ones that serve both public and professionals fairly. This results in the state meeting its goal—more predictable and reliable construction.

More and more, though, owners tend to look for every cost reduction and often make the reductions without commensurate understanding of the consequences. Contract administration is "ripe" fruit for picking when cost reduction is the issue. Although, as noted, it is contained in the basic package of professional services, owners often attempt to delete it, thinking that the costs saved are substantial and the effect inconsequential. Certainly, some owners have staff adequate to perform the contract administration function on their own. Even then, the ongoing presence of the architect/engineer has tremendous intrinsic value, particularly as the person with the most in-depth knowledge of the project. Costs for the replacement of this service may well exceed the apparent reduction. The depth of the professional's knowledge and expertise cannot be replaced, at any cost. Contractually, to the architect, the project is the prime entity; the crux of service, the source of successful contract completion, and the satisfaction of the client. Not to downgrade the owner or owner's input, but the professional is contractually required to remove her/himself from the absolute operative of the owner to the position of agency, where appropriate answers to inquiries or situations are "no," "perhaps," "maybe," and "let's study that." In seeking proper and satisfying completion of the project, the professional must wear several different hats and must carefully exercise the granted authority of each. Cutting off professional service prior to completion (or even implementation) of construction is flatly counterproductive to project success.

PRECEDENCE OF CONTRACT DOCUMENTS

Throughout the normal course of a project, there are instances when parties need to be reminded of various aspects of the work, nuances and changes, and the precedence of the documents. Often this information is crucial to resolving a situation and/or preventing a dispute or claim. Rarely is it specifically noted. Some legal precedents have been established, but specific references are most valuable to an orderly and smooth-running project. Certainly these are preferable to interpretation, dispute, claim, arbitration, and so forth.

Primarily, keep in mind that the contract documents are *complementary*. This means a provision, work item, system, material, and so on shown on any one document is just as binding as it would be if shown on them all. Drawings and specifications are often referred to as "complementary and supplementary" to indicate that they support, relate, and add one to the other, and only consideration of both will provide complete information; not *all* pertinent information is shown on both, individually. Even here there can be discrepancies. Therefore, any gaps, errors, conflicts, discrepancies, ambiguities, or contradictions must be brought to the attention of the design professional, by the contractor, in writing, *before* proceeding with the work involved.

In recent years, the concept of "contract documents" has been reviewed and now is more inclusive and comprehensive. Previously, the definition included merely the contract, the specifications, and the drawings, without explanation or expansion. More recently, the Construction Specifications Institute (CSI), for instance, has offered new insight and an expanded list. The best list is one that reflects the actual on-site situation while the work is in progress. To assist proper interpretation, resolution of situations, and other such issues the following is suggested as being the "proper and best" list of document precedence (based on the in-progress situation):

1. **Change Orders and Work Change Directives.** These take precedence over the agreement since they are approved and included after the agreement is signed, and reflect the work as it is most currently required to be performed.
2. **The Agreement (contract).** This is the Owner-Contractor agreement, signed prior to start of construction, and based on the contract drawings, specifications, and the contractor's bid.

3. **The Specifications,** and the following contents in precedent order:
   - Addenda
   - Contractor's bid/proposal
   - Supplementary general conditions
   - Notice inviting bids
   - Instructions to bidders
   - General conditions of contract
   - Technical specifications
   - Standard specifications
4. **The Drawings,** in precedent order:
   - Numbers over scaled dimensions
   - Details over general/overall drawings
   - Change order drawings over contract drawings
   - Contract drawings over standard drawings
   - Contract drawings over shop drawings (change order required for major deviations in shop drawings)

Figure 3.4 gives a more detailed look at how to interpret contracts, based on the precedence of the documents.

Of course, this precedence and interpretation of the contract documents again points distinctly to the need for education of the owner and elimination of anything that could be considered degradation of the owner's status and protection. Further, it is a necessity for establishing the parameters of the owner-architect relationship over the entire project construction life.

BENEFITS OF CONTRACT ADMINISTRATION TO PROFESSIONAL AND OWNER

An alarming number of professionals too easily dismiss contract administration as being a process that is a nuisance or that adds unnecessarily to their liability exposure. It is crucial that all professionals understand the two primary benefits contract administration provides:

1. A program whereby professionals can protect their liability exposure from unnecessary or unwarranted attack by being active participants who can control much of their own destiny on the project
2. A method of producing a project exactly as planned, designed, and desired by the owner, which is further beneficial to the status and reputation of the professional

These points were recognized by the AIA, and others, in that contract administration was made a part of the basic services provided under the B141 Owner-Architect standard contract form. That certainly removes any doubt an owner may have that this is an added-cost program of service. There must be a conscious move by the professional to explain the extent and terms of the contract and the situation being created, however.

Too often, this entire program is allowed to languish in the contractual framework by lack of appropriate action by the professional and lax enforcement of the contract by the owner. Perhaps this scenario is best described as "an unconscious breach of the contract," which serves neither party. Merely ignoring this basic contract service is unprofessional, and certainly counterproductive to the direction of the contract. What is even more dangerous is the fact that this breach can be the basis of legal action initiated by a dissatisfied owner. Even though the owner was lax in enforcing the contract, the prevailing standard of professional practice may well show the need

> - The contract must be read as a unit (whole). Contract language is presumed to express the intent of the parties. Negotiations can examine and explain, but cannot enlarge the contract.
> - Objective word meanings are preferred over subjective. Ordinary meaning is preferred over technical, but technical controls in a technical context.
> - A lawful interpretation is preferred over an unlawful one; a practical result is preferred over an absurd literal interpretation.
> - Specific terms control over general terms; typewritten controls over printed; and handwritten controls over typewritten.
> - Specifications control over drawings, if in conflict; specifications and drawings are to be read together. They are complementary and supplementary—what is shown or described on one, is as if so done on both.
> - Performance specifications control; design specifications are deemed correct.
> - Trade practice governs contract performance, *but* plain correct language overrides trade usage.
> - Ambiguity is "uncertainty of meaning" and thus is more than simple disagreement; ambiguity will be construed against the drafter.
> - With ambiguity, the contractor's interpretation, if reasonable, controls. *But,* if ambiguity is obvious, the contractor must ask the design professional for clarification, and the professional must clarify.

FIGURE 3.4 Concepts and rules for interpreting contracts.

for appropriate involvement by the professional. That will go a long way, in a courtroom, to showing the shortcomings of the professional. Certainly, one cannot establish a viable defense by saying, "Sure, it is usually a part of the basic services under the contract, but no one really pays any attention to that."

Consequently, contract administration relies on answers to two fundamental questions:

1. Does the owner want to engage the design professional for this service?
2. Does the design professional want to participate in this activity/service?

Although these questions may seem simple, there is a distinct need for resolution of the answers. Certainly, the owner solely retains the option to delete the service from the contract for cause. Then the issue becomes, "Does the owner fully understand all the ramifications that spring from that decision?" In like manner, one wonders if the professional, seeking refuge from added liability, is aware of all of the "downside issues" that could arise from a decision to opt out of contract administration. Even though the expanded legal implications are fairly predictable, does this decision serve the client? Or is this a self-interest determination on the part of the professional? Obviously, some owners more than others are in dire need of this service.

It is more disconcerting that these issues are such an important part of a project and that they occur so early in the process, i.e., as early as selection of the design professional. But society offers other solutions, some of which work in even better fashion for all concerned. Thus, there is a strong rationale for early, in-depth, and complete explanation and understanding of both issues.

IMPORTANCE OF REVIEWING CASE LAW

The case law discussed in Chapter 2, in particular *Hunt v. Ellison & Tanner, Inc.,* sets a course for the professional. (See a fuller discussion of *Hunt* in Chapter 5.) However, the professional's course has many aspects. Any one can be the source of problems (up to and including litigation) for the professional. There is a wealth of case law regarding professional activities.

When these cases are reviewed with others regarding the responsibility of the professional to provide (1) a project suitable to the site and the requirements of the owner, and (2) in full compliance with prevailing building regulations and other applicable laws, the onus on the professional is quite clear—that person is to design the project "properly" and is to ensure that it is executed as designed, approved, and placed under contract.

Following is a list of other cases with brief comments regarding professional design services and legal responsibilities. (Full review of the case decisions is necessary to ascertain all of the circumstances reviewed by the courts.)

Summary of Recent Case Law

In this case the court found that documents and project need not be "perfect" where the contract does not require that:

Seiler v. Osterly, 525 So. 2d 1207 (La. App. 1988)

The professional is "negligent per se" if the building is not designed in conformance with the building code:

Huang v. Garner, 157 Cal. App. 3d 404 (1984)
Burran v. Dambold, 422 F.2d 133 (10th Cir. 1970)

Documents cannot require contractors to "exercise clairvoyance" in noting and resolving ambiguities:

Blount v. United States, 346 F. 2d 962

Forcing bidders to examine "all other applicable codes, etc." to see how they might conflict with the specifications necessitates contractor interpretations that would render the specifications and contract invalid:

Green v. City of New York, 283 App. Div. 485

Work performed in accord with the documents but that proves defective cannot be the basis of a claim against the contractor, who has no authority to alter the documents:

Mac Knight Flintic Stone Co. v. The Mayor, 160 N.Y. 72

Professionals can be held under a shorter term of liability since their "education, training, experience, licensing, and professional stature" will cause them to perform with a higher level of care than contractors:

Zapata v. Burns, 542 A. 2d 700 (Sup. Ct. Conn. 1988)

Basic responsibility for proper preparation of the documents cannot be shifted from the professional to:

the owner—*Simpson Bros. Corp. v. Merrimac Chem. Co.,* 248 Mass. 346, 142 N.E. 9
a building inspector—*Johnson v. Salem Title Co.,* 246, 425 p. 2d 519 (1967)

Chapter 3   The Concept 55

one delegated for such preparation—ibid. at 409

The professional cannot evade basic responsibility for document preparation where the contractor fails to check them or discover the defects:

Chiavernin v. Vail, 61 R.I. 117, 200A. 462 (1938)

Covil v. Robert & Co., 112 Ga. App. 163, 144 S.E. 2d 450 (1965)

The following are other cases that specifically involve issues in the construction phase; for example, a professional failing to condemn and order remedy of defective work:

Skidmore, Owings & Merrill v. Connecticut General Life Ins. Co., 25 Conn. Sup. 76, 197 A.2d 83 (1963)

The professional failing to exercise proper supervisory powers:

Aetna Ins. Co. v. Hellmuth, Obata & Kussabaum, Inc., 392 F.2d 472 (8th Cir. 1968)

Permitting installation of material that was not code approved:

St. Joseph Hospital v. Corbetta Construction Co., 21 Ill. App. 3d 925, 316 N.E. 2d 51 (1974)

Failing to order changes required to make work comply with the codes:

Mississippi Meadows, Inc. v. Hodson, 13 Ill. App. 3d 24, 299 N.E. 2d 359 (1973)

Not consulting a soil tester in conjunction with placing of excess fill:

First Ins. Co. of Hawaii v. Continental Casualty Co., 466 F.2d 807 (9th Cir. 1972)

Allowing work to continue after finding the contractor was using unsafe methods:

Associated Engineers, Inc. v. Job, 370 F.2d 633 (8th Cir. 1966); see Annotation, 59 A.L.R. 3d 869 (1974)

Issuing certificates of payment in a negligent manner:

Aetna Ins. Co. v. Hellmuth, Obata & Kassabaum, Inc., supra

Performing scheduling and coordination in an incompetent manner:

Peter Kiewit Sons' Co. v. Iowa Southern Utilities Co., 355 F. Supp. 376 (S.D. Iowa 1973)

Failing to retain a consultant and check work with the consultant:

Cutlip v. Lucky Stores, Inc., 22 Md. App. 673, 325 A.2d 432 (1974)

Not warning the contractor of general precautions not known in the industry, although the contractor was experienced:

Vonasek v. Hirsch & Stevens, Inc., 65 Wis. 2d 1, 221 N.W. 2d 815 (1974)

Failing to observe deviations from the design when checking shop drawings:

Jaeger v. Hennington, Durham & Richardson, Inc., 714 F.2d 773 (8th Cir. 1983)

Failing to issue a warning notice regarding bankruptcy when paying for material the contractor had in his possession:

Travelers Indem. Co. v. Ewing, Cole, Erdman & Eubank, 711 F.2d 14 (3rd Cir. 1983); not successful

Where there is enough confusion in a public contract that the bidding process may be affected by different bidders reading the contract in different ways, the bidding process is flawed, and the resulting contract cannot stand:

Carbo Construction v. Utilities Authority, 558, A.2d 54 (1989) N.J.

An architect is not liable for errors of judgment, but is liable only when his/her conduct is substandard:

McKeen Homeowners Assn., Inc. v. Oliver, 586 So.2d 679 (1991) La.

As a general rule, when a contractor reasonably relies on the accuracy of specifications when placing a bid, the contractor will be entitled to additional expenses incurred as a result of inaccuracies in the specifications:

Clevco v. Metro, 799 P.2d 1183 (1990) Wa.

The following ruling recognizes verbal change orders. The court rejected a building owner's contention that the fact that the owner had orally instructed a contractor to make changes—which the contractor performed—didn't matter. The court ruled that when alterations have been made with the knowledge of all concerned, the original provisions of the contract may be waived:

Frantz v. Vangunten, 521 N. E. 2d 506 (Ohio App. 1987)

An architectural firm that directed repairs on leaking windows could not rely on the statute of limitations to bar the lawsuit filed against it when the repairs failed:

Sr. Housing, Inc. v. Nakawatase, Rutkowski, Wyns & Yi, Inc., 549 N.E.2d 604 (Ill. App. 1989)

WHEN THE OWNER DOES NOT WANT CONTRACT ADMINISTRATION SERVICES

In analyzing the question as to whether the owner wants to (or should) engage the design professional for contract administration, it is often the case that the owner wants to reduce the overall cost of the project; no other rationale can reasonably be attributed. Where this is done by fairly sophisticated owners, the risk to the professional likely will be reduced. Either the owner has adequate staff to administer the contract as an internal operation, or plans to use some variation of construction management. However, no matter what format may be utilized, there is still degradation of the expertise available to the owner, since the design professional has no input to, nor controlling influence over, the project. The person most familiar with the project concept and interrelationships is excluded from any further participation; chances are portions of the concept and relationships will be changed with no spokesman to properly account for them.

Where the owner is less sophisticated, and/or where cost has been an imposing problem from the outset, the risk is highly escalated. If the owner resists all attempts to retain administration as part of the basic services, one is well advised to see that a clear, firm, and distinct line of contractual demarcation is established in writing. It is necessary to ensure that complete and proper revisions are made to any standardized documents (AIA forms, for example), contracts, conditions, and so on so there is no reference to, or involvement of, the professional in the construction phase of the work. This must be done with extreme care, preferably with legal consultation.

In this event, there must be full and assiduous enforcement of the contractual agreement, in that service to the owner, during construction, is totally withheld. This may seem harsh, but it is here that numerous and severe problems can arise and involve the professional. Even a simple phone call asking for information can result in a situation in which the professional is perceived to have given help, service, approval, or other advice, which was then relied on and used on the project. When problems develop, the professional can easily be drawn in as a contributor to the problem. Of course, it is obvious that such advice, without knowing all of the details involved, is replete with risk from the beginning. This is no place to be a "good guy" to aid the owner. Even answers like, "I don't see anything wrong with that" and "I would think that is okay" can (and will) come back to haunt the professional. The more definitive the statement is, or the more it seems to be part of the professional's expertise, the greater the tendency for the owner to rely on it. The professional must exercise extreme caution in these situations; they are best avoided entirely.

This is not to advocate "blackmailing" an owner into contracting for administration, but it serves as a blatant warning that there is no way in which "bad" contracts (and owners) can be ferreted out and avoided. It is essential that the professional meet the contract obligations, and nothing more. And those obligations should be properly drawn so a contract, fair to both parties, results. With acceptance of each "additional service" there is an inherent professional risk, which requires that a certain professional demeanor should follow. Where that is absent, problems will occur, from irritating misunderstandings on one side to full-blown, destructive litigation on the other.

Two cases show the ease with which problems arise, even with the best of intentions. In the first case, an architectural firm was required to spend several thousands of dollars to have itself removed from a suit regarding a project for which it provided drawing and specifications only. One day the seven-story, highrise apartment building "shed" its outer brick facing. The brick facing was part of a 10-inch cavity wall, which rose the full six stories above the lobby/entrance level. The facing rested on a single ledge angle installed on the end of the second floor slab (the inner course of concrete block rested on the slabs). The brick and block work was so poorly installed that the brick headers stood "free" in voids in the block work behind and provided no masonry tie to the backup.

Evidence showed that the project was thoroughly and properly detailed by the professional firm and had adequate specifications for the masonry work's proper execution. The owner acted as general contractor.

In the second case, a local school board castigated an architectural firm for a project the firm had designed for the board. The board staff was providing the construction services. Professional services had been limited to production of drawings and specifications only.

A casual meeting between the architect and the project superintendent resulted in a caustic "chewing out" (later formalized). Apparently all the dimensions on the drawings were faulty: inappropriate, inaccurate, didn't add up, and so on. The project, now in progress, was in total disarray, and the board was being advised to sue the architects.

Eventually, the problem became clear (at least to the architect). The drawings and specs, prepared in accord with the board's design manual, utilized "modular" sized brick; the school board had selected a "norman" brick for use on the project. This was pointed out to the board's attorney. No suit was filed, and the board did assign other work to the firm in the next few years.

CODE COMPLIANCE

Another problem must be resolved when a client is reluctant to "buy" contract administration, or the professional wants no part of the process. That problem is code compliance.

Granted, this aspect of any project is relatively small, but its demands are eminently imposing; it is the voice of the general public and the law of the jurisdiction. Furthermore, the design professional, in the first instance, carries the responsibility for establishing code compliance and ensuring its full and proper incorporation into the finished project. By making a thorough code search as part of the preliminary design process (programming), the professional ascertains what is regulated and what is required to comply with the regulations. The design function, in part, is to create a project concept in a manner that correctly addresses local law. Case law on this issue is unequivocal.

Both common sense and case law determine that if the professional is responsible for establishing and creating compliance, the professional must exercise proper means to ensure that the finished project contains all that is necessary to comply with applicable codes, standards, regulations, and other such provisions. (See Figure 3.5. In the figure "yardstick" refers to checking or measuring against a requirement, standard, or contract provision.) This general axiom certainly undergirds the inclusion of contract administration as part of basic architectural services; it is a recitation of the legal onus placed on the professional. In addition, most state professional registration laws require that the registered professional practice in a fully lawful manner, and part of that is the "protection of public health, safety and welfare" (incidentally, this is the same charge given to the local building code official or inspector).

It is for the design professional to establish the program of code compliance and then describe it in the contract documents. It is for the contractor to perform the installation of the compliance as part of his/her contractual obligations. However, *both* must ensure that the compliance required is installed in every aspect of the project, or that appropriate alternatives are installed where project conditions change (and this includes the necessary prior approvals from the regulatory agencies).

Shop drawings are defined in AIA document A201 as "submittals" and are not contract documents. They depict how the contractor intends to meet the design professional's design or design concept. Shop drawings are not viable documents for use by code officials in assessing code compliance. The professional is required to produce documents in such detail as to show the method of compliance and the overall scheme of compliance for the project condition. This duty cannot be delegated. It cannot be left open-ended, whereby the specific details are inserted later through the shop drawings of the successful subcontractors or suppliers. That information is

FIGURE 3.5 The professional is responsible for making sure the finished project is in compliance. These terms are part of the process of verification.

VERIFY

V — Variable/view/validate/vouch for/veracity
E — Evaluate/examine/endorse/establish
R — Review/record/reaffirm/remedy/resolve
I — Investigate/involve/indicate/initiate
F — Find out/firm up/finalize
Y — "Yardstick"

Chapter 3   The Concept

made available far too late in the project sequence for compliance review, and will not advance the course of prompt and early building permit issuance.

In the aftermath of the hotel disaster in Kansas City, Missouri (where suspended, bridgelike walkways failed and collapsed), the court in *Duncan v. Missouri Board for Architects, Professional Engineers and Land Surveyors,* 744 S.W.2d 524 (1988), issued four distinct thoughts regarding professional purview and the codes:

> By statute and under the contract, the owner of the building (Kansas City Hyatt Regency Hotel) was entitled to a building structurally sound and safe. . . . The owner did not receive such a building because of the appellants' (the structural engineer) breach of their professional responsibility.
>
> An act which demonstrates a conscious indifference to a professional duty would appear to be a reckless act or more seriously, a willful and wanton abrogation of professional responsibility. . . .
>
> Indifference to the duty is indifference to the harm. . . .
>
> . . . did not meet the design specifications of the Kansas City Building Code. That Code is intended to provide a required level of safety for buildings within the City. It is difficult to conclude that gross failure to comply with the Code can constitute other than conscious indifference to duty. . . .

(Also see the case of *Burran v. Dambold,* mentioned in this chapter and in Chapter 5.)

In this context, problems will arise if the professional is hired for "plans and specs only" and has no contractual obligation for construction-phase service. Who, then, ensures that code compliance will be maintained and fully installed in the work, both as it progresses and in its final configuration? Only part of this situation is within the purview of the government building inspector. The inspector will find the noncompliant work, but will not perform the remedial work required. The owner, who has the bottom-line final, basic responsibility for everything to do with the project, will have to hire some person to both analyze and solve the problem; an added expense both unanticipated and unnecessary! This extra person should be a professional who understands the situation, is able to resolve it, and is willing to accept the responsibility for this isolated work, without knowing every other aspect of the project—a dangerous liability problem.

For the sake of argument, let's assume that a solution is presented and installed. However, later there is a failure of some sort, which has its genesis in the remedial work. Resulting litigation will, in the norm, include the original design professional, who is then in a position of having to defend his design actions, contractual obligations, and so on, at no small cost. The discovery process alone could heavily involve this professional as the attorneys attempt to ferret out "who did what and when." The cost is not something recoverable through normal fee and billing procedures. Obviously, only carefully drawn contract provisions will prevent extensive discovery and involvement.

Included too, for this example, is the owner. Unwittingly a party to the failure, the owner now has no base of support, no scenario that portrays prudent resolution of the problem, unless careful records were kept. That would be unusual when there is no impetus to do so, no professional to do it. This is further aggravated by the fact that having provided no on-site presence via the original design professional, the owner is now suspect in all other actions. In other words, the prime method of ensuring compliance with the codes was removed simply because the owner did not want to pay for construction-phase service by the very person who knows the most about the nuances of the project.

Since the burden of litigation for death or injury is far heavier than for lack of timely completion, dysfunction, or budget overruns, it behooves the professional to aggressively see that contract administration is retained within the basic services contract. The single issue of code compliance is so pervasive throughout the project that both owner and professional are best advised to respect the scenario set forth in AIA standard form B141, Owner-Architect Agreement. And, since it is the design professional who creates the concept, accepts or rejects materials, documents the pattern of compliance, and has intimate knowledge of every detail, it is wisest to retain contract administration even with the construction manager configuration.

The wisdom written into the AIA documents (and those of similar organizations) is undeniable. Part of that wisdom is an attempt to make a legal distinction among terms commonly (and often inappropriately) used. *Inspection,* for instance, is one of those terms.

The Inspection System

Inspect, tr. v. 1. To scrutinize carefully and critically, especially for flaws. 2. To review or observe officially.

Inspection, n. 1. The act of inspecting. 2. Official observation or review.

Inspection—it connotes a problem or some kind of negative experience. The definition itself seems to be directly opposed to the great American axiom of "innocent until proven guilty." It implies that there are flaws or irregularities, that something is not proper.

In construction the inspection system is extremely important to a great many people for varying reasons. All the participants on the project, as well as the public, are direct beneficiaries of inspection. The design professional, the owner, the contractor, and the subcontractor all share an inspection function. All vitally need this function for their own welfare. The government inspectors (building, fire, sanitary, engineering, etc.) provide the necessary safeguards for the public, since the general public is not able to do this for itself; this is a proper function of the political jurisdiction.

Other definitions from more construction-oriented sources soften the basic dictionary definition.

Inspection, n. 1. Examination of work completed or in progress to determine its compliance with contract requirements. 2. Examination of the work by a public official, owner's representative, or others. 3. The process of measuring or checking materials, workmanship, or methods for conformance with quality controls, specifications, and/or standards.

The word *examination* implies a softer approach. It seems to allow for a more cursory look, rather than exacting, critical one. Perhaps the best way to approach construction inspection is to modulate the inspection work between strict inspection and mere examination.

A certain tolerance must be employed in all inspection functions. Nothing is so absolute that it cannot be changed. Compliance with approved documents, manufacturers' instructions, and industry standards is the basic aim of the inspection system. Inspection techniques can be modified to see that this is accomplished, but this does not mean that all inspection must be extremely critical and minutely detailed. Much of the construction material is manufactured and installed with built-in safeguards and factors of safety. The wise inspector knows what is acceptable and molds his/her techniques to be flexible within these safety limits. Examination that is too lenient is just as intolerable as inspection that is too severe.

There is no doubt that the contractors on construction sites do not really want to participate in any kind of inspection program. They feel it is an affront to their egos or an attack on their expertise. They feel sure that they can perform the work in the proper manner, at the proper time, and in abidance with budgetary restrictions. This may very well be true; but with so many interests involved, the inspection process must be constantly functioning on all levels. No participant can allow this responsibility to be ignored because not only his/her best interest but that of the entire project will suffer. An extremely delicate balance must be struck by the inspection system as a whole to allow proper and consistent progress without allowing substandard or noncomplying work. Sometimes inspection can degenerate into an absolute intrusion on the project itself.

Chapter 3   The Concept61

It is important that all the various inspection agencies at least understand one another's province if they cannot come to some kind of coordination and cooperation. Surely, no project or its contractor should be exposed to a sequence of inspections that countermand one another, that demand varying degrees of performance, and that really are counter-productive in the overall view of the project.

There is no doubt that the various interests represented by inspection groups should be active in the project itself. In the end the contractor, subcontractors, and individual workers can rest more easily knowing that the proper safeguards have been taken, knowing that they have produced the project in the best manner possible, and that any repercussions, from simple adjustments to major problems, can be backed up with evidence that the project was built in a prudent, cautious, proper manner.

Without the inspection sequence, no such evidence is available. The project can be suspect, and those who participated in the project can be held liable for the performance or, more properly, the nonperformance of the project.

Inspection can be a constructive element of the project if it is handled properly by the various agencies involved. To know what the other people are doing, to know how far they can go and exactly what their limits are, and to know their proper input into the project can encourage all the inspection agencies to work as a team. In the sports world emphasis is often placed on teamwork. Certainly this type of team effort on a construction project is not only refreshing to all the participants, but is essential to the proper and prompt construction and occupancy of the project.

Unfortunately, the construction "team" never really has time to fully develop and become coordinated. A construction project is built in place and the team must function as best it can while the work is continuing. We do not have the luxury of building, testing, evaluating, and adjusting (repeatedly if necessary) prior to the use or construction of the project. Once the process of construction starts, those involved must proceed with all caution and diligence. Besides, even after the extensive "trial" process, the mere use of the building can downgrade the protection that is built in.

It is essential for all participants in the construction process to understand the necessity for construction inspection. It simply must be done! It is unfortunate that companies and corporations often do things that individuals would not. Our economy is one of "buyer beware"; although this is both philosophical and legal, it becomes quite a problem in the end product. In many instances the buyer can actually oversee the process of the building as the seller is doing the work and building the project. This leads to a situation in which the participants, instead of working as a team, can get into adversary positions.

The adversary condition can be perpetuated from previous projects that were not well run. One contractor, for one reason or another, may have a financial claim, or a lack of respect, or may have had a bad experience with another participant in the project. This is unfortunate; but it is human nature to try to do one's job as one judges it should be done, with a sense of pride and integrity. The variance of judgment can cause problems in cooperation.

No project can successfully tolerate adversary conditions. Whether this is manifested in simple pique ("I'm not talking to you") or open aggression ("I'll get there before he does and do my work—the heck with him"), the overall project climate will be highly charged. This can lead only to frayed nerves and more confrontations. The final project can be affected in a number of adverse ways—not meeting the schedule, low quality work, work stoppage, need to redo work, bad decisions, and so forth. Such a situation must be handled and remedied at the earliest possible time.

It is most important, however, that the inspection sequence be carried forward no matter what the climate of the project, so the end result will reflect the best possible coordinated effort. Whether this is done in a team atmosphere or in an adversary position, the project should be the primary goal of all participants. If it is not, the participants themselves must do some deep soul-searching and discuss problems frankly among themselves to ensure that the project does not suffer because of individual prejudices.

Figure 3.6 specifies the architect's responsibilities and liabilities during the design, contract, and construction phases of a project.

A. What the owner can expect from an architect when engaged or contracted for services:
   1. Perform the services, as set forth in the contract definition.
   2. Changes or modifications to the services, as set forth in the contract definition.
B. Owner must expect changes to the basic concept of architectural services because of current conditions and practices.
C. The responsibilities of the architect (to the owner) are established by:
   1. Contract—written or oral.
   2. If not by contract, by custom and practice of locale.
D. AIA document B141—standard form of agreement between owner and architect (1987):
   1. AIA document B141 does not specifically describe how work will be done.
   2. This is a general professional standard since the architect is dealing with random factors unable to be precisely measured. An architect cannot be certain that a structural design will interact with natural forces as anticipated.
   3. When the standard in the paragraph above is not achieved by the architect, the owner can assert liability on the basis of contract or tort.
   4. Standard imposed by judicial decision—"one who undertakes to render professional services is under a duty to the person for whom the service is to be performed to exercise such care, skill, and diligence as men in that profession ordinarily exercise under like circumstances."
   5. By agreement, this standard may be made more strict or less strict.
E. What responsibilities does B141 impose on the architect? What must the architect do to earn his/her fee? Perform the following basic services:
   1. Prepare schematic design documents and a preliminary cost estimate (Art. 2.2).
   2. Prepare design development documents and an updated cost estimate (Art. 2.3).
   3. Prepare construction documents and update prior cost estimates (Art. 2.4).
   4. Assist owner in bidding and award of contract (Art. 2.5).
   5. Administration of construction contract (Art. 2.6). This administration is performed in conjunction with AIA document A201—General Conditions of the Contract for Construction.
      a. Relationship between B141 and A201 is significant.
F. Difference between design services and construction services performed by architect:
   1. Professional vs. nonprofessional (perhaps).
   2. Active participant vs. acting as observer (Art. 2.6.5).
   3. Control of work vs. noncontrol (Art. 2.6.63).
   4. In the design phase, the architect deals with the owner; in the construction phase, with the owner, the contractor, and the subcontractor.
   5. The design phase deals with concepts; the construction phase deals with the product of the design.

All of the above are areas fertile for the production of disputes, claims, and litigation involving the architect.

Architect's Responsibilities: Construction Phase

A. Significant duties undertaken by the architect in administering contract pursuant to Art. 2.6:
   1. Visit site at <u>appropriate intervals</u> to become familiar with progress and quality of work and determine, <u>in general</u>, if progress indicates completed work will be in accordance with con-

FIGURE 3.6 The architect's responsibilities and liabilities.

tract documents. <u>Exhaustive</u> and <u>continuous</u> on-site inspections <u>not required</u>. On basis of on-site <u>observations</u>, inform owner of progress and quality. <u>Endeavor</u> to guard owner against defects and deficiencies in work (Art. 2.6.5).

    a. What reliance may owner place on this duty?
    b. What reliance may the contractor, subcontractor, or third party place on this duty?

2. Certify contractor's request for payments (Art. 2.6.9).

    a. Compare Art. 2.6.10, what the architect does not represent when he/she certifies.

3. Reject work and require additional inspection (Art. 2.6.11).

4. Review and approve shop drawings, product date, and samples submitted by the contractor (Art. 2.6.12). Note: Review and approval for limited purpose of "Checking for conformance with information given and the design concept expressed in the Contract Documents." Further limitations are set forth in Art. 2.6.12. <u>Pay careful attention to exclusions</u>.

Concept of Liability

A. Liability of the architect:

1. Exposure of the architect to liability is an ever-increasing reality because of:

    a. Type of services rendered.
    b. Expected product or result of services.
    c. Number of people affected by the services.
    d. Amount of money involved.
    e. People's litigious attitude.

2. An architect's exposure to civil liability is based on contract or tort. In contract liability, you must show that the architect breached his/her agreement. In tort liability, you must show that the architect's negligence caused injury/damage to a person or his/her property. Further distinctions:

    a. Contract liability depends on the existence of an agreement and privity. That is, the party asserting a claim must have a direct relationship with the party against whom the claim is made. This is established by determining the parties to the contract or agreement (<u>see Art. 9.7</u>).
    b. Tort liability depends on a negligent act resulting in damage to person or property. The privity requirement is not a factor.

3. Based on the distinction in paragraph A.2 above, <u>architect contract liability</u> involves only the parties to the contract—the owner—whereas <u>architect tort liability</u> may involve the owner and/or third parties.

4. Examples of third parties arising out of the architect/owner relationship:

    a. Contractor and subcontractor.
    b. Workers for contractor and subcontractor.
    c. Lenders and sureties.
    d. Users of the structure.
    e. Passers-by during and after completion of work, etc.

Architect's Liability to Owner During Construction

A. Architect's liability to owner in the administration of the contract:

1. Site visits at appropriate intervals. Art. 2.6.5. limits architect's liability.

2. Architect's liability for contractor's performance is limited—no control over means, methods, techniques, etc. (Art. 2.6.6). This limitation extends to safety precautions.

3. Liability for review and approval by shop drawings only covers checking for conformance with information given and design concept. This limits liability (Art. 2.6.12).

FIGURE 3.6   *Continued*

4. Liable for contract interpretations and decisions if made in bad faith. This limits liability (Art. 2.6.16).

Architect's Liability to Third Parties

A. Third party concept and relationship between architect:
 1. Third parties are all parties not having a contractual relationship with the architect. Hence, third parties lack privity with the architect and cannot hold the architect for a breach of contract.
 2. In the past, this lack of contractual privity was also a defense against third party claims based on negligence.
 3. In 1916, the courts did away with this defense for manufacturers and in the 1950s, architects also lost this defense.
 4. Today, lack of privity is no longer a defense in negligence claims.
 5. When acting in his/her professional capacity, an architect has the duty to exercise such care, skill, and diligence as a professional would exercise under like conditions.
 6. If the above duty (paragraph A.5) is breached and the act of the architect causes bodily injury or property damage to a foreseeable third party, the architect will be held liable under today's tort principles.
 7. Economic loss is not covered by paragraph A.6.

B. Architect's liability to third parties during contract administration:
 1. AIA document B141, Art. 2.6., "Construction Phase—Administration of the Construction Contract," attempts to limit the architect's liability in the construction phase.
 2. Architect's site visits (Art. 2.6.5):
    a. Not exhaustive or continuous.
    b. At appropriate intervals.
    c. Become generally familiar with progress and quality of work (note: no supervision).
 3. Architect has no control over or charge of construction means, methods, techniques, sequences, procedures, or safety precautions (Art. 2.6.6—very broad language).
 4. Shop drawings. Architect's review and approval limited to checking for conformance with information given and design concept (Art. 2.6.12—limited approval given).
 5. Paragraphs 2, 3, and 4 are the areas most used by third parties to impose liability upon the contractor. The architect's defense is that the specific language of the contract relieves him/her from liability and that the areas in question are strictly within the province and expertise of the contractor.
 6. Generally, courts accept the above defense (paragraph 5) provided the architect does not deviate from his/her agreement, for example, supervise or interfere with the contractor's work; does not act in a clearly negligent manner, for example, aware of a serious, unsafe condition and takes no action; or performs one of his duties in a negligent manner, for example, approve shop drawings that describe an inherently dangerous procedure.
 7. When the third party is other than a contractor or a subcontractor, the exculpatory language favoring the architect in B141 and A201 will not be given any weight since the architect and contractor cannot "contract away" another party's rights.
 8. With the present-day expansion of tort liability, the defenses set forth in paragraph 5 should be questioned and the architect prepared to accept greater liability exposure. (Refer to law review article, "Crumbling Tower of Architectural Immunity: Evolution and Expansion of Liability to Third Parties," discussed in Chapter 6.)

FIGURE 3.6 *Continued*

EXERCISES

The following tasks are based on the concepts and information in the chapter. All of the material or information required to complete the tasks may not be contained in the chapter text. Independent research is required. Consult your instructor for guidance.

1. What problems occur most frequently when a design firm supplies only "plans and specs" as opposed to full comprehensive service?
2. Why is contract administration often considered to be an "afterthought" by many owners?
3. Why have the professions included contact administration as part of the basic services package?
4. What is an agent, and what is required of such a person?
5. Discuss the precedence of contract documents and why they are ordered as they are.
6. What is case law? How can it aid or inhibit a contract administrator?
7. Is code compliance a part of contract administration? If so, how?
8. A tractor-trailer load of metal studs and drywall arrives at the site. Upon investigation, the contract administrator sees that the drywall is not a product included in the specifications, and no substitution has been submitted. What action does the contract administrator take?
9. Concrete arrives at the site in the second of what could be an eight-truck sequence. As the concrete begins to be unloaded into the chute, you can see that the mix is extraordinarily dry and not well mixed (segregated). As the contract administrator, how much water do you tell the contractor to add?
10. The professional and the construction manager are both under contract to the owner, but they are not under contract to each other; therefore they are competitors. True or false? Discuss the situation.
11. The professional, via the contract for professional services, is really the "policeman" of the Owner-Contractor Agreement. True or false? Discuss.
12. Is full-time representation on-site part of the normal, comprehensive service contract package offered by the professional?
13. Write a prioritized list of reasons to be given to an owner, citing why on-site representation by the professional is advantageous to the owner.

New Concepts

The past is for inspiration, not imitation; for continuation, not repetition.

—Israel Zangwill, English author

We can say that the contract administration process is one of the last vestiges of the master builder system of biblical and medieval times, if, indeed, that system existed. The "designer" was an intrinsic part of the actual construction process. Old tales of designs drawn on trestle boards or scratched in the sand are still part of our architectural heritage. While these methods may seem ridiculously outmoded, the concept still conveys both the physical and conceptual closeness of the professional to the actual work. And they further note the ability of the professional to ensure that "his" design was executed *exactly* as he conceived it.

This close alliance of design expertise to actual work may have been due to accountability, whether to general law or to the ruler. The designer was "under the gun" to produce. Where the designer was given leave to dream great dreams, we can assume great levels of accountability and the need for direct, close, if not hands-on supervision of the work.

There is little doubt that education, experience, and training varied greatly. The "chosen few" had the ability to dream, to think, and to resolve. Common workers were skilled in narrowly drawn trades. Son followed father, apprentice followed journeyman, seeking to be the "best hands" in the guild (trade), not the greatest mind. Entire generations of families thrived by working on one project, usually with a love and commitment long lost to us. Closeness to the project, then, was both a matter of personal pride and a dire necessity for all concerned.

Unfortunately, over time, many institutions teaching architecture may have contributed (unwittingly or otherwise) to the demise of the master builder concept. Through the axiom "design is everything" the separation has continued to widen. Some advocate innovative design so long as "it looks good—no matter how you build it." A fair question is, "Where does the artist in the architect stop and the engineer begin?" Architecture is a finely delineated blend of art and engineering that produces beautiful buildings, which are fully usable by the owners, safe for all users, and containing all the physical features required to function properly. Yet the academic direction may not deal with all of these issues, not the interface between them. As internationally known architect Michael Graves said, commenting on his undergraduate grounding in architecture at the University of Cincinnati, OH,

> When I went to graduate school at Harvard, I was the only one of 30 classmates who could detail a building. On the other hand, the others knew who Proust was.

Some voices decry the loss of the architect's status as leader of the design team. While not trained in a master builder philosophy, numerous academics and many professionals seek that very position for themselves and their students. Of course, not every project is an award winner, and neither is every building a success for its owners. But when one reads of award-winning buildings that fail, structurally or otherwise, one must wonder what value a design award will have in the litigation that will follow.

A good case in point is a college dormitory that was designed with many innovative features. It was a distinctive and handsome building. However, because of the unique design features, it took political action to build the building. Fourteen refusals were issued by the local building department and the local Board of Building Appeals. The innovative design simply did not address the provisions of the building and other codes. And no compromise could be established. This forestalled construction, until the City Council was approached and eventually ordered that a building permit be issued. Within a decade the building was closed by the city's fire chief, who deemed it "unsafe" functionally and in basic design. It had become a continuing and increasing threat to the hundreds of students it housed. The university shut down the building and sought to remedy the problems. It found that the cost to renovate the structure, to make it safe and code abiding, would exceed the original cost of construction. Shortly thereafter, the building was demolished.

Despite the fact that the architect still has the design award for this dormitory on his wall, was his client properly served? Bad enough the encumbrances of the innovative design, but there was also poor construction. It is evident that the professional's interest lay elsewhere than with his client and with proper construction of his own design concept. There is a deep need for the professional to provide not only comprehensive service, but coordinated, well-executed service, which addresses and resolves every aspect of the project. As Gerald G. Weisbach, FAIA, an architect/attorney in San Francisco, CA, has said,

> . . . the act of architecture is not finished when the design is done, but when the building is built. However, most students see and are taught that a design is the end product.

Too often, the philosophical aspects of architecture contend for attention. Many hold that the "art" of architecture is its primary responsibility. However, unlike a work of fine art, the "product" of architectural endeavor is not merely something to be viewed and appreciated; it is something to be used and occupied; it is the form that supports various activities. It is a three-dimensional mechanism that contains and supports other functions. To think otherwise is both myopic and irrational. That being true, it is difficult to justify current curriculums that supposedly prepare architects for their chosen profession, but do not include the realities thereof. See the 1996 report, *Education of Architects and Engineers for Careers in Facility Design and Construction,* by the U.S. National Research Council, Board of Infrastructure and Constructed Environment. This booklet is an excellent resource for explaining the differences between training and educational programs for the professions and other construction-related personnel. It speaks to the differing titles, academic curriculum content, and overall philosophical tone that set the direction for the graduates. Pointedly, it shows the basic differences between the parties, as noted in "The Context" in this book (see pp. xxi–xxiii), which they bring to the construction project site.

In general, graduates from professional programs are ill-prepared for the process of construction contract administration. Usually they do not have enough basic academic understanding of professional practice, the legal ramifications of the profession, and the technical expertise required to "run work" in the field. Obviously, "after-education" is mandatory for any person or firm seeking to undertake contract administration. The process is far too complex to address casually through on-the-job training. Surprises are commonplace, what with ever-changing construction procedures, methods, and materials. One must be a functionally astute professional who understands design and how design concepts are conceived, developed, and brought to fruition. Respect for the design concept must be carried through every phase of the work. The epitome of good contract administration is faithful execution of the design concept.

In addition, most academic training does not expose the design professional to the legal ramifications of professional practice. There may be passing allusion to legal requirements, but there is no directed, in-depth study of how the law affects professional practice. This lack has led to professionals being quite surprised by some of the suits that befall them. Contract administration, properly undertaken, can greatly simplify this scenario.

Academic training should include the basics of the legal parameters within which the practice functions. Students should review and discuss case law and understand how these decisions can be used as guidelines for professional practice. In addition, the law distinctly sets forth the professional's responsibilities. Without such information, the new professional will embark on a precarious path, one that will more than likely find its way into a courtroom.

For too long, the major emphasis in architectural education has been placed on design, and not commensurately on those other activities that are necessary to bring any design to fruition. But these "other services" will make or break the project, and

> The case of 2314 Lincoln Park West Condominium Assoc. v. Mann, Gin, Ebel, and Frazier, Ltd., 555 N.E. 2d 346 (Ill. 1990), was a landmark decision. The court found that a professional's potential liability is much narrower when third parties (not party to the construction process) are involved. The plaintiffs' contention was that their economic loss (not personal injury) arose from the architects' negligence. The court, however, using the economic loss doctrine, found their loss was one of quality, not safety, in that their grievance came from dissatisfaction with the building design and its failure to meet their expectations. Therefore, the court ruled, the claim would be better resolved through contract law rather than a tort action regarding architectural malpractice for negligence. This established a significant legal precedent that greatly restricts professional liability.
>
> In some jurisdictions this claim would be disallowed since there was no privity (contract) between the parties. The architect's duty would lie with the original owner, not subsequent owners.

FIGURE 4.1 Restricting professional liability.

with it the professional. The failure of schools to instill a well-rounded approach to architecture is most unwise, even if there is a limited budget and limited time.

For example, professionals are rarely exposed to litigation because of "poor design" (see Figure 4.1). Faulty design of elements that fail and cause damage or injury will result in such action, but courts are never asked to scrutinize the aesthetics of a project nor to award damages for incorrect use of design principles. Some may say that this needs to be done, but as long as design is a subjective determination, there will be no standard against which to compare it; it will remain a "meeting of the minds" between client and professional. However, how that meeting of the minds is transformed into a real and occupiable structure is a far different question.

To this end, it would seem advisable that there be some alignment of the curriculum with contractual breakdown, i.e., roughly equating instruction with the percentage of time and fee allotted to each project phase. Some hold that the design process takes more time (hence, more fee) because of the involved and extensive research and conceptualizing it requires, as opposed to the more mundane, standardized processes of drafting and construction. But this is not necessarily true. Also, where is the greatest risk to the professional and the practice?

A realignment of courses would provide for at least minimal attention to project and contract administration. Here, the process of realizing the project in real terms can be explored and analyzed. Although not covering "everything there is to know" about construction and contract administration, this will provide a proper perspective on which the student can build a healthier and more well-rounded philosophy. This is essential to developing a fully trained professional who is cognizant of all aspects of professional practice. When able to understand the project over its entire sequence of development, the professional will tend to approach the work in a more even-handed, expert manner. The end result can only be projects of higher overall quality.

DEFINING RESPONSIBILITIES DURING CONSTRUCTION

It is distressing that contract administration is given so little attention in schools. Perhaps the best impetus toward better training and understanding of this aspect of practice is the motto of the U.S. Navy submarine school in Groton, CT: "If you think training is expensive, try ignorance."

No doubt some young professionals may be able to work successfully in the area of construction contract administration early in their careers. Usually though, it is the more experienced professional, one who understands the technology and detailing of putting a building together and who has field experience, who will find the greater measure of success, and more quickly. This person will be the more effective administrator, the one who understands the value of the system and how it contributes to the success of the project.

Despite what some practitioners think, and despite what some schools present as the "state of the art of architecture," those who detail and document projects and those who see to proper and faithful execution of those documents also serve—and serve well—the profession. Clients talk among themselves; no professional firm can have better advertising than the recommendation of a pleased and satisfied client—one who has received *full* service throughout every aspect of the project. That includes construction contract administration.

Recently, and as a direct result of the Kansas City Hyatt Regency disaster (see Table 4.1), there have been attempts to delineate professional responsibilities on construction projects. Of course, any such undertaking by a particular professional group results in perceptions of biased or misplaced priorities; that is most unfortunate. The complex network of responsibilities on the Hyatt project, and indeed on any project, needs some direction to eliminate gaps, overlaps, confusion, professional pique, and jurisdictional battles. Still, there must be some flexibility, since not every project is produced with the same contractual configuration.

In addition, it must be recognized that quality in the built project is an ongoing process, not merely a singular determination by one party or another. Quality pervades every aspect of work on the project; it is neither a design criterion nor a financial consideration.

There is also an interesting relationship between contract administration and quality in the built project. The *Manual of Professional Practice: Quality in the Constructed Project,* published by the American Society of Civil Engineers (ASCE), defines quality, in part, as *"meeting the requirements."* This refers to the requirements of the owner, the design professional, the constructor, and the public. Overall, it alludes to adequacy, timeliness, fairness, and reasonableness. What is required is a diligent and continuing effort; or, as the design professional may call it, contract administration.

Of course, this program does not necessarily guarantee success, as all of the players must participate fully for that to occur. But it does show distinctly that a quality project is not realized with minimum effort or restricted services. For example, if a decision must be made in a timely manner in the schematic design phase, then a similar decision must be made promptly in the construction phase. Where that does not happen, both success and quality are inhibited, if not lost. The flowchart in Figure 4.2 shows the team-building process that results in a successful project.

Construction creates a commonality of purpose, but it does not provide a system whereby one participant watches out for the interests and rights of the others. A continuing, watchful presence is required to ensure that both the project as a whole and the requirements of the various individuals are fulfilled. A fully successful project demands full participation by all parties throughout the entire project sequence. In this way, true project quality, as defined in the ASCE *Manual,* will be achieved. In the words of Henry Ford,

> Coming together is a beginning;
> Keeping together is progress;
> Working together is success.

Some Sample Contract Definitions

Contract administration is the primary, if not the sole, vehicle available to the design professional to ensure (1) completion of the project in full accord with the approved

TABLE 4.1 Responsibilities of the owner, architect/engineer (A/E), and contractor/construction manager (C/CM). After the Kansas City Hyatt Regency Hotel disaster, where suspended walkways collapsed and hanging bridges fell, the breadth of professional responsibility became a major issue. This table specifies who is responsible for the various tasks: P denotes primary responsibility; S denotes secondary; O denotes none.

Work Item/Task	Responsibility		
	Owner	A/E	C/CM
Preparation/mobilization for construction	O	O	P
Insurance requirements: verify/monitor	P	S	S
Contracts for contractors: prepare/process	O	S	P
Provide full-time, on-site coordination	O	O	P
Prepare schedule for beneficial occupancy	O	O	P
Supervise or observe contractors' work	O	S	P
Assess for compliance with contract documents	O	P	S
Make interpretations of plans and/or specs	O	P	S
Update construction schedule to reflect progress	O	O	P
Process and control shop drawings and samples	O	S	P
Review, check, and approve shop drawings and samples	O	P	S
Certify payment requests from contractors	O	P	S
Disperse payments to contractors	O	O	P
Provide cost control for project	O	O	P
Convene job meetings	O	S	P
Bulletins for contractors: prepare/process	S	P	S
Approve quotations for bulletins	P	S	S
Issue change orders	O	P	P
Oversee and administer: safety program	O	O	P
security program	O	O	P
quality control program	O	S	P
Maintain "as-built" drawings	O	P	P
Coordinate owner occupancy schedule	P	S	P
Prepare "punch list"	O	P	P
Oversee completion of punch list	O	O	P
Certify substantial completion	P	P	P
Secure regulatory agency approvals	O	P	O
Secure certificate of occupancy	O	O	P
Demonstrate operation of systems/equipment	P	S	S
Start-up and recommended maintenance	P	S	P
Submit operations manuals and warranties	O	P	P
Inspect for final compliance	S	P	P
Final accounting, determine final payment	P	P	P
Final review, approval, and acceptance	P		

contract documents, (2) the complete satisfaction of the owner, (3) proper fulfillment of the professional's concept and intent for the project, and (4) equitable resolution of all contractual obligations. In this respect it may be well to consider the following sample statement and the associated definitions that refine the statement. These definitions purposely differ from AIA and other documents; they are presented to clarify other sources.

STATEMENT OF PURPOSE

It is the owner's intention to pursue a contract that will produce the project, as depicted and described in the contract documents.

Unless properly authorized and executed modifications are issued (as provided in the contract documents), the Owner's expectation is that the project will be constructed

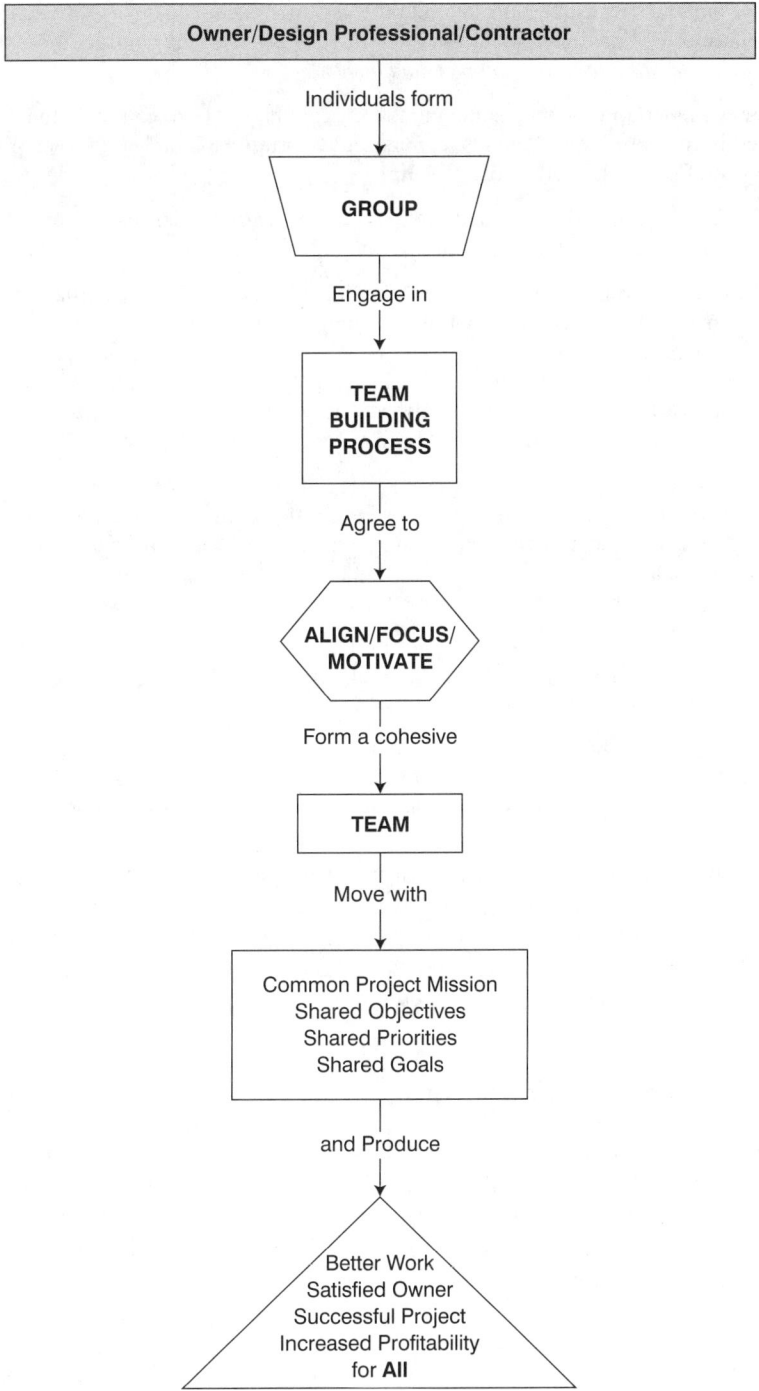

FIGURE 4.2 The team-building process leads to success.

exactly as depicted and as described, and turned over to the Owner, at Final Completion, in a fully finished, fully operational, and "new" condition.

ARTICLE 2—DEFINITIONS

Contract means the bid offer of the contractor, *totally responsive to all of the work*, conditions, responsibilities, and obligations of the Contractor contained in the contract documents, as officially accepted by the Owner, evidenced by the Letter of Intended Acceptance, executed Owner/Contractor agreement or contract, and the other executed finance encumbrance documents.

Contract amount means the sum stated in the contract, including any authorized adjustments thereto; it is the total amount payable by the Owner to the contractor for the *performance of the work under the contract documents.*

Contract completion time means the number of calendar days to complete the work, as specified in the contract documents; compare to Substantial Completion. (See pages 75 and 223 and Figures 4.3, 7.1, 7.26, and 7.30.)

Calendar day means a day of twenty-four hours, measured from midnight to the next midnight, and includes Saturdays, Sundays, and holidays.

Contract documents means *all* of the work included, depicted, described, covered, intended, necessary, or required by the agreement between the Owner and contractor, the payment of a Performance Bond, the General, Supplemental, and Special Conditions, the drawings, the specifications, all addenda, and modifications and Change Orders issued after execution of the contract. The contract documents complement each other; whatever is required by one shall be as binding as if required by all.

Work means and comprises the complete scheme of construction *required by the contract documents.* It includes *all* labor, material, systems, tests, ratings, devices, apparatus, equipment, supplies, tools, adjustments, repairs, expendables, aids, temporary work and/or equipment, superintendency, inspections/approvals, plant, releases, and permissions required to *perform and complete the contract* in an expeditious, orderly, and workerlike manner.

Work order means a written notice from the Owner to the contractor, authorizing the contractor to commence work under the contract and establishing the beginning date from which the contract completion time shall be established.

Change order is the *sole means for authorizing a change* in the work, an adjustment in the contract sum, and/or an adjustment in the contract completion time; it entails a written order to the contractor signed by the owner, issued after the execution of the contract.

Some may say this is stating the obvious. However, there are far too many incidents of taking liberties, when contractors or suppliers make changes or substitutions outside the provisions of the contract documents—at will. In some cases, these changes can force other modifications, which can adversely affect the final project and, perhaps, the owner's operation.

The Importance of Scheduling

There also is a tendency to misunderstand the timing and sequence of substantial and final completion. Obviously where remedial work is required in response to the "punch list," time must be allotted. Usually this occurs between substantial completion (when the punch list inspection is done) and final completion (when everyone is expected to leave the job site pending occupancy by the owner). Too often, if every contractor and subcontractor is given the identical completion time, everyone will work until the last minute. So rougher work will be done after finished work is "finished," only to create more remedial work, and often disputes among contractors. This can be avoided by staggering dates for final completion and allowing the finish trades to finish without the need for further touch-up work. Most of the time the contractors fail to understand this process, leading to a last-minute period of absolute chaos.

Most of this is done in a rather off-handed manner, usually to expedite work and without regard for ancillary impact and contractual obligations. However innocently done, the lack of forethought is of concern. This shows precisely the context of and need for construction contract administration.

Although some of these considerations should be resolved early in the process and should be fully reflected in the contract documents, changes in concept, scope, site conditions, availability of materials and systems, and other extraneous circumstances do continually play on the project's course. Minor changes in the sequence of construction are commonplace and to be expected. Of course, where entire concepts are disrupted, major impact results and extraordinary measures are required to steady the project. In this scenario, it is unwise for the owner to be alone in his/her dealings with the contractors. There is a distinct need for an impartial, third-party voice, i.e., the design professional who knows "everything" about the project, including how the owner's desires were incorporated into the work. Without the professional present, much can be lost, leading to a final result that can prove annoying or unsatisfactory to the owner ("it is not what I expected").

Defining the Professional's Functions

Part of the function of the professional is the ability to compare project conditions with the owner's desires and with the work specified on the documents. No other party has this three-dimensional perspective of the work, nor the ability to assess any condition from various angles. This cannot be emphasized too much, especially to a reluctant client. Further, this requires that a broad range of expertise be brought to the project, which is done almost solely by the design professional; others have some of the same attributes, but none match them exactly.

In the overall analysis of contract administration, one can observe that this program requires knowledge and expertise in many areas, namely, management and administration in:

budgeting	project control
contracts	quality
decision making	resource
information	risk
materials/systems	scheduling
value	safety of the work

Each of these areas appears on every project to some degree (see Figure 4.3). Of course, there is a direct correlation between the size of the project and the scope of each of these project areas.

In addition, it is necessary that these project elements be part of *every* phase of the project—feasibility, programming, schematic design, design development, contract document production, bidding/negotiation/award, construction, occupancy, and warranty/guarantee. There is a crucial mixing of the owner's requirements, their complete evaluation, their incorporation, and their impact on each of the listed project areas.

Although not responsible for scheduling, the professional should monitor progress and make a strong effort to forestall the contractors' tendency to believe that they can work up to the last minute of the last day of the contract term. Work must be scheduled to allow for those who, by necessity, must follow and who must also meet the contract terms. Therefore, the following statements, if made mandatory procedures, can be excellent aids.

The contractors shall plan, schedule, staff, and work the project in such a manner that the only work performed during the period from substantial to final completion is remedial work. Such work shall be done solely to satisfy the list of discrepancies ("punch list") produced by the architect's inspection at substantial completion.

Substantial completion shall occur no less than thirty (30) calendar days prior to final completion (end of contract term/time).

FIGURE 4.3 One of the professional's responsibilities is to coordinate progress on the job.

The design professional is the person responsible for formulating the entire project scheme. Anyone even minimally familiar with the professional's work phases can appreciate the enormous amount of information that must be processed, resolved, and finalized in every area of impact. Schematically expressed this would be as follows:

$$\begin{matrix} \text{OWNER} \\ \text{REQUIREMENT} \end{matrix} + \begin{matrix} \text{WORK} \\ \text{PHASE} \end{matrix} + \begin{matrix} \text{MANAGE/ADMIN.} \\ \text{AREA} \end{matrix} = \begin{matrix} \text{DECISION/DESIRED} \\ \text{RESULT} \end{matrix}$$

This, of course, is a dynamic, ongoing process that cannot be ignored. It is the skeletal framework on which the project is hung. These activities, so necessary to the delivery of a successful project, must be realigned to meet the requirements of project management, design, contracting, construction, construction coordination, and of course contract administration. The design professional is the only person who is active in all this *and* who has the expertise to work through the process.

Hence, design professionals have an interest in the construction of every project they design, not just the production of "plans and specs." Many times, professionals have been dragged into litigation (particularly when a disaster occurs) even though they only produced the documents and had no involvement in, much less control over, the project during construction. The onus on the professional may become much more imposing where contract administration *is not* part of her/his contract. In that case, the documents produced, which then have no professional control, had better be complete, meticulously detailed, and virtually error-free. Any subsequent discovery of "shortcoming" in those documents will invite interpretation and resolution by someone other than the original design professional. The original professional can also become involved when it is necessary to sort out responsibility and liability. All this is a very knotty situation to untangle in a courtroom.

On the other hand, the professional who provides the complete array of basic services has a long-term and acute interest in the project. For both contractual and personal reasons the design professional will carry the burden of the project to the finale. Here, too, the professional is well advised to ascertain all of the legal requirements that surround him/her. Most of the legal coverage is aimed at identifying situations that might occur and, hence, should be avoided. With legal understanding and control of the project per the contract, the professional is in a position to directly influence what action is taken, when, and in what form. To attempt contract administration in any other manner is risky at best and professionally fatal at worst. For success, to the professional, lies beyond merely a complete, fashionable, owner-satisfying project; it also lies in the enhancement and continuation of the professional's skill, reputation, and career.

SUCCESSFUL ADMINISTRATION OF CONSTRUCTION PROJECTS

The concept and, indeed, the successful administration of construction project contracts involves at least three essential elements:

1. A set of well-prepared documents, thorough, accurate, well coordinated, clearly depicted and described
2. A commitment to a proper level of enforcement of the provisions in the documents and that work specified in the documents is expected in every detail
3. An understanding and respect for the trades involved, their training and expertise

This being true, it is clear that a project should not be left unfulfilled for the lack of a comprehensive program of contract administration. Usually only two persons are involved in the project from inception to finale: the owner and the design professional. Obviously, they are the only persons, then, who participate in, formulate, and work through all of the elements noted above. But the professional is *best* qualified for this. Having a feeling for the overall, general "flavor" of the project, the professional is sensitive to every nuance of the project, and this essential information must be conveyed by the professional for the owner/client.

The Professional's Relationship to the Project

This may be the place to review the relationship of the professional to the project. Both by the length of time involved and the intimacy developed by working on "every detail," the professional develops an affinity for the project.

Several conditions, however, play against the provision of full-range architectural service. First, of course, is the glamour and prestige assigned to the designer of the project. This is deeply instilled in the student architect from the very outset of the educational process. Yet being both artist and engineer, the architect must develop a balanced perspective. This should start the first day of class, so each student knows the profession overall. The student should be permitted to observe and understand the range of tasks in the profession.

If, for example, a firm is replete with excellent designers and enjoys a reputation for good design, the firm must also have the ability to execute its designs, providing the client with the intended function, space, and relationships, as programmed. Should any of this falter, the client will look askance upon the best of designs. In the same light, well-built but poorly designed buildings fail. No client wants to be known as the owner of "the ugliest building in town."

The designer, of course, fits all of the program pieces into a coherent functional, and beautiful package, after extensive questioning of the client and analysis of the needs/desire. Once the overall scheme and concept is set forth in the preliminary design, the designer may well be recycled to another project. When the preliminary design is accepted and approved by the owner, the firm's concept has been embraced by the client and the creation can move toward reality.

Some will find the project most challenging when they are charged to "detail" the project. At the same time, they must understand and retain the overall concept of the approved design. This can be a fascinating challenge, particularly where an unusual concept is involved. There can be excitement in maintaining the excellence of the design while providing a safe, substantial, and functional building.

From this juncture, the matter of contract administration rises for resolution. Some professionals will take great delight in the field execution of the design concept and the contract documents. They revel in the hands-on relationship to the "full-size" project, from excavation to final punch list. Their satisfaction lies in the conversion of the paper information into reality. Their work is as vital to the integrity of the project, its design concept and documentation, as any other process in the sequence.

The praise and awards for a project will go to the designer, the principals, the firm as a whole, or perhaps to the project architect. At the same time, the contract administrator can walk away from the finished project with a deep sense of satisfaction and both personal and professional accomplishment. This is not to cast the contract administrator as an unsung hero, but rather as one who has skillfully performed the assigned job, quietly, unobtrusively, and professionally, and has succeeded. The design concept is faithfully executed and exists in the completed project. While perhaps not as flashy as design award plaques or even changes in salary or status, the personal feeling of satisfaction can be a reward.

Not everyone, though, takes the same level of personal pride in the same things. This is most apparent in the production of "plans and specs" projects, where the professional retains no control over anything that happens once the documents go out the door. A valid idea, perhaps. However, the entire design concept can be sacrificed for some reason, with no one there to defend or support it, even to the client. Detailing can become murky and ugly for want of control, discipline, and evaluation. Even a construction manager with proper expertise has a certain level of detachment from the heart of the project; the design professional is not so detached. This simply reflects the more pragmatic approach construction managers take to projects, as opposed to the more "passionate" approach the design professional often takes.

Construction contract administration is a program that often neither professionals nor owners want. The professionals simply do not want the extra involvement and effort, because of the number of problems, and the added liability of trying to deal with those problems and successfully building the project. The owner often misunderstands that the program is not an added service for which an extra fee is required. Hence, the owner seeks to reduce the professional's fee by deleting the administration work from the service contract—the very work that will ensure a successful construction process. Both of these views are shortsighted and for the most part shortchange the very parties who directly benefit.

The professional, for instance, seeks to reduce liability exposure by merely producing projects in the form of "plans and specs." While this is a good way to produce architecture that enhances one's design stature, it does not necessarily produce projects that will be strict manifestations of the design. Rather, it produces projects that are easily sublimated into "non-plus" projects, not examples of design excellence. Too often, other factors play into the decisions of constructing the project and do so outside the original design concept; this is a professional tragedy. But since the professional opts to have no control during construction of the project, it will continue to happen. The professional cannot seek to produce projects as paper exercises and also expect faithful reproduction of the design concept. The chances for disruptive influences and changes are great.

If the professional is truly interested in faithful execution of the design concept and in faithful fulfillment of the contract obligations, he/she must seek, even aggressively endorse and in a sense demand that contract administration be retained as part of the basic services provided, as written in the standard contract (see Figure 4.4). Then the project, professional, and client are served to the highest degree.

Owners seeking to eliminate this program usually have no other motive than to try to reduce the professional fee. However, says Gerald G. Weisbach, FAIA, architect/attorney in San Francisco, CA,

> ... as costs go up, expectations for performance increase, leading to conflict over satisfaction of those expectations.

In place of administration by the professional, owners plan to use themselves, their staff (qualified or not), or some other surrogate. Usually they will not succeed and will wonder why. Others, almost without fail, are too close to the project (from the owner's perspective) or serve a self-interest divergent from the owner's best interest. Only the professional, by training and expertise, is qualified as the independent third party who

FIGURE 4.4 The six essential elements of project delivery.

1. Project management
2. Design and documentation
3. Contracting
4. Construction
5. Construction coordination
6. Contract administration

can operate in an objective, all-encompassing mode, seeing both sides of problems and their correct solutions. One may say that the professional also is self-interested in that he/she will seek to preserve the design as a cherished commodity. This, though, fails to observe the fact that construction of the project proceeds *only* after the client has approved the design concept and has allowed the project to be bid and built. Any "selling" of the concept has already taken place, prior to production of the contract documents. Construction should be viewed as a joint venture by owner/client and the professional. No other party to the project has this relationship with the owner.

Herein is the rationale for contract administration by the design professional. It may well be that the client has other ideas for the execution of the project, but certainly it is worth the effort, on the part of the professional, to remain as an integral part of the design/construction team with well-defined authority, tasks, duties, and responsibilities. Every party to the project will benefit from the presence of the professional. Again, while interests may vary, there are strong elements of prestige, pride, and success in being part of a well-developed project; the professional greatly aids in developing those attributes in the entire project scenario.

CHANGING ATTITUDES AND CONCEPTS

> If you are doing something the same way you have been doing it for ten years, the chances are you are doing it wrong.
>
> —Charles Kettering

Scheduling and Productivity

Since the mid-1960s, there have been more and more formalized efforts to reduce the time between initiation of a project and actual occupancy of the building. In an economy where it is increasingly costly to borrow money and the need for more immediate use of space allocations is becoming urgent, trying to reduce the project time has become perhaps the most vital concern of the construction industry.

Productivity, of course, is most important as far as the individual worker is concerned. He/she is capable of laying so many bricks per day, installing so many panes of glass per day, and so forth. Productivity, however, cannot be increased over and over again, because it simply is not within human capability to become ever more efficient. Scheduling should take into account normal productivity. It is risky to schedule a project based on a greatly increased productivity per person. Workers can be added for increased total productivity, but increased costs will reflect the extra personnel. The proper time should be allowed for each operation, in any event.

This does not suggest that the construction workers should have a slovenly attitude, doing the least amount of work or just enough work to get by. The attitude should ideally be one that strongly encourages maximum productivity, that tries to achieve a truly coordinated effort of combining procedures, taking proper shortcuts, and seeking new methods to reduce overall project time.

Although only recently developed, fast-tracking and construction management have worked extremely well. Combined, they cut down overall project time. Large sums of money have been saved, and pleased owners have gained occupancy months ahead of traditional time schedules.

Fast-Tracking

Fast-tracking is a system whereby certain phases of the construction are actually in process before the complete complement of construction documents is finished. For example, the foundation work may be going on while the architects and engineers are still working on the upper floor plans, elevations, and mechanical systems drawings for the building. While both the field workers and the office forces are working toward the same goal, they are slightly out of phase, with the field processes naturally lagging behind the drawings. This sequence allows for the collapsing of the time to produce documents, and time is not wasted. The project is not standing in limbo while the design professionals complete all the documents before any work is actually started.

Of course, certain risks are involved. New approaches must be taken to initiate fast-tracking. Obviously, if there are major changes in the building after the foundations have been designed and laid, there could be a very serious problem with either underdesigned footings or an inadequately designed superstructure. In most instances the building must be designed in an order contrary to the more traditional one. In years past the design professional would know the total loading of the building before the structure, the footings, and the foundation were designed. Today, this is not always so, and one must rely on carefully calculated guesses in regard to foundation design.

Experience and familiarity with the system have eased the pangs of change. The new method is widely accepted and easily incorporated as standard procedure, and the benefits of the end results far outweigh the changes required of the contractors and design professionals.

Construction Management

Construction management is a new concept in overall control of the project. Although there are still bidding periods and various contractors on the job, the responsibility for the overall management and coordination of the project is in the hands of the construction manager.

The construction manager may have any one of various backgrounds: an architectural or engineering firm or perhaps a general contracting firm. Although formal education and degree programs in construction management are now available, the first ventures were initiated by established firms. They sought additional income by offering expanded services. The design professional could maintain tighter control, and the general contractor could enter a business with far less risk and more reliable income.

Proper management of the project is essential to the end result and to the timing of the project occupancy. The manager, working in concert with both the field operations and the design professionals, must be prepared to influence both the planning and the construction phases. He/she must ensure that all processes progress at proper intervals and in proper sequence.

For instance, the manager may be working with the architect on the interior of the building, moving toward biddable documents for interior partitions, and at the same time, coordinating with the structural engineer and the contractor for the footings and foundations that are being installed in the field. This type of management effort can be achieved only by a person not actively engaged in the actual building process; this is the key to the construction manager system. The manager's role could be compared with that of the old general contractor. The general contractor was concerned not so much with planning and sequencing as with the actual construction of the project. The difference is a simple matter of priority and job description.

THE NEED FOR BETTER UNDERSTANDING AND COOPERATION

The successful project is a blend of many divergent viewpoints into a well-directed, cooperative effort (see Table 4.2). This is true with both the general contract concept and with construction management. And the professional is also part of this effort.

TABLE 4.2 Successful projects blend different views into a cooperative whole.

	PROJECT	
	Successful	Unsuccessful
Scope (3)	Well defined	Badly defined
Planning (2)	Extensive, early	Little, poor, late
Changes	Few, quick response	Large number throughout
Management	Good at all levels	Poor, spotty
Team chemistry (5)	Proper, interactive	Communications breakdown
Control (4)	Whole project concept	Not good, spotty
Client (1)	Positive relationship, involved	Unrealistic scope, budget, timing

Client comments:
1. Owners' roles vary by choice
2. Owners differ on amount
3. Tend to break projects into parts
4. Control schedules, but differently
5. Agree that "people are key"

Where the contractors and their subs are left to operate in an atmosphere of "you just come in and do your work and don't ask any questions," there will be little, if any, mutual effort for the common cause—the project. Each worker today needs to be made part of the team and his/her work a valuable part of the final project. Part of this must come from the contractor-employer, but the concept may be fostered, if not initiated, by the professional and other upper-level project personnel. Often the individual or even "company" cause must give way to the common cause. This is not to say that things must be given away or that extreme personal efforts are required. Rather, it is a simple matter of cooperative understanding and respect for the work of others, one's own work, and the overall end result. Certainly, everyone should be able to walk away from a project satisfied, if not pleased, to have been part of that effort, no matter how remote, how hidden, or how small their work may be.

Consulting with Contractors

To this end, there is a need to open the project to the contractors. This should begin with an educational effort. Here the intricate procedures, the complexities, and the inner workings of the project should be explained. This will bring the contractors on-line, not toeing the line. It will make them feel part of an involved effort, one that is worthwhile and worth doing well.

It must be recognized that each client and, hence, each project presents its own set of peculiar circumstances. These require both attention and adjustments, as well as compliance. Even where standardized documents are used (which set forth certain requirements, conditions, policies and procedures) there are often revisions made to meet the client's situation. When a project is being produced for a large corporation or a government entity, it is likely that the entire array of procedures will require changes to those mandated by the client.

In any event, it is necessary that the pertinent procedures be made clear to the contractors, so compliance can be easily achieved. Since the contractors move from job to job, they will have seen numerous procedures for the same operations, so it is necessary to focus them on those that are to be used for the project at hand. Also, contractors may seek to use their own procedures within their own business operations (the "we always do it this way" syndrome). While they may be able to retain a good portion of their customized procedures, they will still have to adjust their work to the requirements of the project in question. Again, it is necessary that this be made clear to all contractors; their "reward" is a smooth course for all actions, especially payments. Contractors are "doers" and often rebel at the paper exercises in which they are requested to become involved. However, other participants on the project have different perspectives, making paperwork, indeed, good documentation, an absolute necessity.

Although a preconstruction conference is most desirable, more sessions may be necessary. These would deal, in depth, with the details of the project's operations (i.e., paperwork, etc.) and the correct methods, dates, and so forth that will ensure prompt and proper decision making and responses throughout the course of the entire project. Such sessions would be to the benefit of *all* project participants. A better project will result. As Edmund Burke said in another context,

> All that is essential for the triumph of evil is that good [persons] do nothing.

In the past the individual contractor, working in an isolated, artisan-type environment, was left to his own devices. His skills and abilities, while still important to any project, change when they are part of a project being worked on within a "team concept." *Team* immediately suggests tighter coordination, collective willingness to work together, greater individual flexibility, and at times a need to sacrifice for the single effort. A brand-new atmosphere must be created under these circumstances. With the more modern approach and the effort at collapsing the time sequence, the atmosphere on the project site must be studied and understood by all participants.

The Influence of Project Documents

Many factors contribute to the project atmosphere. Among these are the actual construction documents, the drawings and specifications themselves, since they directly affect all participants. How practical the design might be (with regard to ease of construction) and how complete and specific the documents are directly influence the abilities of the various parties. This also has a direct impact on the administration, supervision, and inspection sequences.

If a goodly number of errors, omissions, ambiguities, overlaps, and contradictory information have been included in the documents, all participants must be made aware of these items and how they affect their particular work, and they must work even harder as a team to resolve these situations. For one contractor to go off on his/her own or for one inspector/administrator in a sequence to suddenly become overfastidious and impractical does not serve the project well. All site personnel must work in concert for a solution. At times, some mutual aid or sacrifice is required. Combined thinking, planning, and procedures must be understood and reexamined. The participants must come together to reevaluate this particular phase of the project.

Abilities and Attitudes of Contractors

Another aspect of project atmosphere is the ability of each contractor on the project. If the contractor traditionally keeps his projects very close to the letter of the contract, or if he believes that he has submitted a bid too low—these actions or perceptions have a direct impact on the cooperative effort of the team, on the administration/inspection cycle, and of course, on the project atmosphere and the completed project.

It may well be that the contractor's attitude will have to be dealt with by the other participants. Of course, each contractor brings his own business approaches to the job. He/she should be accepted into the team at face value and should be instructed in the new approaches being taken on the job (if any), not by making the contractor totally acquiesce to the desires of others, but by strongly and firmly attempting to orient her/him toward teamwork.

Chapter 4   New Concepts

The contract format in longest use and considered most traditional is that of the general contractor (GC) with an array of required subcontractors. In addition, there are separate contractors for plumbing, mechanical (HVAC), and electrical work. Bidding is carried out for each of the contracts in separate sequences. This is now also called the "multi-prime" format.

Obviously, this format requires added coordination efforts on the part of the contract administrator. For example, who is responsible for drilling/cutting holes for the running of plumbing pipes—the general contractor or the plumber? This should be resolved in the contract documents to ensure clear and distinct responsibility.

To forestall some of the required coordination effort, another contract scenario was adopted. The "mechanical contracts" are "assigned" to the general contractor. A separate, extra fee is allotted to the general contractor for administration and coordination of the mechanical contracts and subcontractors. Bidding can be done by separate bids with assignment to the GC contract, or the mechanical work can be bid directly to the GC, with normal subcontracts between the GC and the mechanical subs. Either way, if a problem arises, it is the general contractor who is approached for resolution.

In general, unlike design professionals, contractors do not work under a "higher duty" set by legal statute, i.e., the protection of public health, safety, and welfare. The prime regulations to which they must adhere almost exclusively deal with a specific area of work, the conduct of their business, and perhaps a licensing process. Even the latter usually will not require the added burden of public protection.

Obviously, these project participants work with a very different perspective: their own self-interest and the protection of their business. Their depth of commitment to these aspects directly influences how they will function on each project. Where the financial aspect overwhelms all other operational considerations, the company will be on the wrong path. However, no one concerned with the project desires a financially derelict company. Basically, one is looking for a financially sound, responsible contractor who will operate in an expert and cooperative manner, as part of a team, with a valuable expertise. Once under contract, they are not in competition with anyone else and should resolve to perform in the best professional manner known to them, facing their own mistakes, as well as aiding others in resolving theirs. To accomplish this, they must also engage in contract administration!

The actual capabilities of the contractors also affect the project's atmosphere. In a very tight economy where there is not a lot of proposed construction work, contractors often bid on projects for which they lack experience and many times they try to "move up" into areas of construction in which they have never been active before (see Figure 4.5). Although one expects the often-voiced complaint, "How do you get

---

2. Immediately prior to placement of pavement base, proof roll all subgrade sectors in the presence of the Architect and Soils Engineer employed by the Owner. Use standard rubber-tired or flat-wheeled compaction equipment as well as loaded truck similar to type used in paving construction operations. Exercise critical visual observation to subgrade reaction, and any areas which exhibit instability or which react excessively under applied load are to be immediately reworked or undercut as directed by the Architect.

FIGURE 4.5   Excerpt from typical specification. Such broad-ranged, detailed requirements for one phase of the work may be new to some contractors. Note language requiring inspection.

a job without experience, and how do you get experience without a job?" no project is a training ground for contractors or an opportunity to try something new; failure to perform well is a disaster. There is no intentional effort in the industry to exclude "new" contractors, but these contractors should be fully aware of the responsibilities and risks in such new ventures. A tremendous effort, an excellent attitude, and a willingness to learn must be forthcoming. Success, of course, can lead to widened business horizons and a brilliant career. Failure can lead to litigation, financial disaster, and working for someone else. The toughest battle to be fought involves finding an owner and a project team that will give a new contractor a chance. Projects are too dear to allow a frivolous contract award.

All phases of the team and the inspection cycle must somehow come to terms with this type of problem, as it may appear on any project. Care must be taken not to allow the problem to overshadow the work.

Beyond the individual contractor's capability, what is the ability of the entire construction team? Is the team balanced: highly skilled, highly motivated, and well organized? Or is this a combination of streetwise, wily contractors and young, eager, inexperienced contractors? Is there lack of skill? Are resources and experience adequate? Not only does the inspection sequence have to acknowledge these conditions, but the entire team effort must be adjusted.

Duties of the Contract Administrator

When a practice operates nationally, or in several states, and in both urbanized and more rural areas, the contract administrator must recognize the changes in working conditions in everything from the methods and materials used to the personnel's expertise and attitudes. In some places, the work may be highly segmented or unionized; elsewhere it may involve workers who "can do it all" (and do so!). Some projects may be built in a very casual, laid-back atmosphere; others are highly charged and quite organized. The more remote the project to workers, suppliers, and resources, the more problems, no matter the attitudes. Since the contract administrator is prevented from active participation in scheduling and methods of construction, she/he is left to point out needs, discrepancies, lack of proper staffing, gaps, inefficiencies, and so on to the project (contractor's) superintendent or the construction manager.

Observe and Comment on Quality

It is both valid and important that the contract administrator comment, firmly and promptly, when quality or compliance is being sacrificed. Such comment, however, must be within the purview of AIA document B352, which sets out the duties and responsibilities of the on-site representative. For example, if quality of work is being reduced simply to meet a scheduled milestone, comment is valid. The contract administrator must try to induce the construction manager/general contractor and the contractors individually (through the CM/GC) to reassess the situation and ensure ongoing quality work—but no comment should be made regarding the schedule or any schedule revisions, which is outside the authority granted by B352. The contract administrator should make it clear that nothing should compromise contractual compliance. This may seem like a vague, off-handed approach—it is not! B352 limits the administrator's authority, but permits proper and valid comment. The project should never be allowed to move at its "acquired pace." The contract administrator needs to act as a catalyst/facilitator to support the contractor's superintendent or the construction manager in keeping proper progress going. Don't let things languish. Enforce or inquire about promises, deadlines, deliveries, resolutions, submittals, requests, questions, and answers to ensure nothing is lost or overlooked, much less "merely" forgotten.

Every project has its own set of circumstances. The administrator should watch for good and bad signs that may occur during the work. Of course, part of this is contingent on the quality of the documentation (see Figure 4.6 for a list of documentation problems) and the complexity of the work involved. Other aspects have to do with the contractors, subcontractors, materials, and suppliers. Questioning is but one phase of this general assessment by which the administrator can evaluate the prognosis for the work. Too much questioning indicates that the personnel do not understand or have not become properly familiar with the work; too little usually indicates that the workers are proceeding in a manner that may seem correct to them, but that may be faulty. If the contractors do not have a set of documents, problems will occur. If their supervisors or managers merely "run school" and verbally describe the work to the crews, and then leave the workers to their own devices, other problems will occur. If all modification documents (for example, addenda, bulletin drawings, clarification drawings) are not distributed to the proper personnel, on-site, still other problems will take place.

Obviously the administrator cannot (and should not) monitor each crew in all of its operations, but where conditions leading to faulty or improper work occur, comment must be made promptly for the sake of the project. Usually each contractor is responsible for his/her own layout, assignment of work, and methods for meeting all contract requirements. In some cases (far too many, in fact) correct, complete, and pertinent communications are *not* established with field personnel, and many contract requirements and operations are forgotten or circumvented, or ignorance of their existence is claimed. This cannot be allowed to happen. The administrator is not responsible for each layout and other operations, but through continual observation and pertinent comment, corrective/remedial actions, and "preventive" communications, many problems can be averted.

Monitor Job-Site Personnel

On the "people" level, attitudes or methods of operation must be such that the project is well served. Each person working on the project must understand that his/her outlook must give way, in many aspects, to the requirements and processes of the project. No one is exempt from this. Care must be taken that "laid-back" does not become inattentive, nonchalant, negligent, inconsistent, inadequate, or the like. The

---

1. Overdetailing; requiring excessive construction.
2. Crossing, confusing, illegible leader lines from notes to drawings.
3. Cross-referencing to wrong drawing, wrong sheet, or nonexistent drawing.
4. Not coordinating different views to show same items in same way (materials, dimensions, relationships, joints, etc.); inconsistent.
5. Using specification data on drawings only; not including them in project manual.
6. Showing items with no specification or with a different, irrelevant specification text.
7. Specifying items not shown or used; never using or referring to items that are in specs, but nowhere else on contract documents.
8. Showing item on drawing, but with no specification data in project manual.
9. Needlessly verbose notations; "instructional" notes about how to do work, etc.
10. Calling same item by different names in different places.
11. Detailing in isolation and not showing relationship to surrounding work.
12. Showing overly complex construction or requiring excessive/inappropriate field actions or work.

FIGURE 4.6 A list of problems or discrepancies that occur frequently during document preparation.

work needs aggressive handling, but the amount of aggressiveness is the key. A slow-paced, laid-back trade superintendent who does not have a set of documents or all of the necessary tools is an immediately visible problem. In this instance, if problem-solving is "normal" in lieu of problem prevention (by farsighted review and questioning), the project is adversely affected.

At the other extreme is a highly aggressive, needlessly reckless, "pushy" attitude—"get it done, somehow, *now!*" Care is thrown out, usually to gain bonuses (for finishing early) or added profit, despite contractual requirements. Again, the contract administrator cannot tolerate these actions, as the project and client will be short-changed. The contract administrator needs to monitor for good pace, consistent with the schedule and with the contractual requirements. Discussions and adjustments in this area are constant and important. Every circumstance that arises needs to be addressed promptly, completely, and properly, with commensurate review of the impact on the overall project. The contract administrator should be part of this, to the extent of authority granted under AIA document B352 or other contract provisions.

It is interesting that in both of these extreme circumstances (laid-back and highly aggressive), questioning is the key issue. Usually, questions will not be asked in these scenarios. The lack of questions should alert the contract administrator to potential problems. In both instances, the personnel will go along making assumptions and decisions without thinking about the impact of improper work. The workers will merely "do what they have to do" to satisfy their immediate bosses. Errors will occur more often and will then be covered or made up with other faulty work. It can devastate a project. Underlying both of these scenarios is the idea that workers will proceed as they choose until told differently, be they right, wrong, or indifferent. It is the contract administrator who must tell them (within proper channels).

In the construction industry today, it is to no one's credit that we wallow in the mentality of doing the least required, in the shortest amount of time, in a just-adequate manner, mainly to achieve nothing more than the cheapest cost. We claim our technology is far superior to anything past, but our attitude is deficient. We are long past the time when pride in workmanship and examples of great skill provided all the satisfaction needed; today, it is "how fat is the paycheck?"

Once self-esteem came with the satisfaction of a job well done or a simple word of praise, or merely the fact that there was a job. Honor and fulfillment came by doing the right thing in the right way, just as one was taught by the "master" of the trade. Tradition and history were honored and things were done correctly the first time.

Today, we decry those who seek to do things once and correctly. Studied determination, steadiness, drive, reliance on experience, and reusing what worked (in the past) is too often lost in the quest for the almighty buck! Then we wonder why projects run over cost, don't fit needs, don't last or "weather," don't endure and become "rememberable." Today even "signature architects" design buildings with a life expectancy of 30–40 years, as opposed to 75 years minimum in the past. Materials are not easily maintainable, their inherent attributes are poor, and virtually no maintenance is attempted. We continue to create "throwaway" building stock.

This whole scenario affects contract administration. With marginal construction, designing for cost consideration more than anything else, and hiring low-skilled workers, there must be a resolve to get the project done as close to the contract documents as possible. This is the one time when we *must* seek to do the best we can.

In contrast, some personnel do have their documents, review them, plan their work, and question situations unclear to them. Seeking information prior to the start of work is both advisable and prudent—it leads to doing the work once and correctly. This is particularly crucial when the project is complex, on a tight schedule, and/or replete with intricate detailing. The contract administrator should always be alert to crew chiefs/superintendents who do not have or consult their documents, and work ordered by contractor project managers that is at odds with current contract requirements. All deviations, no matter the source, should be challenged and rectified promptly!

Chapter 4 New Concepts

Each region of the country has its own trade atmosphere. A project with a new materials system or an innovative design may encounter problems because of a lack of trade familiarity. In some areas buildings are almost exclusively steel framed, for instance. To introduce a concrete frame project would demand a level of expertise and skill that the local contractors may not be able to supply. Out-of-town firms must then be attracted, which in itself can be a tough job for a single project, especially in times of heavy construction activity. Costs can increase drastically, whether other firms come in or local firms try to deal with unfamiliar systems. In either event, it is obvious the project will suffer. Again, proper reaction is required not only by the inspection sequence, but by the entire design and construction team.

Adjust to Field Conditions

Beyond the project atmosphere are the actual field conditions at the job site itself. Part of the team, particularly the design professionals, can be working in remote air-conditioned offices, well lighted and well heated, with all the resources they need available to them. In this atmosphere they may become complacent and not accurately assess what is happening in the field. Can the work being detailed be executed properly in the field? Are there conditions such as weather, noise, or other environmental factors that may affect the workers and perhaps the work itself?

The inspection team also must work within this environment and must adjust, just as all members of the construction team must take field conditions into account. There is no choice for the inspection system but to be flexible and willing to change its techniques to those that are best for the particular project. Failure to adjust can jeopardize the project.

Work with Owner

In today's marketplace, where the buyer must beware, and with the tremendous amounts of money being spent for construction jobs, owners are demanding that the quality of the job meet the value they place on it. The owner is not in a position to control all the variables—scope, cost, and quality—of the project. An owner is most interested in scope and cost—that is, how much building he/she can build for the money available. With scope and cost being tightly controlled by the owner, the contractors, the entire design team, and the inspection system must be extremely active in the only flexible element left in the program—quality.

There is no doubt that there is a great deal of fluctuation in the quality range on any project. Something can always be found that costs less and still meets the project requirements. However, the inspection system must monitor all substitutions to ensure that no minimum standards are violated.

Because of the time involved in any building project, the owner has a chance to oversee almost every detail of it and therefore is cognizant of all shortcomings that may appear. Most certainly, the owner will look to the inspection system to produce the highest quality from the various elements of the project environment, and will want to achieve the highest quality with a minimum of effort on his/her part. The owner will want excellence, and part of the job of the inspection system is to see that excellence is produced in the best way possible given the project conditions.

Obviously, there are shortcomings in any particular project; 100 percent "sterling" excellence cannot be achieved in any job. The individual worker, though full of pride and skill, still may not be able to produce, for varying reasons, the excellence that the owner expects. But it is up to the inspection system on that particular project to see that the project is produced in the best possible fashion. This means that each member of the inspection system must be active in ensuring that each part of the project within his/her responsibility is executed properly, whether he/she is viewing the basic design or the "extras" that are put into the project or simply seeing that minimum standards are being met.

Many factors can vary on any project. It is not the job of the inspection system to try to force a particular fashion or method of construction or to see that excellence is produced merely by doing the same work over and over again. The inspection team must be fully aware of what level of competence can be achieved. This level of excellence will vary from project to project because of the varying conditions that create the project environment.

Work with Design/Construction Team

The methods of construction, whether techniques of building or of management, old or new, still rely on the inspection system as the basic quality control element. It is important for all members of the design/construction team to understand exactly where and how each element of the inspection system fits into the job. What is the basic criterion? What are the basic areas of responsibility? Until these are understood, members of the team will be vying for position. There might even be open attempts to circumvent or gloss over one area of inspection or another. The inspection system must be accepted and the flexing of ego muscles must cease, so that inspection is in place and active at several levels. All should be given proper due for knowing their jobs, being able to perform properly, and having the integrity to do their best work.

The inspection system in construction is not set up to be obtrusive or to retard progress. The process should be a continual, progressive monitoring of the construction work. If the inspection system does become obtrusive, it is being counterproductive to the project and should not be tolerated.

Neither should any activity that attempts to circumvent any part of the inspection system be tolerated. The system should be open to all participants so that they know exactly what is expected of them. Their work should be adjusted to meet the criteria of the various inspection agencies.

Promote Communication and Feedback

If a project unfortunately breaks down so that some participants in the design/construction team become adversaries (in other words, they simply don't talk to each other), the project is in deep trouble. This element of the project environment is so necessary that it must be established at the very beginning of the project, and it must be nurtured constantly so that it functions continuously and properly. It may be that early in the project some basic meetings will have to be held to eliminate misunderstandings. There may be a need for some simple instruction sheets to be passed out on how to communicate, so that there is no chance for misunderstanding.

Out of small misunderstandings come the big problems, the problems of adversary positions and, worse, noncommunication.

There is also a tremendous need for feedback on a construction project. If a problem has developed in the field and is not fed back through the communication system at the proper level for resolution, the problem will burgeon. The problem will produce more adversary feelings and could eventually strangle the project. This could be manifested as a work slowdown, a full strike, a loss of productive time, or another situation that could drastically affect not only scheduling but quality and budgeting.

Briefly, problems of misunderstanding and feedback should be ironed out with the following guidelines in mind:

1. Approach all situations with an open mind. Try to see how the situation looks through the eyes of the other person.
2. Resolve to solve the problem for the good of the project. Do not reproach, accuse, or seek revenge.

Chapter 4   New Concepts

3. Anticipate anxiety in both yourself and others. Everyone has some concern about his/her position, and this must be overcome in solving problems. Don't deal from emotion at any time. Prevent outbursts of temper. Be tolerant of everyone's viewpoints.
4. Strive to have all situations clearly stated, properly documented, and understood by all. There are times when criticism is necessary, but it should be based on a vision of the whole picture and not be a nit-picking accusation.
5. Do not deal in envy, gossip, trivialities, prejudices, or shows of pride.
6. Concentrate on making things clear at all stages of the project.
7. Debate in a shared, side-by-side inquiry. Don't try to make deals or resolve problems behind the backs of some people. In trying to clarify and still give a firm viewpoint, preface your remarks with the phrase, "It seems to me...."
8. Try to let bygones be bygones. What happened on other projects simply has no relevance to the present one. Show others that this same attitude and the need to compromise are necessary for the project. Practice the Golden Rule. Use constructive action at all times.

In dealing with others and bringing everyone to understanding and good communication, there are times when feedback is absolutely necessary. Feedback should be open, and candor should be the hallmark. Feedback should be pertinent and not a discussion of "war stories." The following list is helpful in determining the quality of feedback and ensuring that it will be useful and for the good of the project:

1. Be specific and don't deal in wide-ranging generalities.
2. Refer to current work and don't bring up past projects.
3. Be sincere and establish trust among the members of the team.
4. Plan your feedback and see how it is best to address the recipients. By the time you are in a situation that needs feedback, you should know the other people, their attitudes, and their behavior patterns pretty well.
5. Study what you will present and show the other person just how this affects you.
6. Be sure that the time is right to discuss a particular situation.
7. Although it may be difficult, be sure that the recipient understands what you are saying.
8. Of course, the recipient must be willing to accept the feedback. Therefore, it is important that the feedback be given in such a manner that the person will not react negatively.
9. Perhaps the most important part is to ensure that the recipient of the feedback is, indeed, able to do something if he/she so chooses. Without this alternative, the feedback becomes valueless because there is no solution to the problem, and it will only lead to frustration if the recipient has no particular action to take.

THE COMPLICATED SYSTEM OF COMMUNICATION

There is no other choice on the modern construction project than to have a communication system that produces clear, concise, complete, and timely communications among participants. This is so important to the success of any project that it simply cannot be emphasized enough. The communication system must be well planned and understood by all participants. It is mandatory that they all fully participate in the communication system as much as possible. Shortcuts cannot be used, nor can partial communications or improper or incomplete distribution lists.

> **SECTION 01001**
> **SPECIAL CONDITIONS**
>
> **ARTICLE 12 — OWNER — ARCHITECT — CONTRACTOR — CONSTRUCTION MANAGER COMMUNICATIONS**
>
> <u>Directives to Contractors shall come through the Construction Manager; directives to Construction Manager shall come through the Architect; directives from the Owner to the Construction Manager shall be issued through the Architect. Contractors shall communicate with the Owner and Architect by way of the Construction Manager</u>

FIGURE 4.7 An excerpt from a recent specification sets forth distinct lines of communication for the project.

The complexity of modern construction is so intricate that if the communication system breaks down even in a small way, the project itself is in jeopardy. Communications within the construction contract administration process should be well thought out and should not be allowed to deteriorate into "paper everyone" exercise. Current and complete distribution lists are vital to the dissemination of information and to the reduction of needless paperwork. Information should be assessed on a "need to know" level. This is not to foster secrecy nor to promote cliques, but rather is an effort to streamline the process by eliminating useless activities. Figure 4.7 gives an example of specified lines of communication.

It is obvious that to administer properly, one must be part of the communications network; one must know, observe, review, analyze, assess, and resolve. This distinctly points to an active *field-oriented* operation, where the main focus is the work in progress. Of course, there is a need to see that all of the administrative (paper) detail is processed and distributed, but this cannot be done without active field observation. The basic-service function provides for "periodic" visits to the site. These, if programmed properly, may provide all of the necessary information to administer the program. However, any use of a quick-fix, superficial observation system will not. To administer, one must be on top of the project, at all times. One must be an active participant, not relying on hearsay, rumor, or assumption, but on substantiated information gathered personally from various reliable sources.

Some feel that "periodic on-site visits" are incompatible with true contract administration; and a good case may be made there. Where full-time, on-site representation is not available, the professional should see that an appropriate program is made available and exercised to the maximum. It can easily be seen that in this dilemma, construction management becomes a more attractive alternative to the owner. There, continual on-site presence is mandated by contract, but not all professional services are available. Further, to turn the project over to a management system leads to interpretations and changes that may redirect the project concept away from what the design professional intended and the owner agreed to and expects. Only professional involvement can forestall this scenario.

From the professional's point of view, internal communication is almost as vital as any other line of communication. It is essential that the professional's field personnel be advised of every situation that affects field operations, be they architectural, mechanical, electrical, or administrative. This applies whether the personnel are full-time, on-site representatives or part-time staffers who rotate between projects. These people are the "lightning rods" on the site for all contractors and become the primary point of contact with the professional. They are the professional's direct, reliable, and continual connection to the site. Their availability is crucial to the suc-

Chapter 4 New Concepts

cess of the administrative process, be it for liaison, determination, interpretation, clarification, or proper communication to solve the problem at hand.

Hence, while certainly not "errand runners," they do convey information both ways. They must be kept current with all the information being handled and distributed, but need not be privy to the high-level, more "political" goings-on that every project encompasses. Certainly, these persons should not and cannot be circumvented or ignored by either their colleagues, their principals, or the contractors. If this occurs, the field representative/contract administrator becomes an unreliable source for the contractors and embarrassingly inadequate (to themselves and to the professionals) to proper administration of the project. This, in turn, frustrates smooth interfaces and communications, particularly on-site. It directly increases the professional's liability exposure, simply by not ensuring that personnel are made aware of the proper information, changes, interpretations, etc., as they occur.

Ideally, the field rep is the most prompt and most reliable contact for the contractor and the professional. This ensures properly inclusive dissemination of information and should foster prompt, accurate information flow. As the professionals' eyes and ears on the project site, the field rep must be an integral part of the communications network.

This is not to advocate that the field rep/contract administrator should be given a preeminent status in administering the construction. The authority of this person remains limited by AIA document B352, and rightly so. However, as the direct, on-site "presence" of the design professional, the administrator should be the preeminent source on site regarding all technical aspects of the work. The best interests of all parties are served when the design professional communicates to the site *through the contract administrator*. The professional, of course, has numerous other lines of communications, not all of which are privy to the contract administrator. Job site and technical project matters need to be seated in the contract administrator/field rep and his/her various activities. Figure 4.8 illustrates communications involving the field rep.

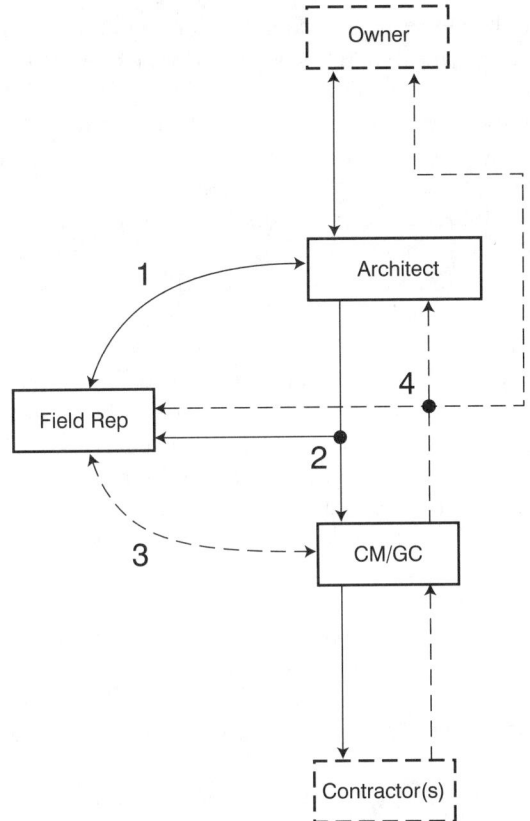

FIGURE 4.8 Communications and the field (project) representative: (1) Architect to/from field rep; (2) architect to CM/GC (copy to field rep); (3) CM/GC to field rep (to architect if necessary); (4) CM/GC to architect (copy to field rep).

It is vital that the on-site representative be proactive, not passive or reactive. This is best seen in a 1991 decision by the U.S. Department of Interior Board of Contract Appeals. *White Buffalo Construction,* IBCA Nos. 2166, 2173, 91-1 VCA, paragraph 23450, addressed the relationship between proper inspection and communication. The issue was whether or not silence on the part of the inspector constitutes acceptance of defective work. The work was properly inspected, but no objection was issued by word or written note. In fact, the inspector made no comment of any sort regarding the work, which was, indeed, defective. Although responsibility is not shifted (according to this board), the contractor cannot rely on silence as acceptance. It is evident, though, that the work is adequate if no comments, notes, or records exist about the condition of the work.

This case, while perhaps extreme, provides an interesting insight into the activities of an inspector (project rep) or a professional engaged in contract administration. Prompt and specific comments, notifications, and/or other records must be made *and* properly distributed so the work will not proceed in a faulty manner. This is not something that can be put off because additional work could well compound the problem and make a later solution more costly and involved; it also could aggravate placement of responsibility, liability, and payment.

No project is immune from the need for communication. A tight inner-city project requires split-second scheduling and manipulation of workers, equipment, and material. An open, wide-ranging project separates functions and personnel to a tremendous degree. Obviously, communication is the necessary common denominator in all projects.

The system of communication begins at the basic level of two workers talking with each other and coming to an understanding about a particular piece of work. From here through a myriad of different combinations of people, the communication builds in intensity to the point that a simple phrase or sentence of instruction given at the wrong level, in an untimely fashion, and not properly distributed to all participants can stymie the progress of the project.

A good deal of time, number of workers, and money is expended on any job simply to keep the lines of communication open and operating. Without this, there is no way to tell just where or when the project will break down. In the event of a breakdown, much more time, money, and personnel could be required not only to find the problem and to solve it, but also to reestablish the lines of communication. Of course, this all works to the detriment of the project.

Prompt communications, especially replies to questions, cannot be emphasized or stressed enough. Contractors do not generally submit frivolous questions. Usually, they will ask about work or material before actual installation; sometimes the inquiry is too late, however. To withhold response, for any reason, is unwise and could affect job progress. This is not to say that hurried, off-the-cuff answers should be given for the sake of a fast reply, but answers should be forthcoming as quickly as practicable. If a reply must be delayed, that information should be made known to the contractor, who can then work around the situation.

If any circumstance arises that prevents any contractor, subcontractor, or supplier from performing *in full compliance with the contract documents,* immediate notice (followed by full written notice) must be made to the contractor *and* the design professional. The notice should explain the reason for possible noncompliance, suggest remedies, and seek direction or approval from the design professional *prior* to executing any work.

Supplying, furnishing, providing, and/or installing systems, equipment, devices, or material is anticipated to be in full compliance with contract obligations. The contract administrator may evaluate such actions differently in view of conditions in the contractual requirements. Added payment for any required remedial work in case of noncompliance, unapproved substitution, error, improper use, or failure should not be entertained. This again points up the need for early review, analysis, and communication on the part of the contractors. Problems will occur, and they must be identified as early as possible.

This is an excellent example of a situation in which the contract administrator must be proactive in ensuring contract compliance. This all takes on added importance when a commitment has been made during the bidding process. As Figure 4–9 shows, many owners require a List of Materials and Equipment at bidding (to eliminate "shopping" for lower prices than those included in the bid; many times the lower costs are not passed on in the form of a reduced project cost). This list forms a contractual obligation, since the bid is part of the contract. Hence, the contracting parties are tied to fixed requirements and cannot merely "freelance" solutions as they see fit. This project control is most valuable to the contract administrator.

The most frequent cause of error in construction is the breakdown of communication. This failure stems from many sources—the distraction of outside worries, pride, inattention, use of words that have different meanings to different people, directions that are misinterpreted, and people's inclination to use their own version

---

ENG. FILE NO. R-742   FORM OF PROPOSAL   ES-185-97
FP-11 of 14

LIST OF MATERIALS AND EQUIPMENT

This List of Materials and Equipment is required by the Owner to be completed by the apparent low bidder within one hour after the close of the official reading of the bids. Failure to submit this list, completed as applicable to the Contractor's bid, may be cause for rejection of the bidder's proposal.

Each item listed shall be clearly identified so the Owner will know definitely what the Bidder proposes to furnish. The use of a dealer's name or stating "per Plans and Specifications" will not be considered sufficient identification.

Where more than one manufacturer is listed for any one item, the Architect reserves the right to select the one to be used.

ITEM   **MANUFACTURER—MODEL/TYPE**

Structural Steel: _____
Metal Roof Deck: _____
Steel Joist and Joist Girders: _____
Architectural Woodwork: _____
Exterior Insulation & Finish System: _____
Membrane Roofing system: _____
    Interfacing with existing warranty: _____
    Discrete with existing warranty: _____
Hollow Metal Doors and Frames: _____
Wood Doors: _____
Aluminum Entrances & Storefront: _____
Skylights: _____
Finish Hardware: _____
    Locksets: _____
    Closers: _____
    Hinges: _____
Drywall Components: _____
Ceramic Tile: _____
Acoustical Ceilings: _____
Resilient Flooring: _____

FIGURE 4.9 An excerpt from the sample Form of Proposal included in a project specification. The length of the form should be adjusted to specific project requirements and the number of items that need to be included for this consideration.

```
Carpet: _____
Paint: _____
Toilet Compartments: _____
Toilet Room Accessories: _____
Louvers: _____
MECHANICAL
Air Handling Units: _____
Exhaust Fans: _____
Return Fans: _____
Pumps: _____
Boiler: _____
Cooling Tower: _____
Unit Heaters: _____
Fin Radiation Convectors: _____
Terminal Boxes: _____
Insulation: _____
Temperature Control: _____
Grilles, Registers, Diffusers: _____
Plumbing Fixtures: _____
    Water closets: _____
    Lavatories: _____
Floor Drains: _____
Roof Drains: _____
Water Heater: _____
Valves: _____
ELECTRICAL
Panelboards & Switchboards: _____
Motor Control Centers: _____
Wiring Devices: _____
Dimming Systems: _____
Fire Alarm System: _____
Sound System: _____
Lighting Fixture Types: _____
    F-1 _____     F-1A _____
    F-2 _____     F-3 _____
            (list all fixtures)
                    END OF MATERIALS AND EQUIPMENT
```

FIGURE 4.9 *Continued*

of instructions. It is important that each person concentrate on using proper wording when speaking to others and on trying to understand the other person's point of view, so that when a communication is given, it can be properly received and put into useful action. The sender of a message must ensure that (1) the receiver is paying attention, (2) the message is clear and in terms the receiver understands, (3) the message is received *and* understood, and (4) the receiver knows what is required and in what form. This outline holds true for both verbal and written messages.

To prevent construction failures entirely may be impossible. Human frailty and material imperfection can lead to construction failure, but miscommunication or the lack of communication should never contribute to it. The following ideas can help ensure that communication is kept open at all times and that everyone can participate:

1. Be sure that you know the person you are dealing with well before you entrust him/her with important orders or directions.

2. Be extremely patient; you may have to discuss, evaluate, or even reword communications at various times.
3. Be sure that both you and your listener are paying attention to the problem at hand and that you are not wandering from the subject.
4. Do not pass on incomplete information, and do not accept incomplete information.
5. If the information you have received is incomplete, not clearly transmitted, or not understood, say so. Be sure that the situation is clarified for everyone's benefit.
6. Do not carry out an obviously foolish interpretation. Check out the information, ask for clarification, and be sure that it makes sense to you before you act on it.
7. Make sure that both you and the person you are dealing with are interpreting words and phrases of construction jargon in exactly the same way.
8. If at any time a picture or a drawing can be used to clarify the situation, be sure to use one. The old saw of "a picture is worth a thousand words" is very apt in the construction industry.

Communication is discussed throughout the other chapters, with particular emphasis on the person involved. Lines of communication in the construction field must be direct and constantly maintained (see Figure 4.10). In the many years since the beginning of the construction industry, vast amounts of money have been lost due to construction failures and time losses that can be attributed directly to a breakdown in communications. It is elementary to say so, but all those on a construction site are, indeed, human beings. The main problem is how to mold these human beings, who come together for a relatively short time, into an open, freely communicating, coordinated team. It is important that this be done—not only that communications are established and understanding is maintained, but that the project is of prime importance. Although everyone involved has various allegiances and may have different alternatives and goals in mind, the inspection function must be "for the project." This sounds trite, but no matter what the allegiance of the inspector, he/she has a firm commitment and obligation: to participate in the entire project in such a manner that does not inhibit or negatively affect it. The inspector's participation and communication should be positive, lasting, and ongoing, so that he/she can contribute in a positive way.

A successful construction project depends greatly on attitude. It is conceivable that a team on a construction project can be made up of several individuals or firms who have had openly hostile relationships in the past. The new situation requires that each person on the project somehow give up these feelings, shed the mantle of hostility, and attempt to do the job properly. According to Dr. Robert H. Schuller, pastor of Crystal Cathedral in Garden Grove, CA,

> Ultimate success is more dependent on human values (honesty, sincerity, helpfulness, etc.) than on professional abilities.

```
2.02  COMMUNICATIONS WITH OWNER

      A. Generally, all communications with the Owner shall be through the
      Architect, _____ (e.g., shop drawings,
      requests for payment, requests for drawings of items furnished by
      Owner, etc.).
```

FIGURE 4.10 Excerpt from typical specification. This shows a definite communications procedure. Although often unwritten, it leaves no room for miscommunication.

It has been said many times that even in a national election one vote can be meaningful. Similarly, if one person on the construction site or one member of the inspection team can work in a positive manner, there is a good chance that he/she can influence the entire team. They may come together and work for the good of the project.

It is fairly accurate to say that the vast majority of professionals on a job site are motivated to do a good job and to produce a successful project. However, motivation alone will not sustain anyone, nor will it produce a successful project. The missing ingredient is information about what is to be done, how it is to be done, who else is participating and in what manner, and so forth. Motivation *and* information produce productivity!

Obviously one should be excited about the work and one's professional involvement, but all this can be for naught where it is directionless or meandering because of lack of instruction, purpose, and a defined goal. Scheduling and timing are also necessary. Professional drive and expertise is "lost" on a project where information is scarce or muddled, where procedures vary, goals and directions vacillate, revising is too common because of the lack of practical determinations, and communications are meager and depend on mood, personality, or shortsighted goals.

There is no doubt that individual communication skills are crucial, not only to "getting one's point across," but to basic understanding and coordination. Certainly specific and properly detailed information is the best facilitator, but in addition the "sender" needs skill in delivering messages. Any public speaking experience, whether in class or before professional groups, is most helpful in developing communication skills. The following list notes the benefits of good communications both in construction and in general.

- Everyone should be able to speak, listen, lead, delegate, and motivate effectively, and to conduct meetings.
- The best of ideas, uncommunicated, contributes nothing, offers no opportunity, is unrealized, and has no real value since it cannot be discussed, improved, or implemented.
- Good communication builds confidence and eases work by eliminating worries about communicating, engaging other people, and allowing fuller participation.
- Improved presentation skills brings respect and admiration from colleagues.
- Success in the workplace is directly proportional to skill in communication.
- Communications creates networks and aids in meeting people from other department and disciplines.
- A by-product of good communications is a wonderful professional camaraderie.
- Good communication builds self-esteem, improves sharing of information, presents opportunities to improve and offer new ideas, and gives one new status as a contributor.

EXERCISES

The following tasks are based on the concepts and information in the chapter. All of the material or information required to complete the tasks may not be contained in the chapter text. Independent research is required. Consult your instructor for guidance.

1. With the concept of the master builder in mind, discuss the development and intent of the terms *craft, trade, union, guild, journeyman, apprentice, skill,* and *specialty.*
2. The project is quite deep (front to back) with long corridors running full length to exits in the rear. Just ahead of the exit doors there is a ceiling height change

down to 7′0″ (under a supporting beam). The wall on either side of the double exit doors is just 4″ wide. The electrician asks where the exit signs should be located so they will be visible.

3. The contract administrator (CA) should actively seek out problems in the documents before the work is done to avoid the problems and not impair work progress. Is this a proper role for the CA?
4. Is it best practice to give a set of documents to the CA on the first day on the site, as work starts?
5. Is it necessary for the CA to be a registered professional?
6. What advantages/disadvantages are there when the CA/project rep is a professional?
7. Explain the difference between mistake, misrepresentation, and fraud.
8. Explain the concept and purpose of "substantial completion."
9. Discuss the difference between substantial and final completion. Cite examples.
10. What significant events occur (per contract) when substantial completion is achieved and certified?
11. List three major impediments to timely completion. Describe them and ways to avoid them.
12. What is a "punch list"? Who writes it? Based on what? Who resolves it? When?

5

The Design Professional

Design professional, n. The professional collectively responsible for the design of the physical environment including architecture, engineering, landscape architecture, urban planning, and similar environment-related professionals.

CONTRACTUAL OBLIGATIONS

The owner, in most instances, has a rather difficult time understanding the position of the design professional and the responsibilities involved. This is understandable. The owner is expending a large sum of money on the project and the design professional largely has direct control over these funds since he/she has direct control of the project itself. The owner often wants the architect to have exactly the same attitude toward the project he has. Often this becomes manifest when the owner, wanting tight, close control over construction operations, desires that the architect be constantly on the site to ensure that the work is done properly, giving the owner full value for the money spent.

The major problem with this attitude is that it puts the design professional in the position of <u>ultimate responsibility</u> and, hence, <u>liability</u>. If the design professional does not perform in the manner and with the attitude the owner wants, he/she has one problem. But the problems are compounded if the professional does adopt the attitude of the owner.

While being the direct agent of the owner, the design professional is a moderating force, taking the desires of the owner and molding them so they can be accommodated by the construction process (see Figure 5.1). Without this understanding, moderation, and accommodation, deep problems can develop.

To take a construction project and overinspect or oversupervise it can be counterproductive to the design professional. Some design professionals prefer a totalitarian approach to projects, taking the attitude that no one can be right except them. Obviously, the team attitude is not present here, and very bad relations often develop. This does not make for a smoothly running project, even though the design professional feels liability would be greatly increased if he/she had taken another tack. This has been reflected recently in standardized contracts and conditions of the contracts from such sources as the American Institute of Architects. Phraseology of the past such as "periodic inspections," "supervision," and the like has been dis-

```
b.   Paragraph 2.2, supplement by adding the following addi-
     tional subparagraph:
     2.2.4.1  The Architect will endeavor to observe and to
              check work but omissions, failures to provide
              proper material, and failure to perform work
              correctly are totally the responsibility of the
              Contractor. The Contractor, not the Architect,
              is responsible for determination that all work
              under his/her contract as it proceeds or as com-
              pleted is performed and installed in accordance
              with the Drawings and Specifications and govern-
              ing regulations. Where laws, codes, standards
              require supervision or inspection of portions of
              the Contractor's work by an Architect, Engineer
              or other competent or qualified person(s), it is
              the Contractor's responsibility to furnish such
              supervision and/or inspection to the satisfac-
              tion of the governing authority and without cost
              to the Owner. Such requirements shall in no way
              be the responsibility of the Architect or
              his/her field representative.
```

FIGURE 5.1 Excerpt from typical specification. A portion of the Supplementary General Conditions states the exact position of the architect. This modifies the basic General Conditions.

carded from these documents because such terms are both misleading to the owner and highly hazardous to the design professional. This may sound as if the design professional is given certain status by the owner, but is either unable or unwilling to perform. However, this is not the case. There should be an open and deep relationship among the architect, engineer, and owner, so that the owner understands exactly what he/she is getting for the fees paid to the design professional.

Many times, the owner will be chagrined to find that the design professional or his/her representative is not constantly on the site. The owner may also find that certain work is installed that meets the requirements of the construction documents, but in a way that is new or unfamiliar to the owner. Having a little knowledge of construction can cause the owner some anguish. His/her experience, usually limited, means that when something new appears, the owner will feel threatened and frustrated. The design professional must address this situation to ensure the owner's understanding and confidence. After all, the owner has made a commitment, hired the professional, and certainly should feel comfortable in placing a large measure of confidence in her/him. However, should the owner want the situation changed or clarified, the owner's contract with the design professional details all the aspects of the relationship. Carefully written contract modifications should be made, where deemed desirable, in the duties of the professional, including those for contract administration.

Historically, the architect has had the status of "team leader" and was usually the person who brought together, coordinated, and led the team from the inception of the project. The architect was the first "project person" brought on board by the owner because of his/her training and expertise over the entire range of project considerations. With this background, the architect was best able to convene the proper personnel for the project team to properly serve the owner. This was seen simply as the way things were done; as normal, good business practice.

In more recent years, for varying reasons too wide-ranging and complex to discuss here, the architect's preeminent status has changed, mainly as owners felt the need for different project information, goals, and configurations. Hence, in the construction industry today numerous options are open to owners regarding the makeup of their design/construction team. Each owner will come to a decision that best suits his/her interests. Of course, nothing demands that every client/owner follow the same path. Any type of personal relationship (pure business considerations may or may not be involved) that the owner chooses to rely on, from old school chums, to church fellows, to neighbors, to golfing buddies can be the basis for selecting one of the many team concepts. The team is now built around client comfort, preference, and confidence. The pragmatic business tradition of the design professional being the first person contacted and put under contract is no longer followed exclusively. A good percentage of work, however, still follows this concept and its primary virtue—overall project conception and feasibility from the "germ" of an idea, through the entire design/construction sequence, to the finale; from site selection considerations through occupancy.

Owners, no matter their personality or business sense, constantly seek several attributes for their construction projects. While always interested in maximizing their influence in the cost/scope/quality quandary, they seek the "largest building, of the highest quality, at the lowest cost." If pressed they will almost always prefer cost and scope to be under their tight purview; the design professional is left to work within the narrow ranges of quality to meet the other parameters.

Beyond this design dilemma, owners seek contractual arrangements that give them better control. However, owners now have different concepts and goals. New ways of doing business, reduced time lines, fast-track construction concepts, and a penchant for establishing single responsibility, for example, have led to owners seeking the project contractual configuration that best suits their needs—tradition and industry "norms" notwithstanding. There has been a movement (more strongly over the last few decades) to find the single contractual arrangement that will produce all of these virtues. It has not yet been isolated.

Contract Options

Today owners have a relatively large palette of options. There is now a proliferation of variations on the traditional design/bid/build format that usually involves four separate construction contracts: general construction (GC), plumbing (P), heating/ventilating/air-conditioning (HVAC), and electrical (E). To this must be added the desire for better and more reliable cost estimating early in the project process and for an analysis of constructability with regard to materials, systems, and manner of detailing the proposed construction. Many different contractual formats and other configurations have been developed by owners, professionals, and constructors. Following are some of these options and variations:

- **traditional** (design/bid/build)

    With five separate contracts—design, GC, HVAC, P, and E

- **tripartite** (single construction contract)

    Separate A/E contract for project design

    GC contract with separately bid mechanicals assigned; or

    GC contract including mechanical trades and all other contracts.

- **multiple primes** (separate contracts)

    A variety of separate contractors (including GC, HVAC, P, and E) bid separately and each is placed under direct contract with the owner

- **design/construction manager**

    Design firm with separate contract for CM functions

    GC acting as CM who hires the design firm

    Professional CM who hires the design firm

- **professional contract management**

    Firm under separate contract directly with owner for traditional CM functions only; or for a variety of construction-related duties (excluding design), as the owner desires

- **design/build**

    Primary firm is usually a contractor; hires a design firm; builds the project with its own forces

    Design/build firm has separate internal departments that do both design and construction

- **turnkey**

    A form of design/build, but with little direction or involvement of the owner (except to establish project goals); uses performance as a guide for a fully self-contained project;

    More closely resembles a "sale" of a project instead of a contract for services; can include requirements for land purchase, financing, equipping and furnishing, on the part of the contractor

- **bridging** ("associated architect")

    Owner hires prime designer for concept only; could be a "signature" architect

    Owner also hires a design/build firm with its own architect for documentation

    Local architect could be hired and then "associated" with a distant designer to document project in accord with local regulations and requirements

- **joint ventures**

    Multidiscpline firms temporarily combined into a single entity and placed under contract for the project at hand (can be design firms only, or may include CM and/or actual construction)

- **phased construction** (fast-tracking)

    Concept usable even on traditional contract projects whereby construction starts prior to completion of the design/contract documents

    Used to shorten time required for a completed project; may reduce loan interest and other fees, but also relies on incomplete information that may cause design omissions, use of inadequate design data, and irregular scheduling and work patterns

- **partnering** (see Figure 5.2)

    New concept to optimize good faith commitments, effectiveness of resources, and harmonious relationships between project participants; "coming together" for a common cause and mutual benefit

    Based on trust, mutual interest/culture, and shared goals, but may be better done with a written charter (an agreement, but not a contact per se)

    Utilized to enhance the relationships, efficiency, innovation, conflict resolution, and quality of service beyond the basic construction documents and agreement, without replacing or superseding those documents

- **teaming agreements**

    Relationship between a prime contractor and subcontractors who aided prime in winning project

    Closely resembles the traditional prime/sub contract, but based on stronger mutual ties and interests

- **build/operate/transfer (BOT)**

    An extended turnkey concept in which a consortium provides financial backing for the project and begins operation before turning the project over to the state/user

    Used for international projects; involves a sovereign state, project sponsors, financiers, and users

All of these options, alternatives, or "opportunities" (reviewed in Figure 5.3) now make it incumbent on the owner to decide on the detailed requirements of the contracts, not only as to content, but also to the configuration of parties that will be involved. Of course, this should be done in the context of the legalities involved, as well as the practical design and construction work. All parties, even though they may

---

The prominent construction weekly, *Engineering News Record,* reported (March 17, 1997, p. 34) a major success for the concept of "partnering." Specifically, the new federal office building under construction in Boston is a unique, complex design. It is a very unwieldy, uncertain, and challenging project for the constructors. The General Services Administration and the general contractor have undertaken seriously the partnering concept, and, despite the "inverted, distorted cone-shaped" facade construction, there have been few contentious or adversarial moments and, most surprisingly, no claims against the project. A most collaborative spirit prevails. Further credit is attributed to numerous meetings and retreats and the inclusion of all stakeholders in the process.

FIGURE 5.2   Example of successful partnering.

CONTRACT OPTIONS					
	Architect/Engineer	Contractor	Contractors	Construction Manager	
Traditional	P(D)	GC	Mechanicals[*]	–	5 separate contracts
Tripartite	P(D)	P(GC)	**	–	2 separate contracts
Multiple Primes	P(D)	GC	Varies	–	Numerous contracts
Design/Build 1	P(D)	GC(S)	–	–	1 contract with A/E
2	JV	JV	–	–	1 contract; 3 parties
3	D(S)	P(GC)	–	–	1 contract with GC
4	–	P(D/GC)	–	–	
Construction					
Management 1	P(D)	–	–	SC	Separate A/E contract for CM
2	D(S)	P(GC)	–	SC	Separate GC contract for CM
3	D(S)	–	–	P	
4	P(D)	–	–	P(CM)	2 separate contracts

P(D) prime contract/design
GC general construction
GC(S) general construction/subcontract
JV joint venture
D(S) design/subcontract
[*]P, HVAC, and E contracts

P(GC) prime/general construction
P(D/GC) prime/design and general construction
SC separate contract
P(CM) prime/construction management
**Mechanical contracts bid separately and assigned to GC, or are direct subcontracts

Number of Contracts
a. single prime contract
b. multiple prime contracts

Contract Types
a. construction management
b. design/build
c. owner/build
d. construction subcontracts

Tripartite: owner/A/E contract; owner/contractor contract
Single construction contract: competitive bid; separate A/E contract
Multiple prime contract: competitive bid; several O/C contracts; separate A/E contract
Construction Management: O/A/E contract with separate contractor and separate CM contracts
Design/Build: O/A/E or O/C contract with appropriate professional sub; also O/administrator/professional contract
Owner/Build: O/A/E contract; numerous subcontracts with owner

FIGURE 5.3 There is a wide variety of contract options.

not be participants in the initial conversations and may not be in a position to influence the owner's decision, must "buy in" to the program that is finally offered. This process should occur early in the project's life (actually in a preprogramming or programming sequence). It is essential to know not only some idea of the general direction of the contracts, but much of the detailed requirements and relationships, even before the first party is brought into the project, no matter who that party may be.

The proper basis on which to shape the contracts is, of course, the best interests of the client/owner and the goals that are set or must be met. Unfortunately it is not very wise to start a project without this determination, or to change the configuration well into the project; this can lead to many complications, both legal and in construction. Most frequently owners seek a single source for liability, responsibility, and accountability. They want the onus for all this in one place, person, or firm. They do not want to have to deal with numerous persons and the mesh of interconnection between them. This is really the mystique and intrigue of the construction industry, but modern owners want no part of it; they want to move through their projects quickly, smoothly, and with minimal added involvement on their part. The owners' point of view, while changing over the last few years, is still rather foreign to the construction industry, where interrelationships abound due to contractual demands.

It is obvious that the owner may have little counsel in making these decisions, other than the legal advisor. Owners may shortchange themselves when they make decisions with no input from design and construction people. Perhaps there are viable (and more attractive) options that the owner and the attorney are overlooking, or of which they have no working knowledge; maybe the project will cost more and be built more slowly because of the owner's final decisions about contract configurations.

No absolute conclusion can be drawn that one option or another is perfectly suited for a particular project. Architects continue to argue that their training lets them provide the best and most wide-ranging expertise to the client and the project. Having at least minimal training in planning, programming, design, construction, structural and mechanical systems, legalities, and all other project aspects, these professionals bring valid experience to any project. Other professions usually lack the fuller range of insight seated in architects. Of course, in certain projects, a narrow expertise may well be better, since it is directed to the core of the project and the solution required by the owner.

Quite often, owners will have a history of modifying or adding to their original facility. Much of this work may be done by merely contacting a contractor, who, working with the owner, will fashion the necessary work (without use of a design professional). Needless to say this works, although there are cases where the work is unsuitable, unsatisfactory, and not in the best interest of the client. But a direction is set, and the owner could follow the same path repeatedly. A good deal of the success of this scenario is because of the close owner/contractor relationship as they work to the owner's time line, mainly under the owner's direction (even including on-site meetings and hands-on direction). Often the owner perceives that this produces better cost control, which is true only to some degree; too often, costs escalate as the owner adds work or makes changes, and the contractor willingly complies without a good discussion about costs. In addition, perhaps hidden from the owner is the loss of time and money in her/his primary business venture, due to excessive involvement in the construction project.

For every successful project using any of the options, there is surely a failure. This only reinforces the fact that there is simply no single configuration good for every project; the configuration is client oriented. Without knowledgeable advice up front, however, owners are left to fend for themselves, leading to nothing more than trial-and-error experiences, or a waste of time and money. There is no forum, except experienced design professionals, available to owners to explain these options and the advantages and disadvantages of each in a objective manner. In their agency relationship, the design professionals must act in the owner's best interest. Too often the personalities involved with the options "pitch" their expertise at the expense of other participants. This, of course, is merely business competition.

What is too easily forgotten or overlooked in this system is the fact that no one has all the answers for a project. "Team effort" is not a catch phrase but is the epitome of striving for success!

As late as November, 1995, magazines such as *Progressive Architecture* still carried correspondence and articles about the creation of form in architecture, the place of technology in architecture, the vying for control between agencies that influence the profession of architecture, and the "dirty work" (technology, construction contract administration, etc.) of the profession. Can we not agree that architecture is not purely an idealistic, theory-oriented, paper exercise? Rather, the accommodation of the client is the true measure of architecture. And architecture cannot exist without a period of implementation known as "construction." Unbuilt "dreams" may be ego-fodder, may please critics, and may provide endless discussion material for academics, but only when the project has run its entire gamut and exists can the determination of good or bad architecture be made.

There must be complete acceptance of the axiom that the design concept (as approved) becomes the preeminent issue. Everything after the client approves the concept must be fully coordinated and aimed at properly and faithfully executing that concept. No other scenario is acceptable to the client nor to the professional. The challenge remains not only conceiving and fashioning good design, but "putting it on the streets" where it can fulfill its true intent: a better environment and a positive influence on the general public and the client/users.

Doesn't "professional" mean to exercise the best knowledge and skill/talent/ability available to one, in a manner that produces the proper solution for the client? Shouldn't excellence be measured over the entire project, not just in the rendering? Shouldn't the project architect and the on-site project representative also share, commensurately, in the kudos when the project is successfully completed and opens? As summarized by Barry D. Yatt, AIA, assistant professor at Catholic University of America and member of the AIA Documents Committee and Education Task Force,

> Perhaps what we all need to accept, if the construction industry at large is to see architects as competent professionals, is the idea that design is a very broad topic, which joins the word "creative" with the words "problem solving," and is as present in the work of outstanding project managers as in the work of outstanding formalists.

The architectural forms created by the designers (and discussed in the trade magazines) fail to exist save for the process called construction. Where that process is flawed or diverted the form will never exist. Those engaged in the pursuit of the project in the field must accept, respect, and faithfully maintain the basic concept of form given to them through the contract documents. The task is to ensure that the correct form rises from the muddy hole and becomes the "unmuddied whole"—that which the client seeks, the designer has designed in response to the program, and the documents require.

With all of these options, the contract administrator in the field must deal with a good variety of attitudes, directions, and intentions. Needless to say, the end result—the finished project in compliance with the contract documents—remains unchanged, no matter what! This must become almost obsessive on the part of the contract administrator.

Differences Among Professional Disciplines

Despite the commonality of purpose in construction—a fully compliant project, profitable to all, and without dispute—the professionals in the various disciplines tend to operate and accomplish their assigned tasks in different ways. This is not necessarily adverse to the work, it is just that differing outlooks, intentions, and attitudes cause different reactions and results. In the main, these differences emanate either from contractual provisions ("what do you want me to do?") or professional philosophy ("this is how we should do our work"). This comes either from academic training or time-tested experience. Some use a hands-on approach, others a more "overview" approach.

Perhaps the best way to discuss the differences among architects, engineers, and construction managers is to look at the work they do and the obligations they are contracted to meet:

Architect

General overall concept and control of project

Detailed selection and documentation of systems and materials

Formulation/control of design team and consultants

"Observation" of work in progress

Technical adviser to owner on the entire project process

Concerned with every aspect of project from aesthetics to mechanical systems

Engineer

Conceptual and detailed control over more utilitarian (and mostly unseen) portions of project

Detailed selection and documentation of selected systems and materials

Coordinates/cooperates within design team

"Observation" and manipulation of work in progress to ensure correct performance

Technical adviser to architect on limited systems

Concerned with proper installation/performance of systems

Construction Manager

Monitors design development; makes cost estimates

Assists/advises on system/material selection

Coordinates between design and construction teams

Administers construction process; solves problems

Technical adviser to owner on costs, time

Concerned with timely/budget-abiding completion

From this matrix each professional selects a "path" of procedures to accomplish the purposes outlined. These vary not only from project to project, but also between persons, offices, and professions. Mainly the variation starts with the item listed first under each profession. For example, what is the extent of the professional's concern? The broader the area of concern, the more the professional tends to deal in a more open and managerial manner. This is not disinterest, but rather a manifestation of the wider responsibility. The architect tends to administer a project in a manner somewhat removed from the up-front, hands-on techniques of the engineer. While fully conversant with scheduling, costing, problem solving, and all the attributes of proper contract administration, the architect tends to oversee the work of others. The engineer takes a more active part in the actual execution of this work and the other functions. Again, this goes back to the charge given each by their contract. Engineering projects are usually larger, more complex, and of a more utilitarian nature (industrial buildings and complexes, public works, etc.). In general, this category of projects requires closer control and coordination, and works to tighter tolerances.

There is still an ongoing dispute regarding what state registration laws allow professionals to do within the scope of their profession. Most permit architects to do "engineering work" within the scope of their projects, and vice versa for engineers. This has no direct bearing on contract administration, but it serves to show that both professions seek to satisfy a narrow range of projects where they function more or less exclusively.

As noted previously, the differences between the professions are a matter of education, training, attitude, experience, direction, and intent. An architect who takes a position with an engineering/design/build firm specializing in industrial work, for example, had better understand that pure architectural considerations will be "minimized" for the most part. The aesthetics of the structures may occupy the lowest priority, but still there is a need for such input. So too with an engineer who joins an architectural firm. Fitting his/her "engineering" into the dominant architecture scheme may prove to be difficult. However, in both of these cases, contract administration *must* be in place to bring the projects to proper, timely, contract-abiding completion. No matter what their training, experience, or professional philosophy, the paramount perspective of *all* on-site administrators must be faithful and successful completion of the project—on time, on budget, and in full accord with *all* contracts!

Two documents currently exist that describe the duties responsibilities, and limitations set upon professionals (and their staffers) engaged in contract administration: AIA document B352 (see Figure 5.4) and Engineers Joint Contracts Documents Committee (EJCDC) document 1910-1-A (EJCDC is a group formed from factions of the National Society of Professional Engineers, the American Consulting Engineers Council, and the American Society of Civil Engineers). Surprisingly, there are some striking differences between these instruments, although they are both intended to cover the same areas of practice. There are variations in the intent and language of the documents, in wording such as "certify" versus "recommend" in regard to payment requests. An in-depth review of the documents is advisable, along with legal scrutiny. In the main, though, professionals rather consistently use the documents promulgated by their respective associations.

No one form of administrative organization can be applied to every construction project without any changes. Usually there is a unique organizational structure for each project. This is done mainly to meet the requirements of the client and the other participants. Office manuals for contract administration should set forth the prescribed program, along with various options and modifications that may be included if desired. In this way, the professional retains control, for the most part, over the administration process. Governmental and corporate projects tend to require more formal organization because of their operating regulations, formats, and various administrative programs. Privately funded projects normally are simpler and have more direct formats.

Documents B352 and 1910-1-A reveal philosophical differences between architects and engineers, as well as the fact that most large, involved projects are usually of an engineering nature. So to best discuss contract administration, first we should review the types of construction projects. Administration is necessary on every project, no matter its attributes, but some demand radically different methods and commitment, as well as added personnel, increased expertise and experience, and innovative methods (and often some plain, old-fashioned common sense).

There are numerous ways to categorize construction, but here we will consider not so much "heavy" and "light" construction, but engineering, industrial, and building construction. The first is comprised of large, even massive projects involving civil and structural engineering practice other than for buildings: roads, bridges, dams, towers, navigational facilities, pipelines, power/water/sewage plants, and some extremely heavy industrial complexes (including some structures and enclosures).

Industrial construction refers almost exclusively to heavy manufacturing and processing plants, i.e., heavy industry facilities. The projects are usually extreme in cost and complexity; automobile assembly plants, for example.

Building construction can be further divided into residential and nonresidential. The former category is divided into single-family and multifamily (unit) dwellings. The latter is divided into:

Commercial	**Institutional**	**Industrial**
residential	financial institutions	light/heavy plants
business operations	colleges/schools	mining operations
restaurants	government buildings	light manufacturing
professional services	hospitals	processing plants
retail sales	long-term care facilities	storage facilities

The document for engineers, an optional/suggestive guide (which needs to be confirmed and specifically implemented), leans toward using an on-site representative who is given wider ranges of duties and deeper responsibilities than his/her architectural counterpart. This reflects the basic differences in the types of construction projects. The language of the document seems to indicate that the engineer should be more proactive in the actual "running" of the work through direct involvement with

Reproduced with permission of The American Institute of Architects under license number #97016. This license expires April 30, 1998. FURTHER REPRODUCTION IS PROHIBITED. Because AIA Documents are revised from time to time, users should ascertain from the AIA the current edition of this document. Copies of the current edition of this AIA document may be purchased from The American Institute of Architects or its local distributors. The text of this document is not "model language" and is not intended for use in other documents without permission of the AIA.

DUTIES, RESPONSIBILITIES AND LIMITATIONS OF AUTHORITY OF THE ARCHITECT'S PROJECT REPRESENTATIVE

AIA DOCUMENT B352

Recommended as an Exhibit When an Architect's Project Representative is Employed

1. GENERAL

1.1 The Architect's Project Representative shall be stationed at the site and shall be responsible for assisting the Architect in the administration of the Contract. Through the observations of the Project Representative, the Architect shall endeavor to provide further protection for the Owner against defects and deficiencies in the Work. Apart from such further protection, the rights, responsibilities and obligations of the Architect as described in the Agreement Between the Owner and Architect shall not be modified by the furnishing of such Project Representative.

1.2 Communications by the Architect's Project Representative relating to administration of the Contract shall in general be restricted to the Architect and Contractor. The Project Representative shall communicate with the Owner and Contractor under the direction of the Architect and with the Architect's full knowledge. The Project Representative shall not communicate with Subcontractors or material suppliers except with the full knowledge and approval of the Contractor and Architect.

2. DUTIES AND RESPONSIBILITIES

The Project Representative shall:

2.1 Perform on-site observations of the progress and quality of the Work as may be reasonably necessary to determine in general if the Work is being performed in a manner indicating that the Work when completed will be in conformance with the Contract Documents. Notify the Architect immediately if, in the Project Representative's opinion, Work does not conform to the Contract Documents or requires special inspection or testing.

2.2 Monitor the Contractor's construction schedules on an ongoing basis and alert the Architect to conditions that may lead to delays in completion of the Work.

2.3 Receive and respond to requests from the Contractor for information and, when authorized by the Architect, provide interpretations of Contract Documents.

2.4 Receive and review requests for changes by the Contractor, and submit them, together with recommendations, to the Architect. If they are accepted, prepare Architect's Supplemental Instructions, incorporating the Architect's Modifications to the Contract Documents.

2.5 Attend meetings as directed by the Architect and report to the Architect on the proceedings.

2.6 Observe tests required by the Contract Documents. Record and report to the Architect on test procedures and test results; verify testing invoices to be paid by the Owner.

2.7 Maintain records at the construction site in an orderly manner. Include correspondence, Contract Documents, Change Orders, Construction Change Directives, reports of site meetings, Shop Drawings, Product Data, and similar submittals; supplementary drawings, color schedules, requests for payment; and names, addresses and telephone numbers of the Contractors, Subcontractors and principal material suppliers.

2.8 Maintain a log book of activities at the site, including weather conditions, nature and location of Work being performed, verbal instructions and interpretations given to the Contractor, and specific observations. Record any occurrence or Work that might result in a claim for a change in Contract Sum or Contract Time. Maintain a list of visitors, their titles, and time and purpose of their visit.

2.9 Assist the Architect in reviewing Shop Drawings, Product Data and Samples. Notify the Architect if any portion of the Work requiring Shop Drawings, Product Data or Samples is commenced before such submittals have been approved by the Architect. Receive and log Samples required at the site, notify the Architect when they are ready for examination, and record the Architect's approval or other action; maintain custody of approved Samples.

2.10 Observe the Contractor's record copy of the Drawings, Specifications, addenda, Change Orders and other Modifications at intervals appropriate to the stage of construction and notify the Architect of any apparent failure by the Contractor to maintain up-to-date records.

2.11 Review Applications for Payment and forward to the Architect with recommendations for disposition.

2.12 Review the list of items to be completed or corrected which is submitted by the Contractor with a request for issuance of a Certificate of Substantial Completion. Review the Work. If the list is accurate, forward it to the Architect for final disposition; if not, so advise the Architect and return the list to the Contractor for correction.

2.13 Assist the Architect in conducting inspections to determine the date or dates of Substantial Completion and the date of final completion.

2.14 Assist the Architect in receipt and transmittal to the Owner of documentation required of the Contractor at completion of the Work.

3. LIMITATIONS OF AUTHORITY

The Architect's Project Representative, in acting on behalf of the Owner, shall not exceed the authority of the Architect under the Agreement Between the Owner and Architect. The Project Representative shall NOT:

3.1 Authorize deviations from the Contract Documents.

3.2 Approve substitute materials or equipment except as authorized in writing by the Architect.

3.3 Personally conduct or participate in tests or third party inspections except as authorized in writing by the Architect.

3.4 Assume any of the responsibilities of the Contractor's superintendent or of Subcontractors.

3.5 Expedite the Work for the Contractor.

3.6 Have control over or charge of or be responsible for construction means, methods, techniques, sequences or procedures, or for safety precautions and programs in connection with the Work.

3.7 Authorize or suggest that the Owner occupy the Project in whole or in part.

3.8 Issue a Certificate for Payment or Certificate of Substantial Completion.

3.9 Prepare or certify to the preparation of a record copy of the Drawings, Specifications, addenda, Change Orders and other Modifications.

3.10 Reject Work or require special inspection or testing except as authorized in writing by the Architect.

3.11 Accept, distribute, or transmit submittals made by the Contractor that are not required by the Contract Documents.

3.12 Order the Contractor to stop the Work or any portion thereof.

AIA DOCUMENT B352 • ARCHITECT'S PROJECT REPRESENTATIVE • AIA® • ©1993 • THE AMERICAN INSTITUTE OF ARCHITECTS, 1735 NEW YORK AVENUE, N.W., WASHINGTON, D.C. 20006-5292 • **WARNING: Unlicensed photocopying violates U.S. copyright laws and will subject the violator to legal prosecution.**

B352—1993

FIGURE 5.4 AIA document B352 enumerates the parameters for the on-site project representative and directs how this person is to function.
Used with permission of the American Institute of Architects.

scheduling, testing, and other administrative operations. The architect can be so involved, but in the norm will be more on the periphery as the operations are "run" by others, primarily the construction manager or the general contractor's on-site personnel. This in no way, however, indicates a more "arms-length" approach by the architect, just a different manner of administering the contract.

Various sources (authors, industry literature, forms, documents) define contract administration in different ways. Some reduce the scope of this program to mere handling of the business relations between parties to the contract (basically viewed as a "paperwork only" administrative function). Others incorporate contract administration as an integral part of construction administration, where it joins the following all-project-related functions:

- conduct of the parties
- relations with the contractors
- communications
- business systems
- establishing and maintaining procedures
- responsibilities/duties/authority of all parties
- documentation requirements
- construction operations
- planning/scheduling of project operations
- coordination at all levels
- materials control
- payment administration
- change orders/other modifications
- dispute/claim handling
- negotiations
- punch list inspections*
- final cleanup*
- administrative closeout*

Perhaps the true measure of contact administration lies between these extremes, incorporating aspects or modified duties of the latter definitions, as follows:

- general contract oversight—relationships and conduct of parties
- observation of in-progress operations/work
- establish and maintain communications
- establish and maintain administration procedures
- defining/monitoring responsibilities/duties/authority of parties
- overall planning
- monitor work progress schedule
- assist/facilitate coordination at all levels
- issue certificates for payment
- issue contract/work modifications
- participate in resolving disputes/claims/negotiations
- participate in all project closeout functions

---

* All project closeout functions.

The outlines above may seem to be very similar. However, the mere change of a word places a different emphasis on some activities and in the last list makes them more appropriate to true contract administration as practiced by most professionals. Obviously, in some instances, these duties and activities change intensity and authority to match the needs of the project.

To aid in analyzing the two styles, the following list notes comparable sections of the EJCDC and AIA documents that address the same or similar topics:

Topic	EJCDC Document 1910-1-A	AIA Document B352
Relationships	Par. A	1.1 General
Observation	Preamble/B5a	2.1
Schedules	B1 partial	2.2
Liaison	B3/B6	2.3
Changes	—	2.4
Meetings	B2	2.5
Testing	B5c	2.6
Records	B8a	2.7
Log/directory	B8b/c	2.8
Submittals	B4	2.9
As-builts	—	2.10
Pay requests	B10	2.11
Occupancy	B12	2.12
Partial occupancy	—	2.13
Final inspection	B12b	2.14
No changes	C13.1	3.2
Test participation	C8	3.3
No contractor work	C3	3.4
No expediting	—	3.5
No safety/advice	C4/C5	3.6
No occupancy	C7	3.7
Limits	—	3.8/3.9/3.10/3.11

The design professional also has a commitment under both the contract and the profession's ethics to perform as required, including periodic site visits and work observations. These functions are not necessarily time oriented. Some construction operations are critical and require on-site review and observation. Placing concrete, for instance, requires that the architect be on the job to see that the proper procedures, equipment, workforce, finishing processes, and so on are involved and executed in accord with the contract documents (see Figure 5.5). This does not, however, require that the architect have ten people on the job all day, overseeing every

---

3A-11   <u>PLACING</u>

a. Place no concrete except when Architect's representative is present unless this requirement is specifically waived by the Architect. Give due notice to the Architect and all Contractors affected before placing concrete. Allow adequate time for installation of all necessary parts and ancillary work.

FIGURE 5.5   Excerpt from typical specification. A firm statement to the contractor, it allows the architect to inspect and control the operation.

> 2.2.10.1 The Architect's field representative does not have the authority to approve materials substitutions, design changes, deviations from the Drawings and Specifications or changes in the Contract amount. Should such considerations arise, authorization for them must be in writing signed by the Project Architect for those items for which the Architect has authority under the Contract or by the Owner.

FIGURE 5.6 Excerpt from typical specification. A procedure that allows the architect to maintain positive control. Decisions are made in the office by a high-level manager, not in the field.

minute detail of the concrete placement, or observing every single nail as it is driven, or other such small operations. By contract, the professional is required to have "general (working) knowledge" of the construction operation.

The design professional is responsible for the overall project. Through experience the professional will know that there are details of construction—some may even be shown on the contract documents—whose execution is left to the expertise of the contractors. This is not to say that the professional ignores the job. There is a generally accepted understanding on a construction job that the contractor is well trained and informed. Therefore, the contractor will perform the work to the best of his/her ability and within the confines of the contract documents. If some improper work is installed, the contractor knows that she/he is under contractual obligation to remove the faulty work and rebuild it properly to the satisfaction of the design professional and the project requirements.

There are further contractual provisions that if the work has progressed and faulty work is covered up, the professional has the authority to have the cover removed to view the underlying work. If that is faulty, the removal, reworking, and recovering is all done at the contractor's cost. A number of similar safeguards are involved (see, for example, Figure 5.6) and the design professional should point these out to the owner. With this understanding, the owner will not be negatively influenced by the decisions of the professional.

As noted earlier, all the complexity of the construction process is carried on within the professional's overall legal obligations and responsibilities. The activity of construction, because of the numerous parties, the close proximity and interface of work, the amount of money at stake, and the administration of varied operations, attitudes, and goals, begets more than its share of entanglements, some so severe they can be solved only through legal action. The accompanying paper discusses some of the specific legal problems that can arise from the work of the professionals during the contract administration phase. The author, who is active in construction litigation, presents a constructive oversight of recent case law. His discussion emphasizes the level of liability and the breadth of situations that can, and do, create problems—often, unfortunately, unresolved except through the court system.

Legal Concerns in Construction Field Administration

Peter C. Halls
Faegre & Benson
Minneapolis, MN

A. Communications and Recommendations

    1. One of the most important tasks is communication

    2. Communicating negative information—defamation risks (libel, slander, defamation, tortious interference with contractual relations)

        a. Examples:

            i. Recommendations against irresponsible bidders

            ii. Disapproval of proposed subcontractors

            iii. Recommendations against claims or change order requests

            iv. Evaluations of work

            v. Removal of personnel

            vi. Recommendations on default or termination

    3. Examining the issue through examples:

From time to time, it is necessary to recommend against awarding a contract to the low bidder because that bidder is not responsible. The architect or construction manager may have had bad experiences or heard bad reports about the contractor. The architect or construction manager then faces the dilemma either of not warning the owner or of warning the owner and facing possible litigation by the rejected contractor for defamation or tortious interference with contract. Although lawsuits are a danger of speaking up, if the architect or construction manager [had] knowledge indicating that a bidder was not responsible and did not inform the owner, there could be liability to the owner for failure to warn should the contractor run into problems and cause the owner damages.

In *Vojak v. Jensen*, 161 N.W.2d 100 (Iowa 1968), an architect objected to the low bidder on a project. After being denied the project, the low bidder was allowed to sue the architect for libel.

WHC, Inc. v. Tri-State Road Boring, Inc., 468 So.2d 764 (La. Ct. App. 1985), involved a road boring subcontractor's claims for libel and slander. The owner's employees had been very critical of the road boring subcontractor's performance and the general contractor terminated the subcontractor. When the general contractor sued the subcontractor for breach of contract, the subcontractor sued the owner for defamation based on allegations that the owner's employees falsely stated that the subcontractor was working in an untimely and unworkmanlike manner. Although the trial court threw the lawsuit out before the trial, in this case the appellate court overruled the trial court and said that the subcontractor's claims could go to trial.

In *Kendrick v. Jaeger*, 436 S.E.2d 92 (Ga. Ct. App. 1993), the City of Gainesville was building its civic center and was taking bids for irrigation and landscape improvements. The landscape architect recommended awarding the

---

Material in this article has been abridged from laws, court decisions, administrative rulings, congressional materials, and other sources. Further details may be necessary for a complete understanding of the statements in this article. For this reason, it is imperative that you consult with legal counsel before acting on any matter discussed in this article.

©1995, Peter C. Halls, Faegre & Benson, Minneapolis, MN. Used with permission.

contract to the third lowest bidder and told reporters that her finding was that the third lowest bidder was really the "lowest responsible bidder based on its record of completing projects of the magnitude of the civic center and based on the quality of work." The low bidder then sued the landscape architect for libel. The court dismissed the lawsuit. The court reasoned that the architect's statements were simply an expression of opinion that one contractor was more suited for a job than another, and that this was an opinion about which people could reasonably differ. An expression of opinion on a subject where people can reasonably differ is not actionable, because it cannot be proved to be false.

Results positive for those in the position of making recommendations also appeared in *Lapar v. Morris,* 501 N.Y.S.2d 82 (N.Y. App. Div. 1986). In that case, a consulting engineer hired to diagnose problems in a wastewater treatment plant issued a report concluding it was the original designing engineer's fault. The designing engineer sued the consulting engineer for defamation, but the suit was dismissed. The court held that the consulting engineer laid out the underlying facts and then expressed his opinion based on those facts. Since the consulting engineer did not refute the underlying facts, the opinions were not actionable.

It occasionally becomes necessary to terminate a contractor. The case of *Victor M. Solis Underground Utility & Paving Co. v. City of Laredo,* 751 S.W.2d 532 (Tex. App. 1988) should provide some assurance to someone in the situation of recommending termination to the owner. In that case, the appellate court rejected the terminated contractor's suit for tortious interference with contract against the architect and project engineer. The court held that, as agents for the city, the designers had an absolute right to recommend termination. The court held that the contractual interest of the architect and engineer with the city was superior to the contractor's contractual interest.

    4. Guidelines to minimizing exposure on negative communications:
       a. Truth is a defense
       b. Qualified immunity—if follow steps
          i. Communicate in good faith
          ii. Interest in subject matter
          iii. Made to one with similar interest
       c. Separate opinions from facts
          i. State underlying facts neutrally and fully
          ii. Document and verify facts
          iii. Identify opinions
          iv. Phrase opinions carefully and without implying untruths
    5. Oral communications—waiver.

The typical construction contract contains great protective clauses requiring modifications to be in writing, claims to be in writing, change orders to be in writing, and on and on. However, even the clause requiring modifications to be in writing can be orally waived. A practice of orally waiving formal written requirements on a case-by-case basis can become a waiver of that entire requirement. *J.A. Moore Construction Co. v. Sussex Associates Ltd. Partnership,* 688 F.Supp. 982 (D. Del. 1988) provides an example of the law in this matter. The contractor contended that the contract which it had signed with the defendant developer was at a price below the actual agreed price as a sham to allow the developer to obtain financing that required that cost be below $10 million in order to qualify for tax-favored status. The defendant allegedly induced the contractor to sign the contract by orally promising to pay the difference between the contract price and the actual construction cost. In addition, after performance began, the defendants requested that the general contractor stop submitting written change orders because the $10 million limit would be reached, and again the defendant promised that the parties would make up for the actual costs of changes and modifications at the end of the project. As can be surmised by the fact that this case had to go to court, the defendant developer then refused to make payment as the project neared completion.

The court held that even if the promise regarding the original lump sum price had been made with the intent to defraud the contractor, the contractor could not establish justifiable reliance sufficient to allow it to support its claim that the contract price was understated. The court held that both parties were sophisticated and that the contract specifically disallowed any prior oral assurances.

However, on the issue of change orders, the court held that claim could go forward. Even though the contract required that change orders be in writing, the law is that a subsequent oral agreement can modify and waive that requirement, even if there is a clause in the contract stating that any modifications must be in writing.

Some states follow this same rule of law; some states say that the requirements for a writing can be waived but only with a higher standard of proof; and a few states enforce the requirement for a writing.

    6.   Recommendations versus orders.

The protective language regarding means and methods being the responsibility of the contractor can be undone by taking actions contrary to that language. Orders by an owner, architect, or construction manager to contractors directing their means and methods can make the owner, architect, or construction manager responsible for means and methods. Questions are a more useful tool and can often get to the same result. The major decisions (not "means and methods," which typically are up to the contractor) should be left up to the owner based on the recommendation of the architect and construction manager.

    7.   Arbiter of contract documents—often a role for designers.
        a.   If interpretation is wrong
        b.   Immunity for good faith decisions
            i.   decisions must be made fairly
- not arbitrary
- comport with customer and trade usage
- allow opportunity to present arguments and support

           ii.   immunity is for acts within scope as arbiter as opposed to acts as designer
           iii.   *E.C. Ernst, Inc. v. Manhattan Construction Co. of Texas,* 551 F.2d 1026 (5th Cir. 1977), *reh'g denied in part, granted in part,* 559 F.2d 268 (5th Cir. 1977), *cert. denied sub. nom. Providence Hospital v. Manhattan Construction Co.,* 434 U.S. 1067 (1968). *See also Kecko Piping Co. v. Town of Monroe,* 172 Conn. 197, 374 A.2d 179 (1977); *see generally* Little, *The Architect's Immunity as Arbiter,* 23 St. Louis U. L.J. 339 (1979); Coulson, *Dispute Management under Modern Construction Systems,* 46 *Law & Contemporary Problems* 127 (1983); Annot., 65 A.L.R.3d 249 (1975).
        c.   Rules of interpretation
   B.   Supervision/Observation/Inspection of Jobsite Activities
       1.   Failure to detect defects.

Despite protective contract language, failure to detect defects during field administration can still impose liability. Even in the best case, one may still be likely to find himself embroiled in a lawsuit.

The architect or construction manager may be liable to the owner for breach of contract in the event that the architect or construction manager is responsible for inspection of the work necessary to complete the project and thereafter permits workmanship and materials of substandard quality to be utilized on the job. This potential liability is illustrated by the decision in *First National Bank of Akron v. Cann,* 503 F. Supp. 419 (N.D. Ohio 1980), *aff'd,* 669 F.2d 415 (6th Cir. 1982). In that case, the owner entered into an agreement with a construction manager who agreed to be responsible for the work necessary to complete the renovation of a bank. During the course of the renovation, granite panels were installed on the building facing. Seven years after completion, the panels began to separate from the building and an investigation revealed that installation of the granite sections had not been performed in accordance with the plans and specifications. The bank, after demand, had remedial work performed and then commenced litigation against its architect and construction manager under breach of contract and negligence theories. The trial court rejected the bank's claim that the construction manager was liable in tort for negligent failure to discover the variances from the plans and specifications, concluding that the bank's cause of action was one bottomed in contract and was, in essence, a claim that the construction manager failed to perform services in a workmanlike manner. However, the court concluded that the construction manager and architect were jointly and severally liable to the owner for breach of contract inasmuch as there were substantial and unapproved deviations from the drawings and specifications governing the work with the consequence that the granite facings as erected were structurally unsound and not of good workmanship or quality.

Some cases show the limited value of clauses exonerating parties from responsibility for construction defects. Architects and engineers have very protective language, yet in *Hunt v. Ellisor & Tanner, Inc.,* 739 S.W.2d 933 (Tex. App. 1987), it did not protect the architect completely. The contract language provided that the architect was not responsible for the contractor's failure to carry out work in accordance with the contract documents. The court held that this favorable language, however, did not exculpate the architect from liability where the contract also obligated the architect to inspect the jobsite. (Even the 1987 edition of B-141 obligates the architect to conduct "inspections." For example, see paragraph 2.6.14 which provides: "The Architect shall conduct inspections to determine the date or dates of Substantial Completion and the date of final completion . . . ") The court explained that the exculpatory language constituted nothing more than an agreement that the architect is not the insurer or guarantor of the general contractor's obligation to carry out the work in accordance with the contract documents. The court ruled against the architect on the basis of additional contract language which imposed a nonconstruction responsibility on the architect, that is, *to visit,* to familiarize, to determine, to inform, and to endeavor to guard. (The 1987 version of B-141 provides in paragraph 2.6.5 as follows:

> The Architect shall visit the site at intervals appropriate to the stage of construction or as otherwise agreed by the Owner and Architect in writing *to become generally familiar* with the progress and quality of the Work completed and *to determine* in general that the Work is being performed in a manner indicating that the Work when completed will be in accordance with the Contract Documents. However, the Architect shall not be required to make exhaustive or continuous on-site inspections to check the quality or the quantity of the work. On the basis of on-site observations as an architect, the Architect *shall keep the Owner informed* of the

> progress and quality of the Work, and *shall endeavor to guard* the Owner against defects and deficiencies in the Work.)

The court held that the architect was not absolved of liability based solely on the language in the contract documents.

In other states, the law is more favorable to designers. In *Moundsview Independent School Dist. No. 621 v. Buetow & Assoc., Inc.,* 253 N.W.2d 836 (Minn. 1977), the Minnesota Supreme Court came to just about the opposite result of *Ellisor & Tanner, Inc.* In *Moundsview Independent School Dist.,* the roof of a building blew off in a windstorm, and the owner sued the architect for failing to discover that the contractor had failed to attach washers or nuts to secure the roof to the south wall of the building as required by the plans and specifications. Based upon the fact that the architect was not a full-time clerk of the works and the standard AIA protective language (the same language ignored by the court in *Ellisor & Tanner, Inc.*), the Minnesota Supreme Court held that the architect was not responsible for the failure of the contractor to perform in accordance with the contract documents and upheld summary judgment in favor of the architect. However, construction managers are on the job full time, or at least more than the architect in the *Buetow* case.

In *Council of Co-owners, Atlantis Condominium, Inc. v. Whiting-Turner Contracting Co.,* 308 Md. 18, 517 A.2d 336 (1986), the court allowed the plaintiffs to state a claim for work allegedly required to correct a latent and unreasonably dangerous condition allegedly caused in part by the architect's failure to supervise the construction project adequately. In *Sweeney Co. of Maryland v. Engineers-Constructors, Inc.,* 823 F.2d 805 (4th Cir. 1987), the court stated that "supervision" is the "function of monitoring and improving construction work." The court stated that a subcontractor in that case in a claim against a contractor could use the certificates of payment approved by the architect as evidence that it had performed work properly.

On the other hand, in *Krieger v. J.E. Greiner Co.,* 282 Md. 50, 382 A.2d 1069 (1978), the court determined that under the contractual arrangements involved in that case, that the duties of the engineer did not include supervision of construction methods or supervision of work for compliance with safety rules and regulations, and therefore denied a claim against the engineer.

Also of some help in reducing liability is the case of *Al Johnson Construction Co. v. United States,* 854 F.2d 467 (Fed. Cir. 1988). In that case, the court held that the contractor could not invoke the doctrine of implied warranty of the adequacy of the design where it neither complied with the defective design nor adequately explained why it thought that the design was defective. The court stated that the implied warranty of design ran only to contractors who had complied with the specifications.

2. Improper supervision.

A construction manager may be liable to a trade contractor for improper supervision of work. In the case of *Gateway Erectors Division of Imoco-Gateway Corp. v. Lutheran General Hospital,* 430 N.E.2d 20 (Ill. App. Ct. 1981), an owner contracted with a construction manager and with specialty trade contractors. Alleging that the work of one of its specialty trade contractors was defective, the owner asserted an action against that trade contractor who in turn alleged that the construction manager negligently directed the trade contractor to proceed with its work and carelessly failed to supervise the work of a subcontractor of the claimant trade contractor. Following dismissal of the trade contractor's allegations of breach of contract and negligence against the construction manager, the Illinois Appeal Court after affirming that the construction manager did not owe a contractual duty to the trade contractor ruled that the trade contractor could maintain a cause of action against the construction manager for alleged negligent direction of the work. The court rejected the construction manager's defense that it was not liable in tort, as it was an agent of the owner and was acting within the scope of its authority, concluding that there is a duty of due care on the part of the construction manager to a trade contractor and thus potential liability on the part of the construction manager to the trade contractor. However, the court affirmed the dismissal of the breach of contract cause of action on the basis that the construction manager was the agent of the disclosed principal and did not agree to become personally liable under the contract between the owner and the trade contractor. Thus, while the construction manager may well breach its agreement with the owner and thus cause a breach of the owner's agreement with a trade contractor, such a breach on the part of the construction manager does not give rise to a breach of contract action by the trade contractor against the construction manager.

A contrary decision was reached in the case of *Harbor Mechanical, Inc. v. Arizona Electric Power Cooperative, Inc.,* 496 F. Supp. 681 (Ariz. 1980). In this case, a trade contractor sued an engineering firm which was retained to supervise and coordinate multiple prime contractors constructing a power plant, alleging delay damages. The court held that the engineer's obligations ran solely to the owner and the lack of contractual relationships defeated any recovery by the trade contractor against the engineer.

As might be presupposed, in the event that an owner, designer, or construction manager causes a trade contractor to incur additional expenses as a result of active interference, they may well be liable to the trade contractor for such damages.

In the decision of *R.S. Noonan, Inc. v. Morrison-Knudsen Company,* 522 F. Supp. 1186 (E.D. La. 1981), an owner and its construction manager were held liable to a trade contractor for $750,000.00 in delay damages and dewatering costs. There, under the terms of the construction management agreement, the construction manager was to coordinate with and advise the architect with respect to various aspects of the design phase, draft proposed bidder lists for the owner, solicit bidders for the owner, prepare bid packages, conduct preaward investigations and orientations, award contracts as agent for the owner, devise a work schedule for the project, and serve as intermediary between the owner and the trade contractors on the project. The court concluded that the defendants failed to provide adequate temporary site drainage and allowed other contractors to discharge water through incomplete storm drains for emergency drainage purposes, thus allowing the work site to become a quagmire. The court held that by failing to restrict and in some instances by affirmatively sanctioning the use of the trade contractors work area as a drainage facility, both defendants interfered with the trade contractor's ability to perform its contractual duties and thus were responsible to the trade contractor for the costs attributable to the delays and for the costs resulting from the removal of the water from the trade contractor's work area.

A similar holding is found in the case of *John E. Green Plumbing and Heating Company v. Turner Construction Company,* 500 F. Supp. 910 (E.D. Mich. 1980). There, the trade contractor under contract with an owner brought suit against the owner's construction manager alleging that the construction manager had continually and intentionally interfered with the work of the trade contractor and also alleging negligence on the part of the construction manager. As noted by the court, the construction manager had assumed the duties in its contract of reviewing plans and specifications so as to eliminate areas of possible conflict among separate contractors, establishing procedures for coordination among the owner and prime contractors and the construction manager, producing a schedule for the project, maintaining a full-time supervisory staff for the coordination of the work, and conducting regular job site meetings to discuss job problems. The court observed that many of these duties had an obvious bearing on the performance of the work by the plaintiff trade contractor.

On the basis that the contract between the trade contractor and the owner contained a no damage for delay clause, the construction manager filed a motion for summary judgment claiming that such clause barred the trade contractor's causes of action. The court held that the no damage for delay clause would not be read to encompass deliberate and willful interference with plaintiff's performance of the contract and therefore the trade contractor was entitled to an opportunity to support those allegations of the complaint. However, the court concluded that causes of action based on negligence would not survive a no damage for delay clause and affirmed the dismissal of that count.

3. Enforcement of test results/hypertechnical inspection.

Unreasonable testing standards can impose liability. If an owner is liable to a contractor, an owner can be expected to turn around and sue the construction manager for indemnification or contribution. Similarly, if a contractor incurs significant additional costs, a contractor could easily include the construction manager in a subsequent lawsuit to collect its costs.

In the case of *B.B. Andersen Construction Co., Inc.,* VABCA No. 2265, 2531, 88-2 B.C.A. (CCH) ¶ 20,630 (Veterans Admin. B.C.A. 1988), the Veterans Administration Board of Contract Appeals held that a concrete slab's deflection was reasonable and the slab was not defective. The contractor was then compensated for testing the slab at the government's request. The concrete slabs had displayed a deflection of 2″ instead of the 1.38″ deflection permitted in the specifications. The contracting officer accordingly ordered additional tests, which showed that the slab could carry the design loads safely.

The architect had anticipated instantaneous deflection of approximately ¾″ and had used a multiplier of 2 to obtain the long-term deflection of 1.38″. The architect then used the 1.38″ as the permissible deflection in the specifications. The board determined that, given the numerous variables, a multiplier of 2.5–3.0 was appropriate when converting from instantaneous to long-term deflection. The architect's multiplier of 2 had been made on the assumption that reshoring would be in place at all times for 28 days. However, the specifications required reshoring only when the amount of weight placed on the slab exceeded a certain minimum. The specifications did not require continuous reshoring. Using the more appropriate multiplier, the permissible deflection ranged up to 2.04″, and the 2″ deflection was within this standard. Therefore, even though the specifications actually called for permissible deflection of only 1.38″, because the actual deflection did not result from a violation of any contractual standards, the contractor was entitled to compensation for costs of the testing.

Construction administrators are not always subjected to liability for requiring testing. Many courts hold that a higher standard of liability is required for certain types of claims. In *Certified Mechanical Contractors, Inc. v. Wight & Co.,* 162 Ill. App. 3d 391, 515 N.E. 2d 1047 (2d Dist. 1987), the Illinois Appellate Court held that the contractor seeking to sue an architect for interference with contract had to establish that the architect acted with actual malice, and therefore could not pursue its claim.

During the course of construction, acting on the architect's advice, the owner ordered the contractor to stop work in order to allow tests on the work to be performed. The contractor alleged that the work stoppage was caused by the architect's attempt to cover up for its own mistakes. The contractor then sued the architect for tortious interference with contract. Although the court acknowledged that an interference claim could be maintained, the court also held that the architect must have induced the owner to breach the contract with malice. The court then went on to add that "actual malice means a positive desire and intention to annoy or injure another person." More than mere ill will must be shown. The defendant must have acted with an intent to harm another unrelated to the interest he was presumably protecting. Thus, only if the architect had induced the breach of contract by the owner not to further the owner's interests, but with the intent to harm the contractor or to further the personal goals of the architect, would the architect be liable for tortious interference with contract.

    C. Shop Drawing and Submittal Review
        1. Liability for approval.

Although construction managers, architects, and engineers have protective contract language, there will always be exposure for approving shop drawings and other submittals. The AIA standard owner-architect agreement provides that architects are to "approve" shop drawings and other submittals, albeit with some limiting language. (Paragraph 2.6.12 of AIA document B141, 1987 edition.) Similarly, the standard AIA owner-construction manager agreement provides that construction managers are to "approve" shop drawings. (Paragraph 2.3.20 of AIA document B801/CMa, 1992 edition.) The construction manager's or architect's defense to a suit for improper approval of submittals may ultimately be proven valid, but there will be an expensive process incurred on the way. Worse, there is a possibility of being proven liable, despite the protective language.

Some responsibility for shop drawing review may belong to the designer no matter what the contract says. The case of *Duncan v. Missouri Board for Architects, Professional Engineers & Land Surveyors,* 744 S.W.2d 524 (Mo. Ct. App. 1988) is an important example of the risks to design professionals. This case involved the appeal following the Missouri licensing board's disciplinary hearing and sanctions imposed following the 1981 collapse of the two walkways in the Hyatt Regency Hotel in Kansas City, Missouri. In that case, the hearing determined that the structural component consisting of hanger rods and box beam hanger rod connections shown on the structural drawings did not meet the design specifications of the Kansas City building code. The commission further found that the structural engineers did not review the shop drawings for compliance for the box beam hanger rod connections with design specifications of the Kansas City building code nor did they review the shop drawings for conformance with the design concept as required by the engineer's contract and specifications. The court defined the term "gross negligence" in the context of licensing as "an act or course of conduct which demonstrates a conscious indifference to a professional duty." The court added that the level of potential danger also helps determine the question of gross negligence. The level of care is directly proportional to the potential harm involved. The court held that "it is difficult to conclude that gross failure to comply with [the building code] can constitute other than conscious indifference to duty by a structural engineer."

Thus, in addition to potential civil liability, the ability of an engineer to be licensed and practice professionally can be at stake.

Another area of concern is turn-around time. Paragraph 2.3.20 of the 1992 edition of B801/CMa covers shop drawing review. It requires that "The Construction Manager's actions shall be taken with such reasonable promptness as to cause no delay in the Work or in the construction of the Owner or Contractors." A similar provision applies to architects. Paragraph 2.6.12, AIA document B141, 1987 edition.

    D. Scheduling

Scheduling can lead to many problems. In preparing or reviewing a schedule, the owner, architect, and construction manager must be careful to determine whether any of the dates create a conflict.

Chapter 5   The Design Professional

One method used by owners, architects, and construction managers to protect themselves is to insert a no damage for delay clause. While these clauses are upheld as valid under the proper circumstances, there are exceptions that destroy the sense of complacency that the clause might otherwise bestow. *See, e.g., Corinno Civetta Construction Corp. v. City of New York*, 67 N.Y. 2d 297, 502 N.Y.S.2d 681, 493 N.E.2d 905 (1986). Generally, delay damages may be recovered despite a no damage for delay clause under the following four conditions:

1. The damage was contemplated, but the owner acted in bad faith;
2. The delay was uncontemplated;
3. The owner abandoned the contract; or
4. The owner breached a fundamental obligation of the contract.

Other courts add as an additional exception an owner or construction manager's active interference with the performance of the contract.

E. Pay Requests
   1. Certification of pay requests can contain liability pitfalls for owners, designers, and construction managers.

In *Magnolia Construction Co. v. Mississippi Gulf South Engineers, Inc.*, 518 So.2d 1194 (Miss. 1988), the Mississippi Supreme Court denied the defendant engineer's request for summary judgment. At the completion of work on a sewer project, the project engineer discovered that the sewer lines were off grade. Accordingly, the owner refused to issue the final payment. The contractor subsequently sued the engineer alleging that the engineer negligently inspected the work. The court ruled that the contractor's claim against the engineer could go forward even though the contract between the owner and the engineer provided that the contractor's only duty in certifying progress payments was to determine that to the best of its knowledge and belief the work was performed according to plan. However, EPA funding was also involved in the project, and the EPA documents required a full-time resident inspector. On top of that, the engineer had an additional 15% override for duties normally undertaken by the city's public works director and the payment certificates contained the word "inspector." Finally, the engineer's on-site representative testified at deposition that, although he inspected only for the quantity of work and not the quality, he would not have certified any payment for work that was not of the proper quality.

Liability for certification of payments might not extend only to the party one contracted with, as was demonstrated in the *Magnolia* case.

Sureties have successfully sued architects. In one case, the architect was not allowed to assert the privity defense when the surety sued it for premature release of retainage. *State use of National Surety Corp. v. Malvaney*, 72 So.2d 424 (Miss. 1954). In *Hall v. Union Indemnity Co.*, 61 F.2d 85 (8th Cir. 1932), *cert. den.* 287 U.S. 663 (1932), the owner sued on a performance bond. The surety was allowed to make the defense that the architect, as the owner's agent, had certified progress payments which overpaid the contractor.

Lenders have also been able to sue architects. It has been held that it is foreseeable that a mortgage will rely on the certification of work completed to issue loan proceeds for payment to the contractor. Accordingly, negligent certification could injure the lender, and the lender injured in those circumstances may sue the architect certifying the payments negligently. *Hobbs v. Florida First National Bank*, 406 So. 2d 63 (Fla. Dist. Ct. App. 1981).

In another twist, an owner has been allowed to sue an architect for negligent certification of payment, even where the architect was hired by the lender rather than by the owner. After the contractor defaulted, the owner sued the architect for certification of work which had not been performed. The court allowed the action to go forward. *Browning v. Maurice B. Levien & Co. P.C.*, 44 N.C. App. 701, 262 S.E.2d 355 (1980).

In the case of *City of Mound Bayou v. Roy Collins Construction Co.*, 499 So.2d 1354 (Miss. 1986), the court allowed a contractor to sue a project engineer who had arbitrarily certified reduced progress payments and thereby violated a duty to act in good faith toward the contractor.

In the case of *Sweeney Co. of Maryland v. Engineers-Constructors, Inc.*, 823 F.2d 805 (4th Cir. 1987), a subcontractor was allowed to use certificates of payment as evidence that it had performed properly in its case against the general contractor.

Occasionally an architect has been exonerated from liability. In *Sheetz, Aiken & Aiken, Inc. v. Spann, Hall, Ritchie, Inc.*, 512 So.2d 99 (Ala. 1987), Sheetz was the project architect and hired Spann to provide the limited services of certifying the contractor's monthly requests for progress payments. After completion, the owner sued Sheetz for not delivering a defect-free building. Sheetz brought a third-party action against Spann. The court upheld summary judgment for Spann. The court reasoned that Sheetz was the inspecting architect and therefore had the duty to verify that the work was in accordance with the plans and specifications. Spann had only the limited role of determining the amount owing to the contractor for direct construction costs.

  2. Some states allow very narrow immunity for an architect's negligence in issuing certificates for payment. *Lundgren v. Freeman*, 307 F.2d 104 (9th Cir. 1962); *Wilder v. Crook*, 250 Ala. 424, 34 So.2d 832 (1948); *Craviolini v. Scholer & Fuller Associated Architects*, 89 Ariz. 24, 357 P.2d 611 (1960).

F. Jobsite Safety

  1. Responsibility for jobsite safety can be explicitly based on contract language.

It can also be by implication often based upon amount of control over the work or by actions taken or duties apparently assumed. Construction managers are often responsible for safety programs, or at a minimum for coordinating them and seeing that they are in place. Such activities can leave construction managers liable for injuries suffered at the construction site. In *Lemmer v. IDS Properties, Inc.*, 304 N.W.2d 864 (Minn. 1980), the "general construction manager" was held liable for injuries to a worker. (The court described this particular defendant alternatively as the "general construction manager" and "general project manager with duties similar to those of a general contractor.") By contrast, the construction manager in this case was responsible for "initiating, maintaining and supervising all safety precautions and programs."

In the case of *Parsons, Brinckerhoff, Quade and Douglas, Inc. v. Johnson*, 288 S.E.2d 320 (Ga. App. 1982), a crane operator and another construction worker employed by a subcontractor to unload and stack steel beams on a site for storage suffered electric shock when the crane hit a nearby power line. There, the construction manager had previously located the site as proper for storage of the steel beams. The court observed that the evidence was undisputed that the construction manager never told the prime contractor that the wires were supposed to be de-energized notwithstanding the fact that the construction manager conducted safety meetings with the prime contractor, advised the prime contractor on safety matters and in fact arranged a conference at which time the prime contractor met with various utility personnel.

Chapter 5 The Design Professional

The agreement between the construction manager and the owner required the construction manager to determine the manner in which the work would be accomplished and required the construction manager to provide advice and assistance to the owner on safety matters as well as develop safety criteria. The court concluded that the construction manager had a duty to consult with the prime contractor with regard to the need to de-energize the power line and that the construction manager failed in the performance of this duty with the foreseeable result that the prime contractor failed to de-energize the power lines and thereby cause injury to the workman.

A construction manager's responsibility to an injured workman was also thoroughly examined in the case of *Hammond v. Bechtel, Inc.*, 606 P.2d 1269 (Alaska 1980) where an employee of an independent trade contractor, injured in a fall through a trailer, brought a lawsuit to recover against the owner and Bechtel as construction manager. The contract between Bechtel and the owner required the construction manager to monitor trade contracts. In addition, the contract required the construction manager to require trade contractors to observe all safety rules and to take any other precautions as were necessary to prevent death or injuries.

In performance of this duty, Bechtel employed job safety inspectors who inspected the equipment of the trade contractor. In reversing a summary judgment in favor of Bechtel, the court held that if Bechtel was independently responsible for the injuries, by failure to prudently exercise control retained over the details of performance of the work at the site or by failure to turn over the premises free of unreasonable safety hazards, that an injured employee could maintain a cause of action. In determining what constituted the requisite control for liability purposes, the court held that there must be such a retention of right of supervision such that the contractor was not entirely free to do the work in its own fashion.

2. Should those with no responsibility for jobsite safety say something or keep quiet if they see something wrong? (Or is it better to be sued and have a good defense or never to be sued at all?)

Construction manager's responsibility can vary from being responsible for the jobsite programs to not being responsible for them. Architects typically avoid safety responsibilities but sometimes become involved in safety programs explicitly or implicitly by actions they take. Architects visit the site and become aware of conditions. Owners have a certain responsibility for the property they own and may become involved in safety programs. In those instances where a party has an extremely limited or nonexistent role in safety, what is his exposure when he knows of an unsafe condition? Generally, modern construction contracts provide that an architect or engineer is not responsible for jobsite safety programs. Where one has no responsibility for safety and no knowledge of the unsafe condition, there is some level of comfort. In *Hamby v. High Steel Structures, Inc.*, 521 N.Y.S.2d 926 (App. Div. 1987), the state's inspection engineer was not responsible for a worker's jobsite death. After the worker was killed, his estate sued the state inspection engineer. The court held that there was no common law duty to provide a safe work site imposed on an architect who does not control the work being performed. In addition, the engineer's agreement with the state did not require the engineer either to supervise or establish safety procedures.

However, where one has knowledge of an unsafe condition, one can have liability exposure despite a lack of contractual responsibility for safety. In *Balagna v. Shawnee County* 668 P.2d 157 (1983), the court held an engineering firm liable for the death of a construction worker caused by an unshored trench. The engineer had no control over the construction operations, but saw the unshored trench, knew it was dangerous, and took no action.

Speaking up will not necessarily create expanded liability. In *Northern Indiana Public Service Co. v. East Chicago Sanitary District*, 590 N.E.2d 1067 (Ind. Ct. App. 1992), the court rejected a suit by workers against the architect and engineer and stated that the A/E had no duty to warn the workers. The A/E's on-site representatives had four unsafe conditions and reported them to the general contractor. This action did not give rise to an assumption of safety duties, and the court stated that the law should encourage those involved in construction to report safety problems without fear of liability for assuming expensive safety obligations.

The case of *Goette v. Press Bar & Cafe, Inc.*, 413 N.W.2d 854 (Minn. Ct. App. 1987) is also favorable for those in construction administration. The plans for this renovation project of adding a second story to the building required through bolting as a method to secure the subframing to the new stucco. The owner decided to use the cheaper lag bolting method instead. The owner also ordered the removal of 8" of brick across the entire front of the building, which was not required by the plans. This removal of brick from the building front removed the support for the parapet above the brick, which accordingly collapsed. The architect was allowed out of the lawsuit on summary judgment. The proximate cause of the collapse was the removal of the brick, and the plans did not call for the removal of the brick and required a different means of securing the subframe from that which was used. The contract did not require the architect to supervise, and there is no general duty to supervise independent of the contract.

However, as a practical matter, when a worker gets injured, every available pocket for recovery is explored. Owners, architects, and construction managers can count on being a defendant in such a lawsuit. One can expect that there will be a strong compulsion for a judge and jury to do everything in their power to assess liability against a party who turns his back on a dangerous jobsite condition. Anything that is said, however, should be said in a way that makes it clear that the party is not assuming any additional responsibility for safety or means and methods.

    3. Contractual means of protecting owners, designers, and construction managers.

Parties can sometimes use indemnification clauses to protect themselves even from their own negligence. Various states have different rules regarding the enforceability of indemnification laws. Some states enforce even those indemnification clauses that require a party to indemnify a second party even for the second party's own negligence. Other states will invalidate any clause that requires indemnification beyond one's own fault. In *Oster v. Medtronic, Inc.*, 428 N.W.2d 116 (Minn. Ct. App. 1988), the court upheld an indemnification clause requiring a prime contractor to indemnify the construction manager, even for its own negligence. Although the jury had found the construction manager 45% responsible, the prime contractor 25% responsible, the owner 10% responsible, and the injured employee of the prime contractor 20% responsible for the injured employee's injuries, because of the indemnification clause, the prime contractor was responsible for all of the damages. Minnesota now has an anti-indemnification statute which did not apply to the contracts in the *Oster* case, but that statute has an exception for clauses which require insurance over the indemnification obligation. Minn. Stat. Chap. 337. The indemnification clause in the Minnesota AGC standard subcontract was held to be an enforceable agreement to insure, and the failure to provide the insurance leaves an indemnification obligation. *Holmes v. Watson-Forsberg Co.*, 488 N.W.2d 473 (Minn. 1992).

    4. OSHA.

        a. There is authority that a construction manager can be cited for exposure of its own employees to OSHA violations.

Chapter 5 The Design Professional **125**

In the case of *Bechtel Power Corp. v. Secretary of Labor,* 548 F.2d 248 (8th Cir. 1977), the court found Bechtel responsible for the exposure of its employees to five nonserious violations of OSHA on a construction site where Bechtel was the construction manager for the building of a power plant. Bechtel attempted to defend on the basis that its employees were not performing the actual work of construction and therefore were not "engaged in construction work" within the meaning of 29 C.F.R. Section 1910.12(a). The court noted that under its contract with the owner, Bechtel was responsible for the administration and coordination of all phases of construction including the safety program and that in fulfilling such duties, Bechtel's employees worked in a managerial or supervisory capacity and did not perform the actual work of construction. Nevertheless, it was held that since Bechtel's functions as construction manager were an integral part of the total construction that it was engaged in construction work within the meaning of the OSHA regulations.

- b. *Reich v. Simpson, Gumpeutz & Heger, Inc.,* 3 F.3d 1 (1st Cir. 1993) (architect had no OSHA duty to protect when had no employees on site).
- c. *Secretary of Labor v. CH2M Hill Central, Inc.,* OSH Review Commission Docket No. 89-1712 (engineering firm acting as project manager was not engaged in construction and therefore not subject to OSHA citation, even if had employees on site).
- d. *Secretary of Labor v. Skidmore, Owings & Merrill,* 1977–1978 O.S.H.D. (CCH) ¶ 22, 101, 5 O.S.H. Cas. (BNA) 1762 (test for being "engaged in construction"—need "substantial supervision" over the "actual construction."

G. Documentation Issues
1. Documentation must be useful for tracking the job as it proceeds, but for litigation purposes, it also must be usable to prove your case later. Generally, you must do what you can in order to make the records accurate, descriptive, and helpful. Don't shoot yourself in the foot in keeping the records.
2. Records usually constitute what lawyers call hearsay. They are allowed into evidence under specific exceptions. The most important exceptions are listed as follows:
    a. Present sense impression. A statement describing or explaining an event or condition made while the declarant was perceiving the event or condition or immediately thereafter. Federal Rules of Evidence, Rule 803(1).
    b. Recorded recollection. A memorandum or record concerning a matter about which a witness once had knowledge but now has insufficient recollection to enable the witness to testify fully and accurately, shown to have been made or adopted by the witness when the matter was fresh in the witness' memory and to reflect that knowledge correctly. If admitted, the memorandum or record may be read into evidence but may not itself be received as an exhibit unless offered by an adverse party. Federal Rules of Evidence, Rule 803(5).
    c. Records of regularly conducted activity. A memorandum, report, record, or data compilation, and any form of acts, events, conditions, opinions, or diagnoses, made at or near the time by, or from information transmitted by, a person with knowledge, if kept in the course of a regularly conducted business activity, and if it was the regular practice of that business

activity to make the memorandum, report, record, or data compilation, all as shown by the testimony of the custodian or other qualified witness, unless the source of information or the method or circumstances of preparation indicate lack of trustworthiness. The term "business" as used in this paragraph includes business, institution, association, profession, occupation, or calling of every kind, whether or not conducted for profit. Federal Rules of Evidence, Rule 803(6).

    d. Absence of an entry in business records. Evidence that a matter is not included in the memoranda, reports, records, or data compilations, in any form, kept in the course of a regularly conducted business activity, to prove the nonoccurrence or nonexistence of the matter, if the matter was of a kind of which a memorandum, report, record, or data compilation was regularly made and preserved, unless the sources of information or other circumstances indicate lack of trustworthiness. (This means if you have a checklist on your reports or space in your reports for certain information, this information must be filled in or the lack of that information can be used against you; e.g., a form that says "contractor problems today.") Federal Rules of Evidence, Rule 803(7).

    e. Statements and ancient documents. Statements in a document in existence for 20 years or more and the authenticity of which is established. Federal Rules of Evidence, Rule 803(16).

    f. Other exceptions. A statement not specifically covered by any of the foregoing exceptions but having equivalent circumstantial guarantees of trustworthiness, if the court determines that (a) the statement is offered as evidence of a material fact; (b) the statement is more probative on the point for which it is offered than any other evidence which the proponent can procure through reasonable efforts; and (c) the general purposes of these rules and the interests of justice will best be served by admission of the statement into evidence. Federal Rules of Evidence, Rule 803(24). (This is the catch-all.)

H. List of Areas in a Project Which Frequently Generate Claims
1. Roofs
2. Structural systems
    a. Change from steel frames to precast or vice versa and failure to re-engineer structural frame (parking garages, hospitals—also lots of utilities for hospitals)
3. Lack of coordination in drawings to reduce or identify interferences
4. HVAC system—lack of adequate system to allow for proper balancing
5. Acoustics/sound transmission
    a. Walls all the way up to slab
    b. Transmission in duct work/piping
    c. Noisy duct work
    d. Inadequate dampers or insulation in duct work
6. Electrical
    a. Not enough outlets
    b. Switches in darkened areas

7. Fire safety controls
    a. Too sensitive or not sensitive enough—false alarms, particularly for hotels and hospitals
8. Fountains and water courses a guaranteed problem
    a. They will leak to the place you don't want them.
9. Skylights
10. Plaza areas above usable space
11. Exterior of building
    a. Brick clad with block or studs and gypsum
        i. Common construction errors because of difficult to follow plans or designs—failure to have flashing carry through from top courses down, not continuously or carelessly installed, or blocked or torn by mortar drippings—omitted or reduced number of wall ties or not galvanized—lack of adequate expansion or contraction at joints or allowance for a lack of adequate allowance for expansion or contraction at joints
    b. Stucco—guaranteed to crack, need plenty of expansion or contraction
    c. Glass-clad buildings—big problems—if no tinted glass or shading, cannot cool building, particularly in hot climates—heat-strengthened glass—nickel sulfide inclusions—oil canning–John Hancock Building—need good wide mullions to keep glass from popping out from wind pressure
12. Parapet wall
    a. Prevent moisture from going through the coping stone into the wall cavity
13. Concrete floors
    a. Lack of adequate expansion or contraction—use true joints, not saw cuts—reinforcement wire mesh needs to be chaired up, doesn't work well on corrugated concrete decking—lack of adequate drainage in storage areas
14. Waterproofing
    a. Need good waterproofing membrane-drain tile (perforated, surrounded with class 5 or other coarse material)—sometimes French drains—positive drainage away from building—gutters and downspouts
15. Painting
    a. Either painter in there too early and has to patch or too late and covers mistakes or doesn't have enough coats to cover mistakes
16. Marble floor tile—vulnerable to cracking, move to thin-set marble increases cracking—marble is not hard—will show wear in high-traffic areas
17. Pressure in highrise apartments and hotels
    a. Need strong pipe and properly based pump and pipe system to provide pressure to high elevations
18. Expansive clays
    a. Heaving

19. Underground—infinite problems. God did not follow the ASTM specification—footings below frost line—sewer and water piping below frost line
20. Compaction
    a. Keep eye on optimum moisture content—clay has a narrow band of moisture content—testing—don't rely on compaction tests from top layer if using sheep's foot
21. Concrete mixes
    a. Dense reinforcement patterns, overvibration or undervibration
    b. Cold joints
    c. Exposed aggregate

CHIEF COORDINATOR FOR THE ENTIRE PROJECT

Since the mid-1960s, there has been a good deal of debate about who is the head of the design/construction team. Traditionally, this role has fallen to the architect or, using the newer term, the design professional. The design professional conceives the project scheme and translates the owner's requirements into reality. Many feel that the planner, the construction manager, or even the contractor, over the years, has assumed the role of the chief of the team.

We feel that the person who is in charge of the project team is the design professional. By virtue of his/her contract with the owner, the professional is the direct agent and the most immediate voice of the owner. The professional has taken the owner's concept and program and translated them into documents that, when executed, will produce the project the owner desires. Given this situation, no one else can be the leader of the team. No one else has this unique position of intermediary between the owner and the project.

Responsibilities of the Design Professional

As the head of the team, the professional has a number of responsibilities that should be carried out, and many of these are not reflected in the contract documents. The professional must make the owner thoroughly aware of what is contained in the contract documents, the scope of the work involved, and the necessary procedures that the owner must perform. This is a major responsibility. The design professional must ensure that the owner understands clearly and in minute detail what his/her responsibilities are and what, exactly, he/she can expect from the design/construction team.

It may well be that some other procedures will be attractive to the owner. Of course, the design professional should participate in the selection of any other procedures and personnel involved. Additional services can be offered, such as additional on-site visits; joint visits by the owner and design professional; full-time representation by the design professional, the owner, or both; or full construction management.

The design professional should provide competent and experienced supervisors from his/her own office. They should be thoroughly familiar with the construction at hand and also able to perform all the intricacies of construction contract administration. It is inappropriate for a person who has been trained in massive projects and heavy construction to try to do a single small building of light construction. And, of course, the reverse is true. In either case the representative must be comfortable with the job. Personnel, construction, materials, methods, and other operations vary, and require a flexible mind and a proper hand at administration. The wise course is to attempt to match expertise with the project as much as possible; far less tension and fewer problems will result.

In the contract documents, the architect will usually provide for the necessary accommodations, equipment, and accoutrements that the project representative will require: job offices, telephones, testing procedures, and the like. Thus, the representative has at hand all the equipment and additional personnel that may be required to serve the project.

It is essential that the design professional establish an on-site "home base." The size of the project dictates the scope of these facilities. The general contractor or construction manager will usually provide the necessary facilities. At times, facilities for the professional will be separate, but those shared with the contractors are adequate. However, the design professional needs a place to review drawings, make telephone calls, hold meetings, and process various documents.

Another important function of the professional is to convene a preconstruction conference just before construction begins. All the various participants are represented at this meeting. Here the ground rules for the project can be set forth, even to the extent of producing and distributing a written manual of procedures to all present (see Figure 5.7). Here each of the participating firms and their on-site representatives can get to know each other, exchange pertinent information, and learn about one another's operations.

To set this meeting before the first spade of ground is turned is essential to getting the project off on the right foot. It shows that everybody is an integral, important part of the project and that the design professional is, in a sense, initiating and soliciting cooperation and coordination. It is important to point out that the professional is interested in opening and maintaining the lines of communication by getting people to sit down at the outset, face to face, to begin the project. The preconstruction meeting is widely used and increasingly important. Documents can be exchanged, distribution lists for communications established, priorities and procedures reviewed, and general thoughts about the project exchanged.

The owner (or an authorized representative) should also be in attendance at this meeting, first to establish her/his identity, but also to indicate the owner's continuing and direct interest in the project. Whether or not this person attends the subsequent job meetings (highly desirable) depends on scheduling, interest, and participation as defined by the owner. At least in this initial meeting the owner is seen as more than just a title.

In setting or assessing the tone of a project, it is well to include the discipline of the project. (The authority of the A/E is discussed in Figure 5.8.) This involves setting goals and priorities, handling problems, resolving similar conditions in the same even-handed manner, and requesting both accurate and reliable communications. This may seem like an odd assortment of topics (and others may also be included) but they are representative of the types of situations that can affect both performance and progress on the project. Also, they establish the atmosphere under which the work will proceed; this develops the attitudes of the various project participants. For example, a good principle is that the contract documents are to be enforced as they were bid and are present on the site, unless there are proper revisions. The expectation is that the finished project will directly reflect that which is depicted and explained in the contract drawings, specifications, and other contractual documents.

The professional in the field must be supported in this effort by a determination in the office that what is required by drawing or specification is what is expected. Taking a theme from code administration, "If you don't intend to enforce, don't regulate." Hence, specifications, in particular, should set the basic operating parameters for the project and should be written with every intention of full enforcement and compliance. Of course, the owner's requirements should also be part of this scenario with an explanation that the contract requirements are set forth and must be met—period. This places a certain onus on the professional to produce documents, and again specifications in particular, that are reasonable, understandable, equitable, workable, and enforceable. Don't make demands on the contractors that cannot be met, for there is no magic in the contract administration sequence that will produce solutions to unworkable or unreasonable requirements.

```
SECTION 01200—PROJECT MEETINGS

PART 1—GENERAL

RELATED DOCUMENTS

Drawings and general provisions of the Contract, including
General and Supplementary Conditions and other Division-1
Specification sections, apply to work of this section.

SUMMARY

This section specifies administrative and procedural require-
ments for project meetings including but not limited to the
preconstruction conference and periodic progress meetings.

PRECONSTRUCTION CONFERENCE

Schedule a preconstruction conference and organizational
meeting at the project site or other convenient location no
later than 10 days after execution of the work order and
prior to commencement of construction activities. Conduct the
meeting to review responsibilities and personnel assignments.

Attendees: The Owner, Architect and their consultants, the
Contractor and its superintendent, all subcontractors, manu-
facturers, suppliers, and other concerned parties shall each
be represented at the conference by persons familiar with and
authorized to conclude matters relating to the work.

Agenda: Discuss items of significance that could affect
progress including such topics as:

     Tentative construction schedule.
     Critical work sequencing.
     Designation of responsible personnel.
     Procedures for processing field decisions and change orders.
     Procedures for processing applications for payment.
     Distribution of Contract Documents.
     Submittal of shop drawings, product data, and samples.
     Preparation of record documents.
     Use of the premises.
     Off-site storage facilities.
     Office, work, and storage areas.
     Equipment deliveries and priorities.
     Safety procedures.
     First aid.
     Security.
     Housekeeping.
     Working hours.
     Schedule of progress meetings.
```

FIGURE 5.7   A specifications excerpt that lists the meetings planned for the project and the general agenda to be taken up. Allows for understanding and preplanning of these sessions.

Responsibilities of the Administrator

It is impossible to understate the need for the on-site administrator to be fully familiar with the legal conditions under which the project will function. The administrator
must know where responsibility lies, what the "pecking order" is, and in general
"how the project will be pursued." Successful contract administration, obviously, is a
function of "doing the correct thing" at the right time and in the right way.

Chapter 5   The Design Professional

PROGRESS MEETINGS

Conduct progress meetings at project site at regularly scheduled intervals established during the preconstruction conference. Notify Owner and Architect of scheduled meeting dates. Coordinate dates of meetings with preparation of payment request.

Attendees: In addition to representatives of the Owner and Architect, each subcontractor, supplier, or other entity concerned with current progress or involved in planning, coordination or performance of future activities shall be represented at these meetings by persons familiar with the project and authorized to conclude matters relating to progress.

> If a subcontractor or material Supplier has been properly notified by Contractor to attend said meeting and it is not attended by a representative of subcontractor or material supplier familiar with this project, payment can be withheld until such a time as another meeting can be held.

Agenda: Review and correct or approve minutes of the previous progress meeting. Review other items of significance that could affect progress. Include topics for discussion as appropriate to the current status of the Project.

Contractor's Construction Schedule: review progress since last meeting. Determine where each activity is in relation to the Contractor's construction schedule, whether on time or ahead or behind schedule. Determine how construction behind schedule will be expedited; secure commitments from parties involved to do so. Discuss whether schedule revisions are required to ensure that current and subsequent activities will be completed within the contract time.

Reporting: No later than 3 days after each progress meeting date, distribute copies of minutes of the meeting to each party present and to other parties who should have been present. Include a brief summary, in narrative form, of progress since the previous meeting and report.

> Schedule Updating: Revise the construction schedule after each progress meeting where revisions to the schedule have been made or recognized. Issue the revised schedule concurrently with the report of each meeting.

SPECIAL CONFERENCES

Conduct conferences required by other section of specifications, including preliminary roofing conference, or as shall be called by special notice as required by conditions.

FIGURE 5.7   (Continued).

While construction is a rather informal operation overall, the legal and administrative functions rely heavily on proper procedure and accurate (full) documentation. This is a logical and necessary form of insurance attesting that everything was done, and done correctly, as mandated by the various contracts. To be an active participant in this process, the administrator must work within proper bounds, on a proactive, not reactive basis.

One of the primary concerns that inspired this text is the lack of an compelling force on the design professions (architects in particular) to engage in contract

Actual	Expressly conferred by owner. Usually limited (not general).
Implied	Incidental to actual authority.
Apparent	Owner's manner leads others to believe A/E has authority beyond what is actually possessed. Result is unwitting expansion of A/E's authority. For example: change order is not signed by owner (but known to owner), but issued by A/E.

FIGURE 5.8 Authority of architect/engineer.

administration, on-site. There is an increasing call from clients for follow-up construction services to ensure proper adherence to the contract obligations. Others have taken over many of the tasks formerly done by the design professions. Where purposeful deletions are made in the contracts, or where the professionals simply do not choose to offer this work, there is still a need, since no one save the design professional has the depth and breadth of knowledge about the project—not even the construction manager, who may be engaged early in the process. Some tasks, concepts, and information remain with the professionals and need full and proper follow-through in every work phase to ensure a project that provides everything the client contracted for.

Of course, it is impossible to discuss every eventuality a contract administrator might encounter. Many may appear to be similar, but are slightly different, which could require an entirely different solution. So we are left to speak in general terms. This is more a "grounding" in principles than in directed activities and actions. Obviously, as one becomes more expert, understanding and flexibility will be enhanced, and so will the ability to deal with any set of circumstances or persons.

Figure 5.9 lists "good practice" principles that seem to work well in contract administration. These thoughts give some basis for approaching the work. Firmness often is required, as is dedication to duty, and loyalty and responsibility to one's employer. At times, it is necessary to stand up for what is right, whoever is involved (be tactful, though, with your employer and your client).

The contract administrator can add to the overall tone of the project by not being strictly an authority figure. The administrator will quickly be recognized by all workers on the site and his/her status known. Simple greetings, pleasantries, or comments to the workers (as they are encountered—do not interrupt, disturb, or distract them) can add greatly to their self-esteem and give them the feeling of being appreciated and of being part of the project. Compliments on good work, extra work, or aid rendered helps morale, productivity, and quality. However, no instructions or comments on "bad" or unsatisfactory work should be passed in this manner; all corrections go to the project superintendent (the administrator's link to the contractors). Every worker seeks to produce decent, if not good work. They also like to talk about their work. Recognition of their effort, comments on conditions, materials, methods (don't require or direct the same), the importance of their work to the project, and other matters (weather, sports, families) can be a refreshing interlude, even if only done in passing. These comments should never be lengthy or distracting. They can be extremely helpful in building a team effort. They should never be adverse to the project work or progress; neither should they become gripe sessions about work, conditions, or supervisors. Be aware of gripes, though, because they are usually symptoms of problems, real or imagined.

It is interesting that the contract administrator or project representative is often cast in the role of educator. Quite often situations that arise during the progress of a project require not only explanation, but instruction by the professional. Since the

- Work by the traditional Golden Rule.
- Know, understand, and work within the limits of your authority.
- Be aware of your professional standard of care; apply its principles carefully.
- There is no shame in asking for help—the end result will be better.
- Quality is both an aesthetic and an economic issue.
- Construction is more art than science, hence, it is imperfect at best.
- No set of documents is perfect. Deviations are common, but need to be controlled (by you).
- Quality assurance is part of your job. It starts at the beginning and needs to be done correctly.
- Be thoroughly familiar with all of the contract documents.
- Approved change orders are the "most current" configuration of the project.
- Anticipate events and plan for them.
- Seek out glitches in the contract documents before work is done as a way to avoid problems, gaps, errors, delays, disputes, claims, and improper work.
- Identifying problems does not always mean you have to solve them too.
- Being friendly does not mean being friends.
- Remember, at least 50% of communication is listening.
- Actively seek to maintain good relationships with all participants.
- Be safety-minded and aware of unsafe conditions, but do not prescribe, direct, suggest, or engage in safety programming, procedures, methods, or systems for the workers.
- "Speak softly and carry a big stick" is not a bad axiom.
- Act in a manner to attract respect, but not personal friendships.
- Griping is a normal on-site activity that you cannot eliminate. Remain unbiased and impartial. Don't perpetuate gossip.
- Make reasonable judgments in strict accord with the contract documents.
- The properly modified contract documents, approved by the regulatory agencies, are the absolute measure of the project.
- Never offer any advice to the contractors regarding methods or means of construction.
- If you don't know, say so—then find out and reply properly.
- You are a professional. Always act like it, but without being overbearing or pompous.
- Being timely does not mean being hasty.
- Always be reliable even in small matters. Keep your word.
- Maintain good records, at a minimum.
- Dress for the conditions.
- Be prepared, but not anxious, to say "no."
- Be firm, but fair.
- Seek to resolve issues in a "win-win" manner whenever possible.
- Be even-handed and even-tempered.
- Be a liaison, convener, conduit, catalyst, facilitator, and a genuine help to the people and the process.
- Be a good example. What kind of person you are often is as important as what you do.
- Don't be overly predictable. Vary your normal routines.
- Make divergent attitudes, talent, expertise, experience, goals, background, and skill work for you and the project.
- The highest quality comes only from a culture (the construction team) that is totally committed to that end. Nurture that concept.
- Remember, satisfaction and achievement do not always jingle in your pocket, but they sure make the heart beat faster and the chest swell with pride!

FIGURE 5.9 Suggestions (not advice) for the contract administrator.

professional works in a realm much broader and more inclusive than the contractors, this educational effort is necessary not so much to "teach" as to inform others about the circumstances that created or contributed to the situation being discussed. Of course, this effort should be made in a gracious, informative manner. More times than not, the other parties will be grateful to have the background information; often it is to their benefit both on the current project and in the future.

The contract administrator should be fully familiar with the authority vested in her/him and should be prepared to exercise that authority (not flaunt it) in a proper and timely manner. Problems are best solved when they occur, while they are still relatively confined and readily resolved. Larger, unresolved problems and those allowed to languish usually wind up affecting too many other items of work or trades. By resolving a problem promptly (not without due consideration) adjustments can be made in the work, and also proper notice can alert others to the need to adjust their work. Most of the time, added costs and claims can be forestalled.

The administrator is not on-site as a whip-cracking disciplinarian, but is there to ensure adherence to the contract provisions and the contract documents (see Figure 5.10). Hence, where situations or problems arise, action should be taken commensurate with the conditions and timely to the schedule of the project. Taking excessive time to resolve problems or answer inquiries never serves the project or the professional (and his/her representative) well.

In all of this, the field personnel have to be ready and able to say "no" decisively and promptly. They must be ready and able to ascertain deviations from the contract requirements and to take the necessary action to have the noncompliance stopped, or remedial action taken to remedy any such work already in place. This must begin at the very outset of the project, if appropriate. One cannot establish a pattern of loose enforcement early and then attempt to create tighter enforcement as the project progresses. Contractors come and go, and all should be treated with the same enforcement techniques, no matter where they fit into the project schedule or work sequence. It is best for the contract administrator to establish a consistent, even-handed, "firm, but fair" course of action from the beginning.

Some professionals simply do not want to "get their hands dirty." They equate strong enforcement with demeaning or unprofessional attributes. So, why produce extensive drawings and well-conceived, meticulous specifications, merely to communicate to a contractor or owner who chooses to ignore them? Again, one might say that the professional is bound (by contract with the owner) to produce the best quality documents possible. Isn't it inconsistent to purposely demur from seeing that those very same documents are fully and properly executed?

Correct enforcement may not make the professional liked, but it is in accord with the authority and direction of all pertinent contracts regarding the work, the project overall, and the services provided. Many in the industry come to fully appreciate and respect this demeanor on the part of the professional, so long as it is consistent in all situations and with all parties. Everyone knows exactly where they stand and what is expected of them. This should not mean an adversarial relationship. "Firm but fair" can be achieved with tact, reason, patience, understanding, and a smile. The contract administrator must establish and nurture good rapport and working relationships with all of the contractors and other project personnel.

It is, though, wise for the professional to remember that rapport with the contractors should never become a relationship that erodes her/his professionalism and status, nor her/his role as the "police officer" over the owner-contractor agreement. It may seem puritanical to stick strictly to the business at hand, within the parameters of the contract, but surely any variance outside those guidelines could be turned against the professional. Everyone likes to be accepted as a human being, but business must be the top priority.

The field representative cannot solve all of the problems, nor prevent them from occurring. However, a representative who reviews the documents in advance of the

FIGURE 5.10 A good working principle for the contract administrator.

Management should be the "lubricant" that aids and allows technology to work better, more timely, and in a more pertinent and fulfilling manner—not vica versa!

work can observe glitches and can move to have them resolved in the office if fitting. Obviously, anything to the benefit of the owner should be presented to, and paid for, by the owner, even if it is an omission on the part of the professional. To try to resolve such items in the field, in an informal manner, is both inadvisable and potentially dangerous. The contract administrator should always function as a conduit for transmitting or developing information and should act in a prompt, far-seeing manner to forestall problems. Where they do develop, the administrator should act to have them described, reviewed, and resolved by others in the quickest possible time frame; this serves all parties well. Further, the administrator remains within the authority vested, i.e., not making changes in the documents and contract requirements.

At times, it becomes necessary for the contract administrator simply to find out the real story. This is not to undercut the contractor or the superintendent, but often situations arise that confound them as well as the administrator. Extra effort may be helpful; a simple telephone call may reinforce the sense of urgency and support the contractor's effort.

For example, the project requires a barrel-vault skylight. The specifications give the impression that this unit will be fabricated in fairly large sections and transported to the site for final placement and assembly. You (the administrator) find out (much too late) that in lieu of a separate subcontractor, the roofing contractor is to provide this unit. It arrives—in "kit" form! Only the curved end sections are fabricated; the other struts and gaskets are loose and the acrylic panels are flat. You find out on your own that the roofer thought the order included both fabrication and installation (as required in the specs). Somehow neither was made part of the subcontractor/material supplier contract.

After a lot of hand wringing and fruitless phones calls, you call the manufacturer. In a few minutes the situation is clarified, much to your chagrin. No installation was included (the roofer, who has no previous experience, is to do it), and the fabrication is what is done "normally" by the manufacturer (despite what the specs require). By securing detailed installation data and requiring close and continuous supervision by the contractor, you make sure the unit is successfully installed and sealed without leakage.

This case points up several routine occurrences. Mainly, this is an example of poor coordination, poor communication, poor understanding, poor contract-making, and less than decisive action. One wonders how long this situation would have lingered, how many meetings might have been held, had the administrator not taken the initiative and found out!

Both contractors and the manufacturer in this example are reliable and produce good work. It seems, though, that their administration techniques need honing; follow-up should become routine. Things don't just happen—they are *allowed* to happen, and taking minimal action often precludes "happenings." As fictional detective Charlie Chan said, "Do not lose *carefulness!*"

In another example, a project was nearly completed. A piece of equipment was to be mounted up in the ceiling void, above the finished ceiling. The unit would be lowered through the finished ceiling and raised when not in use. As the project continued toward completion, the unit never appeared. A number of issues arose and pressed for resolution—How big was the ceiling opening? What kind of support was required for the unit? What was the actual operating size of the unit? How was it connected, electrically? The site superintendent couldn't get the information required, even from his project manager. Finally, a phone call to the manufacturer provided the name of the supplier in the area for the specified product. It was the same supplier who was to furnish other equipment to the project! After a number of calls, it was finally learned that the contractor's project manager was not satisfied with the quoted price, deleted the unit from the order, and never found another source. The unit, needed *now*, was not even ordered. Without the calls, what would the outcome have been? Think of all the work and the other trades involved in closing out this loose end, waiting for simple information that was not coming because no one had a reason to send it!

The very essence of contract administration is that the documents, as bid and made part of the contract, must be enforced. These documents come to the professional's representative in the field. By definition in AIA document B352, no changes are to be made in the field. This is not to say that the documents should be considered perfect—usually they are not. However, any necessary changes or substitutions must be put through the proper procedures and incorporated into revised contract documents. Construction "horse-trading" (informally relieving some requirement or accepting marginal/improper work in exchange for a "favor" or work not included in the contract documents) has been going on for years, but it must be done with extreme care, and minimally! To indicate that the contractor should do some work not included in the contract documents is inadvisable and tends to leave "IOUs" against the professional, which could prove embarrassing and even costly when they are called in. Overuse can easily lead to reduced respect for the professional. It is best not to use this method of operation. The risks far outweigh the benefits.

The AIA, of course, discourages such practices. Instead, that group advocates good project documentation through the development and distribution of numerous standard forms. In regard to major changes in the work, AIA document G701 is available for use (see Chapter 7). However, where work is changed but no change is required in either contract sum or contract time, AIA document G710 should be used (see Figure 5.11; note the similar numbering of the documents). This form is an excellent vehicle for quickly and consistently tracking job changes without elaborate letters, undersized short-form messages, or scraps of paper. The form can be accompanied, where appropriate, by clarification drawings/sketches and, in fact, can act as a transmittal for such graphic work. It provides the necessary information in a format easily distributed to all concerned, including site personnel. Its size permits easy and uniform recognition, filing, tracking, and retention.

Proper use of this form, i.e., concise but complete information describing the change(s) anticipated or ordered, reduces the amount of paper required to facilitate the new work or revisions. This process is highly advisable on projects of all sizes to eliminate confusion and misunderstandings. In addition, it provides a quickly retrievable, logical, and chronological "diary" of the work items, their locations, and their fit in the project sequence. This often proves to be an invaluable set of documents, especially where claims or disputes are involved. It eliminates "word of mouth" agreements, instructions, and informal passage of information, as well as "forgotten" work, lack of coordination, and improper work. It is simply good business practice regarding changes in the work.

Even though most contract documents call for the contractors to provide all necessary items for a complete and finished installation (see Figure 5.12), there are often items that are overlooked, but which are really necessary. In the main, most contractors will cooperate and assist the professional by providing extra items or labor when asked. However, this relationship should never be exploited or abused by the professional, and should never be a point upon which he/she relies. Any time a professional becomes indebted to the contractors, there usually will be a future incident when the "debt" is called to the professional's attention, and an obvious conflict of interest arises (mostly to the contractor's benefit). Voluntary help and cooperation is one thing, but significant added work is beyond that and should be processed via other appropriate and prescribed procedures. Figure 5.11 notes one area where slight adjustments may be required in the work, due to changes or provisions missed in the complex regulatory documents. Most of the time these are minor (requiring little added time or cost), but nonetheless require some modification to achieve compliance.

AIA document A201, section 3.7.3, has a provision requiring the contractor to report any discrepancies in the documents and the process for getting new or corrective information from the professional. This, too, is part of the contract and is used

Reproduced with permission of The American Institute of Architects under license number #97016. This license expires April 30, 1998. FURTHER REPRODUCTION IS PROHIBITED. Because AIA Documents are revised from time to time, users should ascertain from the AIA the current edition of this document. Copies of the current edition of this AIA document may be purchased from The American Institute of Architects or its local distributors. The text of this document is not "model language" and is not intended for use in other documents without permission of the AIA.

ARCHITECT'S SUPPLEMENTAL INSTRUCTIONS

Owner ☐
Architect ☐
Consultant ☐
Contractor ☐
Field ☐
Other ☐

AIA DOCUMENT G710 (Instructions on reverse side)

PROJECT:
(name, address)

ARCHITECT'S SUPPLEMENTAL INSTRUCTION NO:

OWNER:

DATE OF ISSUANCE:

TO:
(Contractor)

ARCHITECT:

CONTRACT FOR:

ARCHITECT'S PROJECT NO:

CONTRACT DATED:

The Work shall be carried out in accordance with the following supplemental instructions issued in accordance with the Contract Documents without change in Contract Sum or Contract Time. Proceeding with the Work in accordance with these instructions indicates your acknowledgement that there will be no change in the Contract Sum or Contract Time.

Description:

Attachments: *(Here insert listing of documents that support description.)*

ISSUED BY:

Architect

AIA   **CAUTION: You should sign an original AIA document which has this caution printed in red. An original assures that changes will not be obscured as may occur when documents are reproduced.**

AIA DOCUMENT G710 • ARCHITECT'S SUPPLEMENTAL INSTRUCTIONS • 1992 EDITION • AIA® • ©1992
THE AMERICAN INSTITUTE OF ARCHITECTS, 1735 NEW YORK AVENUE, N.W., WASHINGTON, D.C. 20006-5209
WARNING: Unlicensed photocopying violates U.S. copyright laws and will subject the violator to legal prosecution.

G710-1992

FIGURE 5.11 AIA document G710 for use where contract/work changes are made that do not involve changes in money or time of project.
Used with permission of the American Institute of Architects.

> APPLICABLE CODES
>
> a. Throughout this specification there is reference to materials and workmanship being in compliance with applicable codes. In this context, applicable codes is to mean: the [State of] Ohio Basic Building Code, the FHA Minimum Property Standards, and any Local, City Building, Zoning, and Fire Zone Standards. All codes shall be the current edition, including all amendments at the time the building permit is issued.
>
> b. Applicable codes shall also include all codes, standards, and testing agencies referenced in any of the codes listed above and thus made a part of those codes, i.e., NFiPA Standards, ASTM, ASHRAE, Underwriters' Laboratories, etc.

FIGURE 5.12 Excerpt from typical specification. This provision sets forth the basis for all work on the project.

by all parties. It is necessary to ensure that the contractors know about this requirement, and that they do not proceed on their own when added or revised information is required. To do so leaves them outside the contract requirements and could lead to even more cost, confusion, disputes, claims, and even litigation.

The contract administrator serves two missions: (1) prevent purposeful noncompliance in the project (i.e., execute the contract documents), and (2) attempt to discover and resolve all discrepancies in the documents before they create undue problems in the work.

Assume Nothing

The contract administrator is responsible for a myriad of things. Overall project coordination is one of these, and the key issue is keeping the project moving in a positive and contract-compliant fashion. In achieving this one essential element cannot be overstressed—make *no* assumptions! There are many negative aspects to assumptions, none of which aid a construction project. The administrator simply must take the initiative to ensure that things are done in a timely and proper manner by the correct parties.

The administrator should always check the drawings, details, references, cross-references, and specifications against each other, project conditions, and the other pertinent documents to discover any gaps, conflicts, inconsistencies, or ambiguities. This is not an attempt to "sharpshoot" the production effort, but is crucial to eliminating nuances, gaps, redundancies, inconsistencies, inaccuracies, missed cross-referencing, inadequate detailing, and so on. To have this done prior to "problem-making" is a vital professional function, which can easily prevent claims and lawsuits.

Most professional offices do not have the time for an objective, in-depth review of the documents before the documents "hit the streets" for bidding. Inevitably, errors and omissions will be included. Since these can become points of contention during the project it is best to find them and resolve them prior to execution of the work. The contract administrator can greatly aid this effort through an insightful, meticulous review of the documents, with particular attention to cross-referencing between documents. Any such problems that are uncovered should be routed back to the project architect for resolution, as this is not a field operation.

Since most firms do not perform the objective review, this is an invaluable service the contract administrator can perform. Indeed, in many instances the administrator may not have been part of the production team, so her/his "new eyes" will see the

Chapter 5 The Design Professional

documentation in a new light and with a new perspective. This is part of the familiarization the administrator must engage to perform well on-site. The problems resulting from the lack of an objective review have been shown to be one of the major producers of legal actions.

Never should there be a conscious decision to allow loose ends in the documents "to be worked out in the field." Enough small inadvertent lapses occur that the field personnel must always be diligent and flexible. Some such items are easily resolved, but care is always necessary to avoid coercion or "stretched" interpretations in achieving the desired result. Obviously, the more care (and expertise) in the preparation of the documents, and the more objective the review, the better. A lower incidence of glitches will result. Perfection, although not achievable, is still a worthy goal!

FEEDBACK

There is yet another aspect to the interplay between field operations and the documentation. This is feedback from the contract administrator to the project architect and perhaps to the office files, particularly the file of standard details and/or specification sections (see Figures 5.13 and 5.14).

This is the last increment of the detail cycle, and provides insight to all concerned with what worked and what did not work on the project, as described in the contract documents. Often this is an ongoing activity throughout the project sequence, but it should also be a follow-up at the end of the project, a debriefing of sorts. Many times, field personnel can lend valuable insight and suggestions that can help avoid problems and ensure better projects in the future. This is a good source of input to

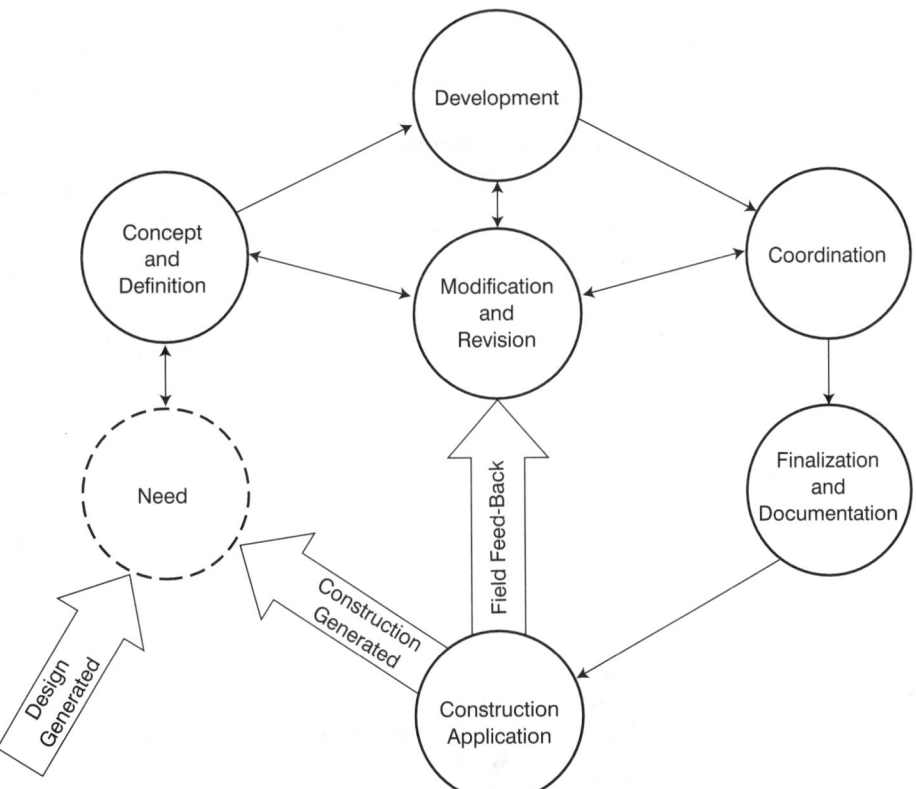

FIGURE 5.13 The chart depicts the detail cycle used on every project. Note that field feedback is an integral part of the cycle, providing valuable information and insight.

```
┌─────────────────────────────────────────────────────────────────┐
│              FIELD FEEDBACK TO DESIGN/SPECIFICATIONS            │
│  PROJECT:                                    DATE:              │
│  PROPOSE CHANGE TO:   (Design)   (Specifications)  (Standard Detail) │
│              (Special Detail)   (Other _____ )         │
│          [Use sketches to illustrate written material in categories below] │
│  ITEM DESCRIPTION:                                              │
│                                                                 │
│                                                                 │
│                                                                 │
│  PROBLEM:                                                       │
│                                                                 │
│                                                                 │
│                                                                 │
│                                                                 │
│  RECOMMENDED REVISION (Attach sketches as necessary):           │
│                                                                 │
│                                                                 │
│                                                                 │
│  Signature _____ , Project Representative  │
│        [Required Distribution; Design Coordinator—2 copies      │
│   Specifications Department; Construction Administration—1 copy—office: 1 copy—field] │
└─────────────────────────────────────────────────────────────────┘
```

FIGURE 5.14 A standard form used to formalize the field feedback and ancillary documents for filing and future use.

the standardized detail bank, since revisions can be made or optional choices included to enhance the design concept on future projects.

PROJECT MEETINGS

Some of this feedback will be developed in the periodic project meetings that the contract administrator or the contractor convenes (see Figures 5.15 and 5.16). Here everyone currently working on the project must participate. It is also a good idea to invite other persons who have left the job or who are anticipating coming to the job. The meeting should be an open forum, with the minutes formalized and distributed to all for the sake of continuity, placing responsibility, and establishing a job history in each participant's file. According to J. Russell Groves, Jr., AIA, associate dean of the School of Architecture, University of Kentucky,

> Conditions depicted in the documents that do not work in the field and those that do should be evaluated as part of an ongoing program for the improvement of contract documents.

1. Cast of Characters
    a. Introductions
    b. Identify the representatives from each team
    c. Who will prepare project directory?
2. Review Contract Requirements
    a. Notice to Proceed issued by owner? Date?
    b. Formal dates of construction contract—start and completion dates
    c. Does contract address liquidated damages? Review
    d. All insurance certificates complete?
    e. Performance and payment bonds
3. Correspondence
    a. Owner's, architect's, and contractor's address and contact
    b. Who is to receive copies of correspondence between parties?
    c. Proper routing of correspondence
4. Agency Approvals
    a. Have all necessary approvals been obtained?
        i) State agencies
        ii) Building permit
        iii) Sanitary district
        iv) Park and planning
        v) Historic preservation
5. Testing Services (Specifications identify requirements)
    a. Identified in spec. section _____
    b. Soil and compaction testing and inspection
    c. Concrete
    d. Steel and welding
    e. Roofing
    f. Paving
    g. Specialty: Security? X-ray?
    h. Test reports
6. Subcontractor and Supplier Approval
    a. Identified in spec. section _____
    b. Preliminary list from contractor (date)
7. Shop Drawings and Samples
    a. Identified in spec. section _____
    b. Contractor's initial approval
    c. Processing
    d. Shop drawing stamp
    e. Substitutions
8. Progress Schedule
    a. Identified in spec. section _____
    b. Submittal dates
        i. Preliminary for review
        ii. Final
    c. Periodic update
9. Construction Photos
    a. Identified in spec. section _____
    b. Submit monthly with the requisition

FIGURE 5.15 Outline for preconstruction conference.

10. Payment Requisition
    a. Requisition form
    b. Schedule of values (submit before first payment request)
    c. Date for request
    d. Review with architect
    e. Retainer

11. Contract Changes
    a. Identified in spec. section _____
    b. Authority to change work
    c. Processing procedure

12. Progress Meetings
    a. Job meetings—contractor/subcontractor
        i. Frequency
    b. Job Meetings—owner/architect/contractor
        i. Frequency
        ii. Location
        iii. Day and time (should follow subcontractor meeting)
        iv. Personnel attending
        v. Meeting conducted by _____
        vi. Responsibility for meeting minutes
        vii. Agenda

13. Construction Inspections
    a. Frequency
    b. Architect
    c. Owner's rep
    d. Other representative

14. Site Organization
    a. Office location
    b. Construction documents, record sets, files
    c. Phones
    d. Security
    e. Conference area

15. Record Drawings and Data
    a. Identified in spec. section _____
    b. Record drawings
    c. Operating and maintenance manuals
    d. Warranties

16. Project Closeout
    a. Punch list
    b. Substantial completion
    c. As-built drawings and manuals
    d. "Attic" stock
    e. Close-out documents

17. General Comments by Owner, Architect, and Contractor
    a. Questions about documents

FIGURE 5.15   *(Continued)*.

> The following format should be used as a standard agenda for meetings and minutes of all construction progress meetings.
>
> Sheet Format
>
>   From: (Minutes author)
>   Subject: (Project name)
>   (Progress meeting number)
>   Project No.:   (No. and file)
>   Date of Meeting:
>   Date of Minutes:
>   Participants: (Names and company represented)
>
>   I. Review of Previous Minutes
> II. Job Status
>   A. Schedule
>   B. Progress to date (since last meeting)
>   C. Submittals
>   D. RFIs
>   E. As-builts
>   F. Upcoming activities (projected next 2 weeks)
> III. New Business
> IV. Change Orders
>   A. Proposal requests (proposed and working)
>   B. Change orders (those issued since last meeting)
>
> Meetings are numbered sequentially 1, 2, 3, etc., and each item under new business is also sequentially numbered using that meeting number as a prefix, i.e., at meeting number 23, business items would be listed as 23.1, 23.2, 23.3, etc.
>
> Items are carried in the minutes under that number (for tracking) until resolved.
>
> Right-hand side of minutes carries an "Action" column. Each open item is to identify who has the action responsibility for a response. In the description for that item identify a projected date for that response.
>
> At the end of the minutes note that the contractor, architect, owner, and anyone else who participated toured the site either prior to the meeting or following the meeting. Site-related comments should be addressed in a separate site visit report.
>
> State the date, time, and place for the next meeting.
>
> All meeting minutes should close with the statement:
>
> "This report states our understanding of the matters discussed and the decisions reached. Each person receiving a copy of this report is asked to review it promptly and to notify (name) of (company), in writing, of any errors or omissions."
>
> cc: All participants and any pertinent parties not in attendance.

FIGURE 5.16   Outline/agenda for construction progress meetings.

The professional and the administrator should each establish an orderly system for all documentation of the job. Correspondence, forms, reports, shop drawings, samples, color selections, approvals, interpretations, memorandums, decisions, and meeting minutes should be accurately produced, properly distributed, and filed. In this way, both office and field personnel can keep the owner, the various construction inspectors, contractors, subcontractors, and consultants fully informed about all aspects of the job.

Each participant is interested in keeping a complete job file on the project for his/her own use. Of course, this file should be as complete and accurate as possible.

It is certainly no waste of paper to send pieces of documentation to a participant who is not directly involved with the memorandum. By virtue of this "nice to know" information, one can maintain a feel for the project and see how things are running.

One of the design professional's principal jobs on the site is to check on the contractors' schedule of values and in-place work. It is, of course, the owner's contractual responsibility to pay the contractors' invoices. This payment should, however, be made only over the certification of the design professional and his/her on-site representative.

The design professional, by contract, is required to police the owner-contractor contract. He/she must be fully aware of the in-place work before signing a certificate of payment. This in effect requires the owner to pay the contractors for work to date. Neither overpayment nor underpayment is proper.

As discussed earlier, on-site observations of the work are critical to its proper progress. Perhaps the key phrase is "timely observation of the work" rather than periodic observation. (See Figures 5.17 and 5.18.) Timely observations are necessary so that a proper standard of acceptability can be established. By being on the job at the right time—the beginning of a procedure, for instance—the design professional's representative can point out exactly what is acceptable and what is not. To do this after the fact or after the work is completed is not in the best interests of the project. This can become an irritant and a problem to the operatives on the project.

Any time an interpretation or a decision or an expansion of information is required, it should be done promptly. If the contractors and workers know that the architect or engineer will be on the job at specific times or within certain time frames, they can plan to approach him/her with some reliability and get the informa-

---

```
d)   All fire code drywall shall be installed in accordance with the sys-
     tem used for the approved applicable fire rating tests obtained by
     the drywall manufacturer and accepted by the state of Ohio and all
     other governing agencies. Installation of fire-rated drywall will not
     be accepted unless installed in accordance with tested assembly
     instructions. . . .

        walls shall have one (1) layer of 5/8" fire code drywall on
        both sides for a one (1) hour rating.

     5. First and second floor ceilings shall have one (1) layer of 5/8"
        fire code drywall for a one (1) hour rating. Third floor ceiling
        shall have a 1/2" fire code for a 3/4 hour rating.

3.03 SCHEDULE
     See Drawings for types and locations of exterior wall insulation,
     partitions and ceiling assemblies and systems:
     a. Typical Exterior Walls: Apply single layer foil-backed gypsum
        drywall over wall insulation blankets and Z-furring channels.
        (Note: Insulation shall extend above gypsum wallboard height or
        in some conditions insulation is applied only to walls; see
        Drawings.)
     b. Other Interior Partitions: Apply single layer gypsum drywall both
        sides (thickness indicated) to metal studs.
     c. Fire-rated assemblies:
        1. Two (2) hour rated wall shall comply with U.L. Design U411.
        2. Two (2) hour rated ceilings shall comply with U.L. Design G503.
```

FIGURE 5.17 Excerpt from typical specification. The specification provision for drywall is vague and inadequate; it does not firmly denote what is required. Section 3.03 states exactly what assemblies are to be used, showing all the detail required for installation and providing an excellent basis for inspection.

Chapter 5 The Design Professional

    1.03   QUALITY ASSURANCE:

    A.   Codes and Standards:

          Perform foundation drainage work in compliance with applicable
          requirements of governing authorities having jurisdiction.

FIGURE 5.18   Excerpt from typical specification. This represents an inadequate specification requirement. It is vague and shows that the design professional is uncertain about what standards apply. This unfairly shifts the design burden to the contractor, and the professional loses control.

tion they need. All information should, of course, be directed through the proper channels. The design professional should never talk to the subcontractors, vendors, or suppliers directly, but should talk to the manager, the contractor, or the people who have the proper lines of responsibility. Any communication, of course, should be backed up with written documentation and be properly distributed to keep all interested parties informed.

ON-SITE TECHNIQUES

The design professional's on-site representation can vary as widely as the clientele, their projects, and the sum of money available. Basically, the representation can involve the design professional in varying degrees, from none to constant on-site representation by a number of people (see Figure 5.19). Some contracts awarded to design professionals do not include provisions for extensive on-site inspections. This is most unfortunate because an outsider unfamiliar with the intricacies of the design concept and its development must take the documents and attempt to learn the project so he/she can properly inspect it during construction.

The outsider, whether an individual, a team of individuals, or a separate company, loses the flavor of the project and can see only the technical mechanisms that are combined to form the project. In some instances this can be a tremendous drawback, not only to the project itself, but to the relationships that are established either by contract or through individual contact on the project. For example, an architectural firm designed a new school for a local board of education. The board chose to have the design professional prepare only the contract documents and made no provision in the contract for on-site observation.

As discussed in Chapter 2, the professional can easily come under attack from the client, or others, where complete insight is lost. This can be personally hurtful, but the professional must ensure that the complaints are not widely voiced or distributed. Even though erroneous, once given to others this type of information is difficult to dispel or refute; it could have long-term implications for the professional's reputation. How easily one telephone call could have eased the consternation of all parties to this incident!

This shows that on-site contract administration by someone other than the design professional can lack insight and full understanding; it can be inadequate, misdirected, or inappropriate. Today, when lawsuits are frequent, the professional can be forced to pay a large sum of money simply to show that a problem that developed during construction was not his/her fault or direct responsibility. It has to be shown that the drawings and specifications (if any) were properly conceived and documented (for example, see the standards specified in Figure 5.20), and that it was during the construction phase, over which the professional had no control, that the

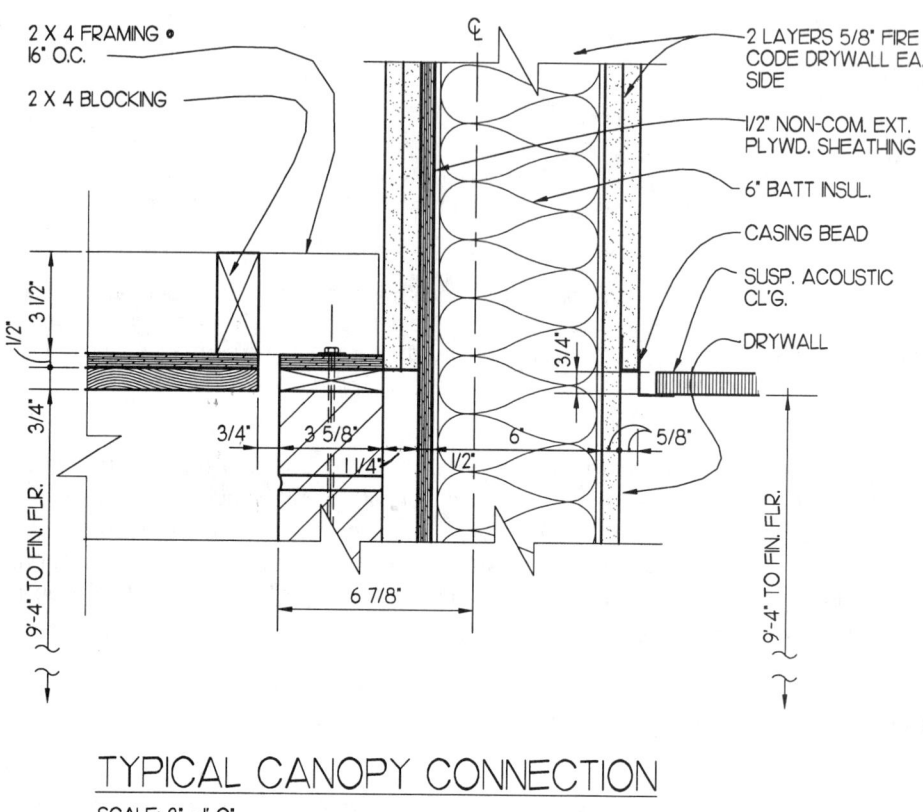

FIGURE 5.19 This architectural detail is for a limited area, but shows several items involved. This is a complicated piece of work for an inspector. Several people will have to view this work when in place.

problem situation developed. The professional has to show that had the documents been faithfully executed in the actual work, the project would not have experienced the problem. But to do this, the professional might have to expend his/her own time and money; often this is not covered by professional liability insurance, despite the amount (it may be less than the policy's deductible). According to Professor Justin Sweet of the University of California, Berkeley, College of Law,

> Americans today are less willing to accept their grievances silently . . . one element in the rise of consumerism in the 1960s. Claimants usually do not assert legal action against persons . . . unable to pay for court judgment or are not insured . . . design professionals [who] carry [professional liability] insurance, or sufficient resources to respond to court judgments. As a result, claims increase. . . . One out of every three practicing architects is likely to find herself [/himself] in litigation . . . as defendant or plaintiff.

Many design professionals take a very hard-line approach regarding contract administration to ensure that they do have at least some on-site control and observation capabilities. They want to be sure that their documents are properly followed. Numerous, costly instances could be cited where the professional was required to provide a defense against litigation because her/his documents were used, but there was no oversight by the professional during construction. Figure 5.21 is an excellent example of the type of information design professionals use to help ensure that their interests and those of the clients are properly protected. Even in situations where only plans and specifications are produced, it is advisable that those documents include the various reference standards and other regulations that apply to the work. The listing in Figure 5.21 must be adjusted for each project. This is direct protection for the professional, because it shows correct concepts and procedures were used. This aids in mitigating any challenge to the professional. Unfortunately, with lawsuits drawn to include everyone involved, some must pay dearly merely to defend their responsibilities and their limits.

Traditionally, the basic service of contract administration has revolved around the use of one person from the professional's office who is familiar with the project, or perhaps several current projects. Usually this person, by touring around on a daily basis, can observe and administer all the projects under her/his control. This system works well for the small- to moderate-sized office with projects of commensurate size.

1.05 QUALITY ASSURANCE:
A. Standards:
    The following standards are hereby made a part of these specifications:

    1) "Building Code Requirements for Reinforced Concrete," American Concrete Institute (ACI 318-71).
    2) "Recommended Practice for Measuring, Mixing, and Placing Concrete," American Concrete Institute (ACI 614-59).
    3) "ACI Standard Recommended Practice for Selecting Proportions for Concrete," American Concrete Institute (ACI 613-54).
    4) "Recommended Practice for Cold Weather Concreting," American Concrete Institute, (ACI 306-66).
    5) "Recommended Practice for Hot Weather Concreting," American Concrete Institute (ACI 605-59).
    6) "Manual of Standard Practice for Detailing Reinforced Concrete Structures," American Concrete Institute (ACI 347-68).
    7) Testing as per current ASTM standards.
    8) Building code of the city and state in which this development is located.

B. The requirements of "Specifications for Structural Concrete for Buildings," (ACI 301-72) are a part of these specifications. These project specifications define materials, additions, changes or omissions from ACI 301-72 and govern over them (items are numbered according to related chapters of ACI 301-72).

A. Codes and Standards:
    1. Comply with the provisions of the following codes, specifications and standards, except where more stringent requirements are shown or specified:

        a. ACI 301, "Specifications for Structural Concrete for Buildings"
        b. ACI 311, "Recommended Practice for Concrete Inspection"
        c. ACI 318, "Building Code Requirements for Reinforced Concrete"
        d. ACI 347, "Recommended Practice for Concrete Formwork"
        e. ACI 304, "Recommended Practice for Measuring, Mixing, Transporting and Placing Concrete"
        f. Concrete Reinforcing Steel Institute, "Manual of Standard Practice"

FIGURE 5.20 Excerpt from typical specification. This lists the standards to which the concrete work must conform. The use of dates in the first two sections may conflict with dates of the standards in the code. The third section states standards and implies the latest editions, which will usually agree with code provisions. Confusion can occur when various dated editions change requirements.

ACI	American Concrete Institute
AGA	American Gas Association
AIA	American Institute of Architects
AIEE	American Institute of Electrical Engineers
AISC	American Institute of Steel Construction
ARI	Air-Conditioning & Refrigeration Institute
ASA	American Standards Association
ASHRAE	American Society of Heating, Refrigeration & Air-Conditioning Engineers
ASLA	American Society of Landscape Architects
ASME	American Society of Mechanical Engineers
ASTM	American Society of Testing Materials
AWSC	American Welding Society Code
BOCA	Building Officials and Code Administrators, International
CSI	Construction Specifications Institute
FS	Federal Specification
IES	Illuminating Engineers Society
NAAMM	National Association of Architectural Metal Manufacturing
NBFU	National Board of Fire Underwriters
NBS	National Bureau of Standards
NEC	National Electric Code
OBBC	Ohio Basic Building Code
ODHS	Ohio Department of Highways Specifications
SJI	Steel Joist Institute
SPR	Simplified Practice Recommendations
UL	Underwriters' Laboratories, Inc.

FIGURE 5.21 Sample of specifications section referring to technical groups and their documents by using their acronyms for identification.

As an office begins to grow, not only in the number of projects but in the scope of projects as well, the one-person observation system begins to prove inadequate. A very large project, for example, may demand so much of this person's time that he/she is forced to delay or give cursory service to the smaller projects. Of course, this is both a disservice to the clients and a quick method for increasing professional liability exposure.

When a firm is in this position, it must evaluate what steps to take and what direction it wants to pursue. It may find that the one-person effort should become a team, or even a separate department devoted exclusively to contract administration. With an entire department, field observation may be augmented with responsibilities for shop drawing review, change orders, and modifications, and other administrative matters could be added. Extra personnel may have to be assigned to this group to assist with the clerical and/or technical aspects of the work.

Although architects and engineers are not specifically or narrowly educated in just observation and contract administration, they often find themselves (by choice or direction) "pigeon-holed" during their careers. It is very hard for someone who enjoys field work and the intricate detail of a job in progress to be drawn back into the office and a design situation. Sheer personality, experience, and the people in the field may steer this person to be purely a field-oriented professional.

It is important that every design professional's office have a well-trained, well-qualified, knowledgeable, flexible individual as its field operative. The person must fully understand and respect the design process through the entire progression of the project. At the same time, the person must have the flexibility to understand the execution of the details of the drawings, so all work is done to support the basic design concept.

Chapter 5 The Design Professional

There certainly is no stigma for the persons involved in field operations. Certainly it is worth a high-quality, if not professional effort for the firm to ensure that none of the actual construction is shabby or shoddy. Any person who adds effort, expertise, and dedication to such a task is not only an invaluable part of the team effort, but could be the "point" effort and the best insurance against noncompliance. In addition if it were not for this contract administration effort, a nasty legal/contractual situation could develop.

Many projects, by scope or complexity, will require, or perhaps the owner will demand, that the professional have an "all day every day" (i.e., full-time) on-site representation. This type of service is available through the use of the system traditionally known as "the clerk of the works." More recently, the person in this position has been designated as the architect's project representative or the resident project representative (for engineers; see Figure 5.22). This person, in many cases, is a registered professional or a construction-experienced individual (usually with supervisory skills and wide-ranging experience) hired by the design professional. Since this is an additional service (over and above the basic services the contract provides) this person reports to and through the professional, but his/her salary and expenses are paid by the professional and reimbursed by the owner. Proper modifications must be made in the standard (AIA or EJCDC) contracts to incorporate this service into the project scheme. These minor nuances should not prevent the use of this program, which provides greater overview, control, observation, communication, and evaluation of the project, since the project representative will be on the job site, performing a variety of duties, eight hours a day, every day. If there is any problem, the other operatives on the site know that the architect/professional has a representative there available at all times. To a large degree, this person is able to solve problems promptly, on-site, or with quick and direct consultation with the professional's office. This is especially helpful when a project is complex and/or fast-paced, involves a large number of workers, or where the construction is so extensive and involved that only continual observation can service the project adequately. An expanded staff of personnel may be necessary on some extraordinary projects.

With the wider use of construction management, many architectural and engineering firms are offering their services as full-scale construction managers. Here, the

---

The Resident Project Representative must be an expert in engineering, architecture, construction methods, labor relations, barroom brawls, public relations, and should have an extensive vocabulary in graphic and colorful construction terms. He or she must be able to work closely with contractors, subcontractors, engineers, architects, and owners, and agree with frequent directives that make no sense from people who know less than he or she does; he or she must be able to understand, then ignore environmental impact reports and regulations without the Sierra Club finding out, and he or she must cooperate with construction superintendents and contractor quality control representatives, and avoid backing his or her automobile into the privy while they are inside.

The resident inspector must have skin like an alligator, the stomach of a billy goat, the temperament of a Presbyterian minister, nerves of chrome-molybdenum steel, the fortitude of Job, and the physical strength to take care of a situation when all else fails. He or she should enjoy his or her job, and will probably love the working conditions, except for the dust, numbing cold, searing heat, knee-deep mud, mosquitos, rattlesnakes, scorpions, muck-covered office trailers, questionable toilet facilities, and physical assault by frustrated foremen.

FIGURE 5.22   The preferred nature of the project representative or contract administrator. From *Western Construction*, Gene Sheley, editor.

> This checklist is provided as a quick review of the items that require attention during the providing of construction phase services by the design professional. It is not all-inclusive and should not be construed as the ultimate answer to construction situations and problems.
>
> 1. Ensure that the client/owner is aware of and fully understands that the design professional <u>cannot be and is not</u> the guarantor of the performance of the contractors.
> 2. Be certain that the client/owner realizes, understands, and embraces the concept that integrity and competence on the contractors' part is the greatest protection against faulty workmanship and projects.
> 3. Purposely ensure that the design professional has *no* responsibility, in any form or language, for job-site safety and personnel. Include *no* safety program information in the project manual or technical specifications.
> 4. Provide all instructions to the contractor *in writing*. Always communicate matters concerning subcontractors through or with the general contractor (or construction manager) and not directly.
> 5. Provide only factual and objective information about any contractors, suppliers, etc. which may be the subject of disapproval.
> 6. Ensure that all on-site and office personnel of the various participants fully understand the procedure for handling shop drawings.
> 7. Only "review" shop drawings, and then just to ensure compliance with the general design concept, the drawings, and the specifications.
> 8. Instruct all employees of the design professionals to carefully avoid all hazardous conditions when making on-site visits. Allow the contractors full control and responsibility for providing a safe and healthful workplace for their employees.
> 9. Be thoroughly familiar with all terms, rights, duties, authority, responsibilities, and obligations assigned to you and your firm in all contracts, conditions, and so on.
> 10. Administer, facilitate, assist, and aid, properly *within* granted authority, the required construction in keeping with current precedence of the contract documents.
> 11. Maintain continuously careful, complete, clear, and easily retrievable files of all documents, correspondence, forms, and all other project documentation; include careful notes of changes, both minor and major; have as-built drawings and documents kept as required by contract.
> 12. Ensure proper written records of all changes, directions, orders, and decisions affecting the configuration, scope, design, and/or cost of the project.
> 13. Provide *all* construction services required by contract.
> 14. Ensure that all consultants perform and service the project in accord with their contracts and the progress of the work.

FIGURE 5.23   Summary of precautions; construction phase services.

design firm expands its purview and offers the added services normally offered by construction management. Usually this involves a wider range of efforts than normally performed by the professional. Often this entails more depth in bidding, coordinating bidders, awarding contracts, scheduling work, expediting, and so on. In some design firms, construction management has become a separate department or even an auxiliary firm within the overall practice structure. Care must be taken to ensure no conflict of interest occurs. The professionals are still working on their own project and attempting to provide the owner with the extra services, while utilizing some augmentation of their same basic cadre of personnel. When the construction manager is a separate adjunct firm with its own staff, the management firm can serve projects produced by other design offices and can provide other ancillary professional services, all adding low-risk (since a specific fee is set) income to the parent firm.

The design professional must be fully aware that all her/his consultants must also have representatives on the job site, from time to time. Whether these consultants are dealing with landscaping, acoustics, or mechanical systems, for example, they need to be represented, at least periodically; at job meetings, for example, and other times pertinent to the work in progress. They must make their observations, share their views, assess and evaluate work progress (for billing/payment certification), give their input/corrections/modifications, and give feedback to the various other operatives. The design professional should ensure that all the consultants are under contract to the professional, in such a way that this service is provided at no added cost, but as part of the basic services package.

Figure 5.23 summarizes some reminders and precautions regarding the design professional's duties during the construction phase of the project.

EXERCISES

The following tasks are based on the concepts and information in the chapter. All of the material or information required to complete the tasks may not be contained in the chapter text. Independent research is required. Consult your instructor for guidance.

1. Write a short paper explaining the benefits of partnering to the client.
2. Discuss the difference between turnkey and design-build projects.
3. Write a brief discussion of fast-track construction, its advantages and disadvantages.
4. A professional in design/build contracts can be in any of three roles. List them and discuss the relationships and responsibilities of the professional in each case.
5. Discuss how an owner's need for design responses is affected by the design-build contract.
6. Define and give examples of: separate contractor; phased construction; cash allowance; design-build drawings.
7. Explain the concepts, pro and con, in using design-build drawings in a fixed-price contract.
8. The way to ensure proper work is to give instructions directly to the workers on the site as the work progresses. True or false? Defend your position.
9. Is AIA document B352 viable and proper as the work outline for the project rep?

6

Obligation, Education, and Liability

Liability, n. 1. The state of being legally obligated. 2. Susceptible.

Why is contract administration included as one of the basic professional services in AIA document B141–1987's (owner-architect agreement) Section 2.6? The drafters of the document intended that the architect would provide comprehensive service, commensurate with the progress of the work, throughout the entire sequence of the project, including construction. As Figure 6.1 shows, Section 2.6.1 of this document is quite specific regarding the scope of contract administration. Section 2.6.5 outlines the services the architect is to provide during construction. Here these are stated explicitly; they are ambiguous only in some aspects.

HOW CONTRACT TEXT DEVELOPED

Some review provides insight into this contract text. For several years leading up to 1987 near chaos in the professional liability (errors and omissions) insurance industry and generally increased concern on the part of design professionals created an atmosphere of consternation. Professional practices were being plagued with the same litigious attitudes that were prevalent in other areas of life. It was obvious from the number of claims and lawsuits filed that there were problems that needed attention. These entailed the fundamental manner of running a practice. One of the first issues for professionals was to reduce their liability exposure in an effort to avoid those areas of practice that caused the greater number of claims and other legal problems. The choice was obvious and easy. Contract administration (or construction services) was, by far, the phase of professional work most attacked by claims and other disputes. For it is here that the glitches of the contract documents come to light.

Hence, professionals sought to transfer or otherwise avoid responsibility and liability. First came the deletion of contract administration, sometimes by default (nonenforcement) and sometimes by well-documented contract changes. Changes were made in most areas of work, even in the documentation phase.

What were once professional-produced drawings became specification items; shop drawings were the replacements for the professional drawings. For example, in lieu of layouts and details of roof trusses and fire sprinkler systems, specifications called for these drawings to be produced by the manufacturer or installer, not by the professional. The responsibility was shifted away from the professional to the other party. In addition, the professional would only "review" these documents for compliance with the "general design" of the project, again avoiding the ominous responsibility of "approval." One small aspect of this scenario erupted when building permits were held up because detailed engineering drawings (required, in our example, for trusses and sprinklers) were not available to the code official prior to approval and issuance of the building permit (see Figure 6.2). Of course, structural design and fire protection are two sensitive issues in the permit process. This points out that unilateral action to avoid liability caused other problems to both the professional and the client. Something had to be done to resolve this and similar issues regarding avoidance or transfer of responsibility/liability.

Many individuals, seminars, papers, and articles extolled the virtues of contract administration in an attempt to win over design professionals to renew use of the program. Of course, the idea of added liability during the construction phase lay at the heart of the issue; reduce liability exposure, and you reduce the cost of insurance and practicing. This led, however, to an increased use of "plans and specs" contracts.

Although statistics are scant, it can be reasonably inferred that with fewer claims now being filed (with insurance carriers, at least) and more modest claims being paid directly by the professionals, contract administration is still a positive contributing factor. Certainly, to ensure faithful execution of the contract documents' provisions, design professionals have realized that proactive participation, control, review, and oversight are critical throughout the *entire* project sequence—hence, increased use of contract administration programs. Still, reevaluation and discussion continues.

THE AMERICAN INSTITUTE OF ARCHITECTS

AIA Document B141

Standard Form of Agreement Between Owner and Architect

1987 EDITION

THIS DOCUMENT HAS IMPORTANT LEGAL CONSEQUENCES; CONSULTATION WITH AN ATTORNEY IS ENCOURAGED WITH RESPECT TO ITS COMPLETION OR MODIFICATION.

SAMPLE

AGREEMENT

made as of the   day of   in the year of
Nineteen Hundred and

BETWEEN the Owner:
(Name and address)

and the Architect:
(Name and address)

For the following Project:
(Include detailed description of Project, location, address and scope)

Reproduced with permission of The American Institute of Architects under license number #97016. This license expires April 30, 1998. FURTHER REPRODUCTION IS PROHIBITED. Because AIA Documents are revised from time to time, users should ascertain from the AIA the current edition of this document. Copies of the current edition of this AIA document may be purchased from The American Institute of Architects or its local distributors. The text of this document is not "model language" and is not intended for use in other documents without permission of the AIA.

The Owner and Architect agree as set forth below.

Copyright 1917, 1926, 1948, 1951, 1953, 1958, 1961, 1963, 1966, 1967, 1970, 1974, 1977, ©1987 by The American Institute of Architects, 1735 New York Avenue, N.W., Washington, D.C. 20006. Reproduction of the material herein or substantial quotation of its provisions without written permission of the AIA violates the copyright laws of the United States and will be subject to legal prosecution.

AIA DOCUMENT B141 • OWNER-ARCHITECT AGREEMENT • FOURTEENTH EDITION • AIA® • ©1987
THE AMERICAN INSTITUTE OF ARCHITECTS, 1735 NEW YORK AVENUE, N.W., WASHINGTON, D.C. 20006   B141-1987 1

FIGURE 6.1   AIA document B141 details the owner-architect agreement.

TERMS AND CONDITIONS OF AGREEMENT BETWEEN OWNER AND ARCHITECT

ARTICLE 1
ARCHITECT'S RESPONSIBILITIES

1.1 ARCHITECT'S SERVICES

1.1.1 The Architect's services consist of those services performed by the Architect, Architect's employees and Architect's consultants as enumerated in Articles 2 and 3 of this Agreement and any other services included in Article 12.

1.1.2 The Architect's services shall be performed as expeditiously as is consistent with professional skill and care and the orderly progress of the Work. Upon request of the Owner, the Architect shall submit for the Owner's approval a schedule for the performance of the Architect's services which may be adjusted as the Project proceeds, and shall include allowances for periods of time required for the Owner's review and for approval of submissions by authorities having jurisdiction over the Project. Time limits established by this schedule approved by the Owner shall not, except for reasonable cause, be exceeded by the Architect or Owner.

1.1.3 The services covered by this Agreement are subject to the time limitations contained in Subparagraph 11.5.1.

ARTICLE 2
SCOPE OF ARCHITECT'S BASIC SERVICES

2.1 DEFINITION

2.1.1 The Architect's Basic Services consist of those described in Paragraphs 2.2 through 2.6 and any other services identified in Article 12 as part of Basic Services, and include normal structural, mechanical and electrical engineering services.

2.2 SCHEMATIC DESIGN PHASE

2.2.1 The Architect shall review the program furnished by the Owner to ascertain the requirements of the Project and shall arrive at a mutual understanding of such requirements with the Owner.

2.2.2 The Architect shall provide a preliminary evaluation of the Owner's program, schedule and construction budget requirements, each in terms of the other, subject to the limitations set forth in Subparagraph 5.2.1.

2.2.3 The Architect shall review with the Owner alternative approaches to design and construction of the Project.

2.2.4 Based on the mutually agreed-upon program, schedule and construction budget requirements, the Architect shall prepare, for approval by the Owner, Schematic Design Documents consisting of drawings and other documents illustrating the scale and relationship of Project components.

2.2.5 The Architect shall submit to the Owner a preliminary estimate of Construction Cost based on current area, volume or other unit costs.

2.3 DESIGN DEVELOPMENT PHASE

2.3.1 Based on the approved Schematic Design Documents and any adjustments authorized by the Owner in the program, schedule or construction budget, the Architect shall prepare, for approval by the Owner, Design Development Documents consisting of drawings and other documents to fix and describe the size and character of the Project as to architectural, structural, mechanical and electrical systems, materials and such other elements as may be appropriate.

2.3.2 The Architect shall advise the Owner of any adjustments to the preliminary estimate of Construction Cost.

2.4 CONSTRUCTION DOCUMENTS PHASE

2.4.1 Based on the approved Design Development Documents and any further adjustments in the scope or quality of the Project or in the construction budget authorized by the Owner, the Architect shall prepare, for approval by the Owner, Construction Documents consisting of Drawings and Specifications setting forth in detail the requirements for the construction of the Project.

2.4.2 The Architect shall assist the Owner in the preparation of the necessary bidding information, bidding forms, the Conditions of the Contract, and the form of Agreement between the Owner and Contractor.

2.4.3 The Architect shall advise the Owner of any adjustments to previous preliminary estimates of Construction Cost indicated by changes in requirements or general market conditions.

2.4.4 The Architect shall assist the Owner in connection with the Owner's responsibility for filing documents required for the approval of governmental authorities having jurisdiction over the Project.

2.5 BIDDING OR NEGOTIATION PHASE

2.5.1 The Architect, following the Owner's approval of the Construction Documents and of the latest preliminary estimate of Construction Cost, shall assist the Owner in obtaining bids or negotiated proposals and assist in awarding and preparing contracts for construction.

2.6 CONSTRUCTION PHASE—ADMINISTRATION OF THE CONSTRUCTION CONTRACT

2.6.1 The Architect's responsibility to provide Basic Services for the Construction Phase under this Agreement commences with the award of the Contract for Construction and terminates at the earlier of the issuance to the Owner of the final Certificate for Payment or 60 days after the date of Substantial Completion of the Work, unless extended under the terms of Subparagraph 10.3.3.

2.6.2 The Architect shall provide administration of the Contract for Construction as set forth below and in the edition of AIA Document A201, General Conditions of the Contract for Construction, current as of the date of this Agreement, unless otherwise provided in this Agreement.

2.6.3 Duties, responsibilities and limitations of authority of the Architect shall not be restricted, modified or extended without written agreement of the Owner and Architect with consent of the Contractor, which consent shall not be unreasonably withheld.

FIGURE 6.1 *(Continued).*

2.6.4 The Architect shall be a representative of and shall advise and consult with the Owner (1) during construction until final payment to the Contractor is due, and (2) as an Additional Service at the Owner's direction from time to time during the correction period described in the Contract for Construction. The Architect shall have authority to act on behalf of the Owner only to the extent provided in this Agreement unless otherwise modified by written instrument.

2.6.5 The Architect shall visit the site at intervals appropriate to the stage of construction or as otherwise agreed by the Owner and Architect in writing to become generally familiar with the progress and quality of the Work completed and to determine in general if the Work is being performed in a manner indicating that the Work when completed will be in accordance with the Contract Documents. However, the Architect shall not be required to make exhaustive or continuous on-site inspections to check the quality or quantity of the Work. On the basis of on-site observations as an architect, the Architect shall keep the Owner informed of the progress and quality of the Work, and shall endeavor to guard the Owner against defects and deficiencies in the Work. *(More extensive site representation may be agreed to as an Additional Service, as described in Paragraph 3.2.)*

2.6.6 The Architect shall not have control over or charge of and shall not be responsible for construction means, methods, techniques, sequences or procedures, or for safety precautions and programs in connection with the Work, since these are solely the Contractor's responsibility under the Contract for Construction. The Architect shall not be responsible for the Contractor's schedules or failure to carry out the Work in accordance with the Contract Documents. The Architect shall not have control over or charge of acts or omissions of the Contractor, Subcontractors, or their agents or employees, or of any other persons performing portions of the Work.

2.6.7 The Architect shall at all times have access to the Work wherever it is in preparation or progress.

2.6.8 Except as may otherwise be provided in the Contract Documents or when direct communications have been specially authorized, the Owner and Contractor shall communicate through the Architect. Communications by and with the Architect's consultants shall be through the Architect.

2.6.9 Based on the Architect's observations and evaluations of the Contractor's Applications for Payment, the Architect shall review and certify the amounts due the Contractor.

2.6.10 The Architect's certification for payment shall constitute a representation to the Owner, based on the Architect's observations at the site as provided in Subparagraph 2.6.5 and on the data comprising the Contractor's Application for Payment, that the Work has progressed to the point indicated and that, to the best of the Architect's knowledge, information and belief, quality of the Work is in accordance with the Contract Documents. The foregoing representations are subject to an evaluation of the Work for conformance with the Contract Documents upon Substantial Completion, to results of subsequent tests and inspections, to minor deviations from the Contract Documents correctable prior to completion and to specific qualifications expressed by the Architect. The issuance of a Certificate for Payment shall further constitute a representation that the Contractor is entitled to payment in the amount certified. However, the issuance of a Certificate for Payment shall not be a representation that the Architect has (1) made exhaustive or continuous on-site inspections to check the quality or quantity of the Work, (2) reviewed construction means, methods, techniques, sequences or procedures, (3) reviewed copies of requisitions received from Subcontractors and material suppliers and other data requested by the Owner to substantiate the Contractor's right to payment or (4) ascertained how or for what purpose the Contractor has used money previously paid on account of the Contract Sum.

2.6.11 The Architect shall have authority to reject Work which does not conform to the Contract Documents. Whenever the Architect considers it necessary or advisable for implementation of the intent of the Contract Documents, the Architect will have authority to require additional inspection or testing of the Work in accordance with the provisions of the Contract Documents, whether or not such Work is fabricated, installed or completed. However, neither this authority of the Architect nor a decision made in good faith either to exercise or not to exercise such authority shall give rise to a duty or responsibility of the Architect to the Contractor, Subcontractors, material and equipment suppliers, their agents or employees or other persons performing portions of the Work.

2.6.12 The Architect shall review and approve or take other appropriate action upon Contractor's submittals such as Shop Drawings, Product Data and Samples, but only for the limited purpose of checking for conformance with information given and the design concept expressed in the Contract Documents. The Architect's action shall be taken with such reasonable promptness as to cause no delay in the Work or in the construction of the Owner or of separate contractors, while allowing sufficient time in the Architect's professional judgment to permit adequate review. Review of such submittals is not conducted for the purpose of determining the accuracy and completeness of other details such as dimensions and quantities or for substantiating instructions for installation or performance of equipment or systems designed by the Contractor, all of which remain the responsibility of the Contractor to the extent required by the Contract Documents. The Architect's review shall not constitute approval of safety precautions or, unless otherwise specifically stated by the Architect, of construction means, methods, techniques, sequences or procedures. The Architect's approval of a specific item shall not indicate approval of an assembly of which the item is a component. When professional certification of performance characteristics of materials, systems or equipment is required by the Contract Documents, the Architect shall be entitled to rely upon such certification to establish that the materials, systems or equipment will meet the performance criteria required by the Contract Documents.

2.6.13 The Architect shall prepare Change Orders and Construction Change Directives, with supporting documentation and data if deemed necessary by the Architect as provided in Subparagraphs 3.1.1 and 3.3.3, for the Owner's approval and execution in accordance with the Contract Documents, and may authorize minor changes in the Work not involving an adjustment in the Contract Sum or an extension of the Contract Time which are not inconsistent with the intent of the Contract Documents.

2.6.14 The Architect shall conduct inspections to determine the date or dates of Substantial Completion and the date of final completion, shall receive and forward to the Owner for the Owner's review and records written warranties and related documents required by the Contract Documents and assembled by the Contractor, and shall issue a final Certificate for Payment upon compliance with the requirements of the Contract Documents.

FIGURE 6.1 *(Continued).*

2.6.15 The Architect shall interpret and decide matters concerning performance of the Owner and Contractor under the requirements of the Contract Documents on written request of either the Owner or Contractor. The Architect's response to such requests shall be made with reasonable promptness and within any time limits agreed upon.

2.6.16 Interpretations and decisions of the Architect shall be consistent with the intent of and reasonably inferable from the Contract Documents and shall be in writing or in the form of drawings. When making such interpretations and initial decisions, the Architect shall endeavor to secure faithful performance by both Owner and Contractor, shall not show partiality to either, and shall not be liable for results of interpretations or decisions so rendered in good faith.

2.6.17 The Architect's decisions on matters relating to aesthetic effect shall be final if consistent with the intent expressed in the Contract Documents.

2.6.18 The Architect shall render written decisions within a reasonable time on all claims, disputes or other matters in question between the Owner and Contractor relating to the execution or progress of the Work as provided in the Contract Documents.

2.6.19 The Architect's decisions on claims, disputes or other matters, including those in question between the Owner and Contractor, except for those relating to aesthetic effect as provided in Subparagraph 2.6.17, shall be subject to arbitration as provided in this Agreement and in the Contract Documents.

ARTICLE 3
ADDITIONAL SERVICES

3.1 GENERAL

3.1.1 The services described in this Article 3 are not included in Basic Services unless so identified in Article 12, and they shall be paid for by the Owner as provided in this Agreement, in addition to the compensation for Basic Services. The services described under Paragraphs 3.2 and 3.4 shall only be provided if authorized or confirmed in writing by the Owner. If services described under Contingent Additional Services in Paragraph 3.3 are required due to circumstances beyond the Architect's control, the Architect shall notify the Owner prior to commencing such services. If the Owner deems that such services described under Paragraph 3.3 are not required, the Owner shall give prompt written notice to the Architect. If the Owner indicates in writing that all or part of such Contingent Additional Services are not required, the Architect shall have no obligation to provide those services.

3.2 PROJECT REPRESENTATION BEYOND BASIC SERVICES

3.2.1 If more extensive representation at the site than is described in Subparagraph 2.6.5 is required, the Architect shall provide one or more Project Representatives to assist in carrying out such additional on-site responsibilities.

3.2.2 Project Representatives shall be selected, employed and directed by the Architect, and the Architect shall be compensated therefor as agreed by the Owner and Architect. The duties, responsibilities and limitations of authority of Project Representatives shall be as described in the edition of AIA Document B352 current as of the date of this Agreement, unless otherwise agreed.

3.2.3 Through the observations by such Project Representatives, the Architect shall endeavor to provide further protection for the Owner against defects and deficiencies in the Work, but the furnishing of such project representation shall not modify the rights, responsibilities or obligations of the Architect as described elsewhere in this Agreement.

3.3 CONTINGENT ADDITIONAL SERVICES

3.3.1 Making revisions in Drawings, Specifications or other documents when such revisions are:

.1 inconsistent with approvals or instructions previously given by the Owner, including revisions made necessary by adjustments in the Owner's program or Project budget;

.2 required by the enactment or revision of codes, laws or regulations subsequent to the preparation of such documents; or

.3 due to changes required as a result of the Owner's failure to render decisions in a timely manner.

3.3.2 Providing services required because of significant changes in the Project including, but not limited to, size, quality, complexity, the Owner's schedule, or the method of bidding or negotiating and contracting for construction, except for services required under Subparagraph 5.2.5.

3.3.3 Preparing Drawings, Specifications and other documentation and supporting data, evaluating Contractor's proposals, and providing other services in connection with Change Orders and Construction Change Directives.

3.3.4 Providing services in connection with evaluating substitutions proposed by the Contractor and making subsequent revisions to Drawings, Specifications and other documentation resulting therefrom.

3.3.5 Providing consultation concerning replacement of Work damaged by fire or other cause during construction, and furnishing services required in connection with the replacement of such Work.

3.3.6 Providing services made necessary by the default of the Contractor, by major defects or deficiencies in the Work of the Contractor, or by failure of performance of either the Owner or Contractor under the Contract for Construction.

3.3.7 Providing services in evaluating an extensive number of claims submitted by the Contractor or others in connection with the Work.

3.3.8 Providing services in connection with a public hearing, arbitration proceeding or legal proceeding except where the Architect is party thereto.

3.3.9 Preparing documents for alternate, separate or sequential bids or providing services in connection with bidding, negotiation or construction prior to the completion of the Construction Documents Phase.

3.4 OPTIONAL ADDITIONAL SERVICES

3.4.1 Providing analyses of the Owner's needs and programming the requirements of the Project.

3.4.2 Providing financial feasibility or other special studies.

3.4.3 Providing planning surveys, site evaluations or comparative studies of prospective sites.

FIGURE 6.1 *(Continued).*

3.4.4 Providing special surveys, environmental studies and submissions required for approvals of governmental authorities or others having jurisdiction over the Project.

3.4.5 Providing services relative to future facilities, systems and equipment.

3.4.6 Providing services to investigate existing conditions or facilities or to make measured drawings thereof.

3.4.7 Providing services to verify the accuracy of drawings or other information furnished by the Owner.

3.4.8 Providing coordination of construction performed by separate contractors or by the Owner's own forces and coordination of services required in connection with construction performed and equipment supplied by the Owner.

3.4.9 Providing services in connection with the work of a construction manager or separate consultants retained by the Owner.

3.4.10 Providing detailed estimates of Construction Cost.

3.4.11 Providing detailed quantity surveys or inventories of material, equipment and labor.

3.4.12 Providing analyses of owning and operating costs.

3.4.13 Providing interior design and other similar services required for or in connection with the selection, procurement or installation of furniture, furnishings and related equipment.

3.4.14 Providing services for planning tenant or rental spaces.

3.4.15 Making investigations, inventories of materials or equipment, or valuations and detailed appraisals of existing facilities.

3.4.16 Preparing a set of reproducible record drawings showing significant changes in the Work made during construction based on marked-up prints, drawings and other data furnished by the Contractor to the Architect.

3.4.17 Providing assistance in the utilization of equipment or systems such as testing, adjusting and balancing, preparation of operation and maintenance manuals, training personnel for operation and maintenance, and consultation during operation.

3.4.18 Providing services after issuance to the Owner of the final Certificate for Payment, or in the absence of a final Certificate for Payment, more than 60 days after the date of Substantial Completion of the Work.

3.4.19 Providing services of consultants for other than architectural, structural, mechanical and electrical engineering portions of the Project provided as a part of Basic Services.

3.4.20 Providing any other services not otherwise included in this Agreement or not customarily furnished in accordance with generally accepted architectural practice.

ARTICLE 4
OWNER'S RESPONSIBILITIES

4.1 The Owner shall provide full information regarding requirements for the Project, including a program which shall set forth the Owner's objectives, schedule, constraints and criteria, including space requirements and relationships, flexibility, expandability, special equipment, systems and site requirements.

4.2 The Owner shall establish and update an overall budget for the Project, including the Construction Cost, the Owner's other costs and reasonable contingencies related to all of these costs.

4.3 If requested by the Architect, the Owner shall furnish evidence that financial arrangements have been made to fulfill the Owner's obligations under this Agreement.

4.4 The Owner shall designate a representative authorized to act on the Owner's behalf with respect to the Project. The Owner or such authorized representative shall render decisions in a timely manner pertaining to documents submitted by the Architect in order to avoid unreasonable delay in the orderly and sequential progress of the Architect's services.

4.5 The Owner shall furnish surveys describing physical characteristics, legal limitations and utility locations for the site of the Project, and a written legal description of the site. The surveys and legal information shall include, as applicable, grades and lines of streets, alleys, pavements and adjoining property and structures; adjacent drainage; rights-of-way, restrictions, easements, encroachments, zoning, deed restrictions, boundaries and contours of the site; locations, dimensions and necessary data pertaining to existing buildings, other improvements and trees; and information concerning available utility services and lines, both public and private, above and below grade, including inverts and depths. All the information on the survey shall be referenced to a project benchmark.

4.6 The Owner shall furnish the services of geotechnical engineers when such services are requested by the Architect. Such services may include but are not limited to test borings, test pits, determinations of soil bearing values, percolation tests, evaluations of hazardous materials, ground corrosion and resistivity tests, including necessary operations for anticipating subsoil conditions, with reports and appropriate professional recommendations.

4.6.1 The Owner shall furnish the services of other consultants when such services are reasonably required by the scope of the Project and are requested by the Architect.

4.7 The Owner shall furnish structural, mechanical, chemical, air and water pollution tests, tests for hazardous materials, and other laboratory and environmental tests, inspections and reports required by law or the Contract Documents.

4.8 The Owner shall furnish all legal, accounting and insurance counseling services as may be necessary at any time for the Project, including auditing services the Owner may require to verify the Contractor's Applications for Payment or to ascertain how or for what purposes the Contractor has used the money paid by or on behalf of the Owner.

4.9 The services, information, surveys and reports required by Paragraphs 4.5 through 4.8 shall be furnished at the Owner's expense, and the Architect shall be entitled to rely upon the accuracy and completeness thereof.

4.10 Prompt written notice shall be given by the Owner to the Architect if the Owner becomes aware of any fault or defect in the Project or nonconformance with the Contract Documents.

4.11 The proposed language of certificates or certifications requested of the Architect or Architect's consultants shall be submitted to the Architect for review and approval at least 14 days prior to execution. The Owner shall not request certifications that would require knowledge or services beyond the scope of this Agreement.

FIGURE 6.1 *(Continued).*

ARTICLE 5
CONSTRUCTION COST

5.1 DEFINITION

5.1.1 The Construction Cost shall be the total cost or estimated cost to the Owner of all elements of the Project designed or specified by the Architect.

5.1.2 The Construction Cost shall include the cost at current market rates of labor and materials furnished by the Owner and equipment designed, specified, selected or specially provided for by the Architect, plus a reasonable allowance for the Contractor's overhead and profit. In addition, a reasonable allowance for contingencies shall be included for market conditions at the time of bidding and for changes in the Work during construction.

5.1.3 Construction Cost does not include the compensation of the Architect and Architect's consultants, the costs of the land, rights-of-way, financing or other costs which are the responsibility of the Owner as provided in Article 4.

5.2 RESPONSIBILITY FOR CONSTRUCTION COST

5.2.1 Evaluations of the Owner's Project budget, preliminary estimates of Construction Cost and detailed estimates of Construction Cost, if any, prepared by the Architect, represent the Architect's best judgment as a design professional familiar with the construction industry. It is recognized, however, that neither the Architect nor the Owner has control over the cost of labor, materials or equipment, over the Contractor's methods of determining bid prices, or over competitive bidding, market or negotiating conditions. Accordingly, the Architect cannot and does not warrant or represent that bids or negotiated prices will not vary from the Owner's Project budget or from any estimate of Construction Cost or evaluation prepared or agreed to by the Architect.

5.2.2 No fixed limit of Construction Cost shall be established as a condition of this Agreement by the furnishing, proposal or establishment of a Project budget, unless such fixed limit has been agreed upon in writing and signed by the parties hereto. If such a fixed limit has been established, the Architect shall be permitted to include contingencies for design, bidding and price escalation, to determine what materials, equipment, component systems and types of construction are to be included in the Contract Documents, to make reasonable adjustments in the scope of the Project and to include in the Contract Documents alternate bids to adjust the Construction Cost to the fixed limit. Fixed limits, if any, shall be increased in the amount of an increase in the Contract Sum occurring after execution of the Contract for Construction.

5.2.3 If the Bidding or Negotiation Phase has not commenced within 90 days after the Architect submits the Construction Documents to the Owner, any Project budget or fixed limit of Construction Cost shall be adjusted to reflect changes in the general level of prices in the construction industry between the date of submission of the Construction Documents to the Owner and the date on which proposals are sought.

5.2.4 If a fixed limit of Construction Cost (adjusted as provided in Subparagraph 5.2.3) is exceeded by the lowest bona fide bid or negotiated proposal, the Owner shall:

.1 give written approval of an increase in such fixed limit;

.2 authorize rebidding or renegotiating of the Project within a reasonable time;

.3 if the Project is abandoned, terminate in accordance with Paragraph 8.3; or

.4 cooperate in revising the Project scope and quality as required to reduce the Construction Cost.

5.2.5 If the Owner chooses to proceed under Clause 5.2.4.4, the Architect, without additional charge, shall modify the Contract Documents as necessary to comply with the fixed limit, if established as a condition of this Agreement. The modification of Contract Documents shall be the limit of the Architect's responsibility arising out of the establishment of a fixed limit. The Architect shall be entitled to compensation in accordance with this Agreement for all services performed whether or not the Construction Phase is commenced.

ARTICLE 6
USE OF ARCHITECT'S DRAWINGS, SPECIFICATIONS AND OTHER DOCUMENTS

6.1 The Drawings, Specifications and other documents prepared by the Architect for this Project are instruments of the Architect's service for use solely with respect to this Project and, unless otherwise provided, the Architect shall be deemed the author of these documents and shall retain all common law, statutory and other reserved rights, including the copyright. The Owner shall be permitted to retain copies, including reproducible copies, of the Architect's Drawings, Specifications and other documents for information and reference in connection with the Owner's use and occupancy of the Project. The Architect's Drawings, Specifications or other documents shall not be used by the Owner or others on other projects, for additions to this Project or for completion of this Project by others, unless the Architect is adjudged to be in default under this Agreement, except by agreement in writing and with appropriate compensation to the Architect.

6.2 Submission or distribution of documents to meet official regulatory requirements or for similar purposes in connection with the Project is not to be construed as publication in derogation of the Architect's reserved rights.

ARTICLE 7
ARBITRATION

7.1 Claims, disputes or other matters in question between the parties to this Agreement arising out of or relating to this Agreement or breach thereof shall be subject to and decided by arbitration in accordance with the Construction Industry Arbitration Rules of the American Arbitration Association currently in effect unless the parties mutually agree otherwise.

7.2 Demand for arbitration shall be filed in writing with the other party to this Agreement and with the American Arbitration Association. A demand for arbitration shall be made within a reasonable time after the claim, dispute or other matter in question has arisen. In no event shall the demand for arbitration be made after the date when institution of legal or equitable proceedings based on such claim, dispute or other matter in question would be barred by the applicable statutes of limitations.

7.3 No arbitration arising out of or relating to this Agreement shall include, by consolidation, joinder or in any other manner, an additional person or entity not a party to this Agreement,

FIGURE 6.1 *(Continued).*

except by written consent containing a specific reference to this Agreement signed by the Owner, Architect, and any other person or entity sought to be joined. Consent to arbitration involving an additional person or entity shall not constitute consent to arbitration of any claim, dispute or other matter in question not described in the written consent or with a person or entity not named or described therein. The foregoing agreement to arbitrate and other agreements to arbitrate with an additional person or entity duly consented to by the parties to this Agreement shall be specifically enforceable in accordance with applicable law in any court having jurisdiction thereof.

7.4 The award rendered by the arbitrator or arbitrators shall be final, and judgment may be entered upon it in accordance with applicable law in any court having jurisdiction thereof.

ARTICLE 8
TERMINATION, SUSPENSION OR ABANDONMENT

8.1 This Agreement may be terminated by either party upon not less than seven days' written notice should the other party fail substantially to perform in accordance with the terms of this Agreement through no fault of the party initiating the termination.

8.2 If the Project is suspended by the Owner for more than 30 consecutive days, the Architect shall be compensated for services performed prior to notice of such suspension. When the Project is resumed, the Architect's compensation shall be equitably adjusted to provide for expenses incurred in the interruption and resumption of the Architect's services.

8.3 This Agreement may be terminated by the Owner upon not less than seven days' written notice to the Architect in the event that the Project is permanently abandoned. If the Project is abandoned by the Owner for more than 90 consecutive days, the Architect may terminate this Agreement by giving written notice.

8.4 Failure of the Owner to make payments to the Architect in accordance with this Agreement shall be considered substantial nonperformance and cause for termination.

8.5 If the Owner fails to make payment when due the Architect for services and expenses, the Architect may, upon seven days' written notice to the Owner, suspend performance of services under this Agreement. Unless payment in full is received by the Architect within seven days of the date of the notice, the suspension shall take effect without further notice. In the event of a suspension of services, the Architect shall have no liability to the Owner for delay or damage caused the Owner because of such suspension of services.

8.6 In the event of termination not the fault of the Architect, the Architect shall be compensated for services performed prior to termination, together with Reimbursable Expenses then due and all Termination Expenses as defined in Paragraph 8.7.

8.7 Termination Expenses are in addition to compensation for Basic and Additional Services, and include expenses which are directly attributable to termination. Termination Expenses shall be computed as a percentage of the total compensation for Basic Services and Additional Services earned to the time of termination, as follows:

   .1 Twenty percent of the total compensation for Basic and Additional Services earned to date if termination occurs before or during the predesign, site analysis, or Schematic Design Phases; or

   .2 Ten percent of the total compensation for Basic and Additional Services earned to date if termination occurs during the Design Development Phase; or

   .3 Five percent of the total compensation for Basic and Additional Services earned to date if termination occurs during any subsequent phase.

ARTICLE 9
MISCELLANEOUS PROVISIONS

9.1 Unless otherwise provided, this Agreement shall be governed by the law of the principal place of business of the Architect.

9.2 Terms in this Agreement shall have the same meaning as those in AIA Document A201, General Conditions of the Contract for Construction, current as of the date of this Agreement.

9.3 Causes of action between the parties to this Agreement pertaining to acts or failures to act shall be deemed to have accrued and the applicable statutes of limitations shall commence to run not later than either the date of Substantial Completion for acts or failures to act occurring prior to Substantial Completion, or the date of issuance of the final Certificate for Payment for acts or failures to act occurring after Substantial Completion.

9.4 The Owner and Architect waive all rights against each other and against the contractors, consultants, agents and employees of the other for damages, but only to the extent covered by property insurance during construction, except such rights as they may have to the proceeds of such insurance as set forth in the edition of AIA Document A201, General Conditions of the Contract for Construction, current as of the date of this Agreement. The Owner and Architect each shall require similar waivers from their contractors, consultants and agents.

9.5 The Owner and Architect, respectively, bind themselves, their partners, successors, assigns and legal representatives to the other party to this Agreement and to the partners, successors, assigns and legal representatives of such other party with respect to all covenants of this Agreement. Neither Owner nor Architect shall assign this Agreement without the written consent of the other.

9.6 This Agreement represents the entire and integrated agreement between the Owner and Architect and supersedes all prior negotiations, representations or agreements, either written or oral. This Agreement may be amended only by written instrument signed by both Owner and Architect.

9.7 Nothing contained in this Agreement shall create a contractual relationship with or a cause of action in favor of a third party against either the Owner or Architect.

9.8 Unless otherwise provided in this Agreement, the Architect and Architect's consultants shall have no responsibility for the discovery, presence, handling, removal or disposal of or exposure of persons to hazardous materials in any form at the Project site, including but not limited to asbestos, asbestos products, polychlorinated biphenyl (PCB) or other toxic substances.

9.9 The Architect shall have the right to include representations of the design of the Project, including photographs of the exterior and interior, among the Architect's promotional and professional materials. The Architect's materials shall not include the Owner's confidential or proprietary information if the Owner has previously advised the Architect in writing of

FIGURE 6.1 *(Continued).*

the specific information considered by the Owner to be confidential or proprietary. The Owner shall provide professional credit for the Architect on the construction sign and in the promotional materials for the Project.

ARTICLE 10
PAYMENTS TO THE ARCHITECT

10.1 DIRECT PERSONNEL EXPENSE

10.1.1 Direct Personnel Expense is defined as the direct salaries of the Architect's personnel engaged on the Project and the portion of the cost of their mandatory and customary contributions and benefits related thereto, such as employment taxes and other statutory employee benefits, insurance, sick leave, holidays, vacations, pensions and similar contributions and benefits.

10.2 REIMBURSABLE EXPENSES

10.2.1 Reimbursable Expenses are in addition to compensation for Basic and Additional Services and include expenses incurred by the Architect and Architect's employees and consultants in the interest of the Project, as identified in the following Clauses.

10.2.1.1 Expense of transportation in connection with the Project; expenses in connection with authorized out-of-town travel; long-distance communications; and fees paid for securing approval of authorities having jurisdiction over the Project.

10.2.1.2 Expense of reproductions, postage and handling of Drawings, Specifications and other documents.

10.2.1.3 If authorized in advance by the Owner, expense of overtime work requiring higher than regular rates.

10.2.1.4 Expense of renderings, models and mock-ups requested by the Owner.

10.2.1.5 Expense of additional insurance coverage or limits, including professional liability insurance, requested by the Owner in excess of that normally carried by the Architect and Architect's consultants.

10.2.1.6 Expense of computer-aided design and drafting equipment time when used in connection with the Project.

10.3 PAYMENTS ON ACCOUNT OF BASIC SERVICES

10.3.1 An initial payment as set forth in Paragraph 11.1 is the minimum payment under this Agreement.

10.3.2 Subsequent payments for Basic Services shall be made monthly and, where applicable, shall be in proportion to services performed within each phase of service, on the basis set forth in Subparagraph 11.2.2.

10.3.3 If and to the extent that the time initially established in Subparagraph 11.5.1 of this Agreement is exceeded or extended through no fault of the Architect, compensation for any services rendered during the additional period of time shall be computed in the manner set forth in Subparagraph 11.3.2.

10.3.4 When compensation is based on a percentage of Construction Cost and any portions of the Project are deleted or otherwise not constructed, compensation for those portions of the Project shall be payable to the extent services are performed on those portions, in accordance with the schedule set forth in Subparagraph 11.2.2, based on (1) the lowest bona fide bid or negotiated proposal, or (2) if no such bid or proposal is received, the most recent preliminary estimate of Construction Cost or detailed estimate of Construction Cost for such portions of the Project.

10.4 PAYMENTS ON ACCOUNT OF ADDITIONAL SERVICES

10.4.1 Payments on account of the Architect's Additional Services and for Reimbursable Expenses shall be made monthly upon presentation of the Architect's statement of services rendered or expenses incurred.

10.5 PAYMENTS WITHHELD

10.5.1 No deductions shall be made from the Architect's compensation on account of penalty, liquidated damages or other sums withheld from payments to contractors, or on account of the cost of changes in the Work other than those for which the Architect has been found to be liable.

10.6 ARCHITECT'S ACCOUNTING RECORDS

10.6.1 Records of Reimbursable Expenses and expenses pertaining to Additional Services and services performed on the basis of a multiple of Direct Personnel Expense shall be available to the Owner or the Owner's authorized representative at mutually convenient times.

ARTICLE 11
BASIS OF COMPENSATION

The Owner shall compensate the Architect as follows:

11.1 AN INITIAL PAYMENT of   Dollars ($)
shall be made upon execution of this Agreement and credited to the Owner's account at final payment.

11.2 BASIC COMPENSATION

11.2.1 FOR BASIC SERVICES, as described in Article 2, and any other services included in Article 12 as part of Basic Services, Basic Compensation shall be computed as follows:
(Insert basis of compensation, including stipulated sums, multiples or percentages, and identify phases to which particular methods of compensation apply, if necessary.)

FIGURE 6.1 *(Continued).*

11.2.2 Where compensation is based on a stipulated sum or percentage of Construction Cost, progress payments for **Basic Services** in each phase shall total the following percentages of the total Basic Compensation payable:

(Insert additional phases as appropriate.)

Schematic Design Phase:	percent (%)
Design Development Phase:	percent (%)
Construction Documents Phase:	percent (%)
Bidding or Negotiation Phase:	percent (%)
Construction Phase:	percent (%)
Total Basic Compensation:	one hundred percent (100%)

11.3 COMPENSATION FOR ADDITIONAL SERVICES

11.3.1 FOR PROJECT REPRESENTATION BEYOND BASIC SERVICES, as described in Paragraph 3.2, compensation shall be computed as follows:

11.3.2 FOR ADDITIONAL SERVICES OF THE ARCHITECT, as described in Articles 3 and 12, other than (1) Additional Project Representation, as described in Paragraph 3.2, and (2) services included in Article 12 as part of Additional Services, but excluding services of consultants, compensation shall be computed as follows:

(Insert basis of compensation, including rates and/or multiples of Direct Personnel Expense for Principals and employees, and identify Principals and classify employees, if required. Identify specific services to which particular methods of compensation apply, if necessary.)

11.3.3 FOR ADDITIONAL SERVICES OF CONSULTANTS, including additional structural, mechanical and electrical engineering services and those provided under Subparagraph 3.4.19 or identified in Article 12 as part of Additional Services, a multiple of ( ) times the amounts billed to the Architect for such services.

(Identify specific types of consultants in Article 12, if required.)

11.4 REIMBURSABLE EXPENSES

11.4.1 FOR REIMBURSABLE EXPENSES, as described in Paragraph 10.2, and any other items included in Article 12 as Reimbursable Expenses, a multiple of ( ) times the expenses incurred by the Architect, the Architect's employees and consultants in the interest of the Project.

11.5 ADDITIONAL PROVISIONS

11.5.1 IF THE BASIC SERVICES covered by this Agreement have not been completed within ( ) months of the date hereof, through no fault of the Architect, extension of the Architect's services beyond that time shall be compensated as provided in Subparagraphs 10.3.3 and 11.3.2.

11.5.2 Payments are due and payable ( ) days from the date of the Architect's invoice. Amounts unpaid ( ) days after the invoice date shall bear interest at the rate entered below, or in the absence thereof at the legal rate prevailing from time to time at the principal place of business of the Architect.

(Insert rate of interest agreed upon.)

(Usury laws and requirements under the Federal Truth in Lending Act, similar state and local consumer credit laws and other regulations at the Owner's and Architect's principal places of business, the location of the Project and elsewhere may affect the validity of this provision. Specific legal advice should be obtained with respect to deletions or modifications, and also regarding requirements such as written disclosures or waivers.)

FIGURE 6.1 *(Continued).*

11.5.3 The rates and multiples set forth for Additional Services shall be annually adjusted in accordance with normal salary review practices of the Architect.

ARTICLE 12
OTHER CONDITIONS OR SERVICES

(Insert descriptions of other services, identify Additional Services included within Basic Compensation and modifications to the payment and compensation terms included in this Agreement.)

This Agreement entered into as of the day and year first written above.

OWNER   ARCHITECT

_____   _____
(Signature)   *(Signature)*

_____   _____
(Printed name and title)   *(Printed name and title)*

AIA DOCUMENT B141 • OWNER-ARCHITECT AGREEMENT • FOURTEENTH EDITION • AIA® • ©1987
THE AMERICAN INSTITUTE OF ARCHITECTS, 1735 NEW YORK AVENUE, N.W., WASHINGTON, D.C. 20006   **B141-1987 10**

FIGURE 6.1 *(Continued).*
Used with permission of the American Institute of Architects.

Chapter 6 Obligation, Education, and Liability

This is especially true of the construction phase. Control is requisite to ensuring that reality transforms the design concept, with its theoretical considerations, into a finished, fully functional structure. Without contract administration, both the owner and the professional become victims, the owner holding unrealistic expectations and the professional finding unrealized expectations. The owner may have the idea that certain things will happen or certain conditions will be resolved eventually (and more than likely in the last work phase, construction). The professional anticipates

the work and equipment under consideration shall be prepared in conformity with rule 4101:2-1-20 of the Administrative Code and be submitted to the building official for inspection.

Exceptions:

(A) No plans need be filed for minor repairs; and

(B) No site preparation or plot plans need be filed with the division of workshops and factories for industrialized units used exclusively as one-, two-, or three-family dwellings.

HISTORY: Eff. 7-1-95
1979-80 OMR 4-712 (A), eff. 5-1-80; 1978-79 OMR 4-216 (E), eff. 7-1-79; 1978-79 OMR 4-212 (R), eff. 7-1-79; prior BB-1-17

CROSS REFERENCES
Ohio Revised Code, 3791.04, Submission of plans

4101:2-1-18 PLANS, WHEN AND WHERE TO FILE

(A) Pursuant to section 3791.04 of the Revised Code, before entering into a contract for, or beginning the construction erection or manufacture of any building for which plans are required under rule 4101:2-1-17 of the Administrative Code including all industrialized units, **the owner thereof shall submit plans** which shall indicate thereon the portions that have been approved pursuant to section 3781.12 of the Revised Code, for which no further approval shall be required, to the building official for approval.

(B) Plans shall be submitted in triplicate.

HISTORY: 1981-82 OMR 683 (A), eff. 7-1-82
1980-81 OMR 175 (A), eff. 1-1-81; 1978-79 OMR 4-216 (E), eff. 7-1-79; 1978-79 OMR 4-212 (R), eff. 7-1-79; prior BB-1-18

CROSS REFERENCES
Ohio Revised Code, 3791.04, Submission of plans

4101:2-1-19 PLANS TO BE ADEQUATE

(A) Plans required under rule 4101:2-1-18 of the Administrative Code shall be drawn to scale and shall be sufficiently clear, comprehensive, detailed, and legible when submitted to the building official so that, together with any accompanying specifications and data, a person who is competent in such matters can determine whether or not the proposed building addition, or alteration, and all proposed building equipment will conform in safety and sanitation to all applicable provisions of OBBC.

(B) If substantive changes to the building are contemplated after first plan submission, or during construction, those changes must be submitted to the building official for review and approval prior to those changes being executed. The building official may waive this requirement in the instance of an emergency repair, or similar instance.

(C) Plans for all buildings shall designate the occupancy, type of construction, and the fire-resistance rating of all structural elements as required by this code. The plans and specifications shall include all documentation or supporting data substantiating all required fire-resistance ratings.

(D) Plans shall indicate how penetrations will be made for electrical, mechanical, plumbing, and communication conduits, pipes, and systems, and shall also indicate the materials and methods for maintaining the required structural integrity, fire-resistance rating, and firestopping.

(E) Plans, when submitted to the building official for review shall be in standard multiples of eight and one-half inches by eleven inches or nine inches by twelve inches in size, and shall include:

(1) An index of drawings located on the first sheet;

(2) A plot plan showing street location; the location of the proposed building and all existing buildings on the site including setback and sideyard dimensions; distance between all buildings; and location and sizes of all utility lines;

(3) Floor plans, including plans of full or partial basement or cellars and full or partial attics or penthouses. Floor plans must show all relevant information such as door swings, stairs and ramps, windows, shafts, etc., and must be sufficiently dimensioned to describe all relevant space sizes. Wall materials must be described by cross-hatching (with explanatory key), by notation, or by other clearly understandable method. Spaces must be identified by code appellation, i.e., an "auditorium" may not be identified as "meeting room" if its size and function dictates that it is an auditorium;

(4) All elevations necessary to completely describe the exterior of the building including floor to floor dimensions;

(5) Cross sections, wall sections and detail sections, to scale, as may be required to describe the general building construction including wall, ceiling, floor and roof materials and construction, and details which may be necessary to describe typical connections, etc.;

(6) Complete structural description of the building on the above drawings or on separate drawings including size and location of all principal structural elements and table of live loads used in the design of the building and computations, stress diagrams and other data sufficient to show correctness of plans;

(7) Complete description of the mechanical and electrical systems of the building on the above drawings or on separate drawings, including plumbing schematics and principal plumbing, heating, ventilation and air conditioning duct and piping layouts and lighting and power equipment layouts; and

(8) Additional graphic or text information as may be reasonably required by the building official to allow him to review special or extraordinary construction methods or equipment.

(F) Upon application for plan approval for buildings or portions thereof constructed of industrialized units authorized pursuant to section 3781.12 of the Revised Code, the building official shall be provided with a copy of the industrialized unit manufacturer's "Letter of Authorization" in addition to documents as required by this rule for on-site construction and documentation required by rule 4101:2-1-18 of the Administrative Code.

(G) Before industrialized unit(s) are set or installed on the site of intended use, the building official shall be provided with the following:

(1) A copy of the plans approved by the board;

(2) Details pertaining to on-site interconnection of modules or assemblies; and

OBBC—Building Code

FIGURE 6.2 Building code excerpts that specify the types of documents required and the schedule to be followed for their submittal.

903.0 Building Code

either in the unit or obtained at the point of installation (see Section 919.0).

Sprinkler: A device, connected to a water supply system, that discharges water in a specific pattern for extinguishment or control of fire (see Section 906.0).

Sprinkler system, automatic: A sprinkler system, for fire protection purposes, is an integrated system of underground or overhead piping designed in accordance with fire protection engineering standards. The system includes a suitable water supply. The portion of the system above the ground is a network of specially or hydraulically designed piping installed in a building, structure or area, generally overhead, and to which automatic sprinklers are connected in a systematic pattern. The system is usually activated by heat from a fire and discharges water over the fire area (see Section 906.0).

Sprinkler system, limited area: An automatic sprinkler system consisting of not more than 20 sprinklers within a fire area (see Section 907.0).

Standpipe system: A standpipe system is a fire protection system consisting of an arrangement of piping, valves, hose outlets and allied equipment installed in a building or structure (see Section 914.0).

Supervisory device: An initiating device used to monitor the conditions that are essential for the proper operation of automatic fire suppression systems (i.e., switches used to monitor the position of gate valves, a low air-pressure switch on a dry-pipe sprinkler system, etc.) (see Section 923.0).

Voice/alarm signaling system: A system that provides, to the occupants of a building, dedicated manual or automatic facilities, or both, for originating and distributing voice instructions, as well as alert and evacuation signals that pertain to a fire emergency (see Section 917.0).

Water supply, automatic: A water supply that is not dependent on any manual operation, such as making connections, operating valves or starting pumps (see Section 914.5).

HISTORY: Eff. 7-1-95

Section 903.0 CONSTRUCTION DOCUMENTS

903.1 Required: Construction documents or shop drawings, or both, for the installation of fire protection systems shall be submitted to indicate conformance to this code and shall be reviewed by the department prior to issuance of the permit.

Note: Since the fire department is responsible for inspecting for the proper maintenance of fire protection systems in buildings, the administrative authority shall cooperate with the fire department in the discharge of responsibility to enforce this chapter.

903.2 Construction documents: The construction documents and shop drawings submitted to the department shall contain sufficient detail as outlined herein to evaluate the protected hazard and the effectiveness of the system.

903.2.1 Information: Construction documents for fire protection systems shall be submitted with the construction documents for the construction permit. Included shall be information on the contents, the occupancy, the location and arrangement of the structure and the contents involved, the exposure to any hazard, the extent of the system coverage, the suppression system design criteria, the supply and extinguishing agents, the location of any standpipes, and the location and method of operation of detection and alarm devices.

903.2.2 Shop drawings: Shop drawings for the installation of fire protection systems shall be submitted for review and approval prior to the installation of a fire protection system. Included on the shop drawings shall be information showing the basis for compliance with the design density, the specific arrangement of the system, the devices and their method(s) of operation, and the suppression agent. The details on the construction documents or shop drawings for the fire protection system shall include design considerations, spacing and arrangement of fire protection devices, protection agent supply and discharge requirements, calculations with sizes and equivalent lengths of pipe and fittings, and protection agent source. Sufficient information shall be included to identify the apparatus and devices utilized and other information as required by this code.

HISTORY: Eff. 7-1-95

Section 904.0 [4101:2-9-04] FIRE SUPPRESSION SYSTEMS

904.1 Where required: Automatic fire suppression systems shall be installed where required by this code, and in the locations indicated in Sections 904.2 through 904.11.

Exceptions

1. An automatic fire suppression system shall not be required in portions of buildings that comply with Section 406.0 for open parking structures.

2. In telecommunications equipment buildings, an automatic fire suppression system shall not be required in those spaces or areas occupied exclusively for telecommunications equipment, associated electrical power distribution equipment, batteries and standby engines, provided that those spaces or areas are equipped throughout with an automatic fire detection system in accordance with Section 918.0 and are separated from the remainder of the building with fire separation assemblies consisting of 1-hour fireresistance rated walls and 2-hour fire-resistance rated floor/ceiling assemblies.

904.2 Use Groups A-1, A-3 and A-4: Where a Use Group A-1, A-3 or A-4 fire area exceeds 12,000 square feet (1116 m^2) in area, an automatic fire suppression system shall be provided as follows:

1. Throughout the entire story or floor level where the A-1, A-3 or A-4 Use Group is located;

2. Throughout all stories and floor levels below the A-1, A-3 or A-4 Use Group; and

3. Throughout all intervening stories and floor levels between the A-1, A-3 or A-4 Use Group and the highest level of exit discharge that serves Use Group A-1, A-3 or A-4 fire areas, including the highest level of exit discharge.

Exceptions

1. Auditorium areas of Use Group A-1 or A-3 where the main auditorium floor is at the level of exit discharge of the main entrance.

2. Naves and chancels of Use Group A-4 where the main floor of the nave or chancel is at the level of exit discharge of the main entrance.

FIGURE 6.2 *(Continued).*

that the approved design concept will be fully and properly executed, hence a successful project will result. Without contract administration, neither party has any way of ensuring what the project will become—or will *not* become! Their collective, direct, insightful control is gone.

Obviously, every professional firm has the right to establish a business policy decision regarding contract administration. Each organization, no matter its size, must come to grips with the reality of liability and the cost of adequate protection. Much depends on the individual choices of the principals of the firm and their philosophy of how they want to do business. Some are perfectly content with a limited practice, wherein they take a small range of projects and/or provide only design and documentation services; plans and specs only, no contract administration. It is, however, interesting that many firms have seen fit to include contract administration as an integral element of their total philosophy of practice. For instance, contract administration is frequently deemed by formal declaration to be *equally* important as the planning and design of a project, as the only scenario that best serves the client.

Despite the traditional breakdown of professional work phases, percentage of fees earned, time budget, number of personnel, and other measures of practice, an underlying philosophical determination that "equal emphasis *will* be given planning, design, *and* contract administration" usually proves to be more successful and most beneficial to all contractual parties. It is an enlightened position. Success for the client, the project, and the professional is achieved in this way, and achieved with a good deal of consistency. In some firms, ranking associates and even principals are specifically charged with overseeing and directing a specific area of practice, including contract administration. This tends to maximize the integration, utilization, coordination, impact, and function of contract administration throughout the entire firm and its project process. It greatly enhances project success. As architect Robert A. M. Stern said,

> Creativity, in architects, is in large part a manipulation of circumstances. This leads architecture to better fitting its individual context—its being!

Optimal use of contract administration would find it continually influencing the decisions, selections, and detailing in the design development phase. This influence is invaluable as it blends field experience with the new design concept. It seeks to avoid problems encountered in previous, similar conditions. It is an aid to the process. The development process is enhanced where field information is fed back into the design and development sequences. Also, it can be a form of ongoing training. Design and other office personnel can benefit from actual field occurrences and learn how to avoid bad situations in future designs. It can create new or improved standard details, procedures, policies, and precautions, as well as being an ongoing, dynamic resource of crucial information (see Figure 6.3). In addition, contract administration can be a direct deterrent to disputes, claims, and even litigation, not only in its field operations, but in the office.

Some years ago it was suggested that each office could benefit from a "quality control coordinator." This would establish a format to forestall problems in operations that directly contributed to claims. One can see the value of at least minimal input from contract administration to resolve the following problem areas:

- Inadequate communications between personnel of the various disciplines involved with a project
- Errors on the part of inexperienced personnel
- Problems directly attributable to very rapid growth of the firm and lack of time for training

FIGURE 6.3 Mastery involves more than merely performing well in a new format.

> Knowing what does not work, as well as what does work, and the appropriate and useful adaptations and modifications between them, makes one a master of any chore, task, job, project, trade, or profession.

- Lack of coordination in the design office
- Junior staffers who have no understanding of how their work relates to the construction and the entire project;
- Lack of insight into the construction process
- Staffers who have little understanding of how their two-dimensional work is converted into three-dimensional reality during construction
- Insufficient time to objectively review, process, and check both original documents and subsequent changes
- Inadequately defined duties and responsibilities

Solutions to these problems were suggested as the duties for the coordinator. As amended to include the contract administrator, they are:

- Select and train employees for specific projects (and for contract administration, if their interests lie there).
- Orient the design team to the total project, with initial and periodic feedback from the field personnel.
- Coordinate all professional efforts for the project in all disciplines. Encourage consultants to review field operations.
- Review the firm's design effort with suggestions from the field and the "construction perspective."
- Distribute loss prevention materials and ideas (include field operations).

Where the provisions for contract administration remain in the contract, strict enforcement of the contract by the owner will mandate that this work be performed. Where this is not performed, a contractual breach will occur. Quite often, however, the owner is unaware of these requirements, purposely chooses to ignore them, or through lax enforcement simply allows the work to go unperformed by default. Similarly, a professional who writes good, well-founded specifications and then allows them to be un- or underenforced is not meeting the demands of the contract. If you are going to control by making requirements and demands, you *must* enforce! Any failure here on the part of the professional is twofold: breach of the owner-architect agreement for a full array of services (now left unfulfilled), and permitting a breach of the owner-contractor contract by not ensuring that the full range of services and products required by the contract are forthcoming to the owner. In both events, the professional (unbeknownst to the owner) could be paid in full accord with the agreed-on fee, supposedly for providing the full array of basic contract services, but without having done so. Not only is that illegal, it is unethical and highly unprofessional.

One professional author wrote that professionals "are faced with the necessity of performing those services which they have contracted to do during, and in anticipation of, the construction phase of the typical architectural contract." What a dispirited comment! First, if you don't care to do something, it is absolutely essential that you don't contract to do it. Certainly that is a more professional approach than offering a half-hearted effort to meet "a necessity."

Oddly enough the same author later notes that "administration of the contract provides the architect the opportunity of observing the construction as it progresses." Absolutely! Not in the sense of an onlooker, but as one who can influence progress, correct construction, and fulfill the client's desires. That is the direction of this effort. We heartily advocate full and continued activity by the design professional in all phases of all projects. Remember that architecture is about real buildings, not only about good design concepts and the "paper exercises" that lead to construction. The reality of the new buildings is the true test of the professional effort.

Surely, too, in any professional design organization, there is an element of the staff that finds the construction effort stimulating. This is a good part of the construction contract administration impetus and does, in the end, carry over to the entire staff. To pass this off as "meeting a contractual necessity" is doing the process a true injustice.

If the service of contract administration is not going to be performed by purposeful *and mutual* decision between professional and owner, the contract provisions themselves should be drawn (or modified, in the case of a standard AIA document) to delete the appropriate text of Section 2.6. It is essential, though, that both parties to the contract understand any such deletion, its consequences, the changes in their responsibilities/liability, and the impact on the project work. This can have a direct (and perhaps adverse) effect on the final outcome of the project, physically, functionally, and/or financially. This is the direct manifestation of the risk involved in deleting the service. It can also result from nonperformance of the service.

If the service requirement is retained, there is a distinct need for the professional and the owner to engage in explanation, clarification, discussion, and definition of what this requirement truly entails. The parties must be careful to review the provisions of Section 2.6 (AIA document B141) and to note every instance of ambiguous language. For example, in Section 2.6.5, what does "visit the site at intervals appropriate to the stage of construction" mean? Who makes the judgment or defines what it means? Is the architect the sole interpreter of this contract and all its provisions? What impact on the service and fee does "or as otherwise agreed" have? What does "generally familiar" mean to each party? How can the architect inform the owner of the "quality of the Work" when he/she is "not . . . required . . . to check the quality or quantity of the Work"? It is essential, and beneficial to both parties, that they fully understand the parameters of the service to be provided. The entire flow of the project and perhaps its ultimate success depend on this cooperative effort. This is even more important in the contract administration phase. Substantial case law has been developed over these very issues.

The basic elements of any contract are (1) an offer, (2) an acceptance, and (3) the exchange of compensation. These are fairly self-explanatory. However, another underlying element, often evaluated by the courts, is "meeting of the minds." Essentially, this entails ascertaining whether or not all parties to the contract have the same understanding, intentions, expectations, and general agreement on what the contract is about and what is to be produced/delivered. Without this element, even the best of contracts can prove inadequate or faulty and become the direct issue of litigation.

It is widely documented that extensive negotiations ensued before the AIA (and other professional societies) published their standard contract forms. The drafters, both professional and legal, went to extensive lengths to include verbiage that was acceptable to all parties and that reflected the work to be done. However, the words used were selected and phrased to pose the least liability exposure for the professionals. Thus, two sets of people are left to determine what, exactly, is intended and what is to be done: the professionals and the judges. Certainly this creates a situation that every practitioner should thoroughly resolve through education, discussion, legal advice, and individual understanding.

THE INFLUENCE OF CASE LAW

Perhaps the most helpful and definitive information can be gathered from case law. This information can serve as enlightening guidelines for professionals. Case law is one of the best and most expeditious ways to study the legal aspects of contract administration. These short reviews of court cases give great insight into the principles involved. Some excellent sources of such material are listed later in this chapter.

Some general principles follow regarding the use and understanding of case law:

1. It is a legal precedent or indication of how another court ruled.
2. It is not mandatory upon another court, even if the circumstances of the cases are identical.
3. It is a valuable aid to the court (by showing the research, rationale, and precedents used by previous courts in similar circumstances) and to the professional (as to advisable practices, pitfalls to avoid, limits of authority, and impact of court findings).
4. It is best to review a series of case law results to see the variations in the circumstances of the cases and the rationale of each court.

Remember that each case was decided on a specific set of circumstances, which may not be present in other cases. Also, some courts disagree with others and overturn previous decisions. Law in the different states also produces differing opinions.

Remember also that contract administration includes code administration. Since professionals are responsible for establishing code compliance, they are also responsible for ensuring that such compliance is achieved in the finished project. Here, case law citations are numerous. Some may argue that if the professional is retained just for plans and specs, the resulting project is none of the professional's concern or responsibility. While this viewpoint may be valid, it is a circumstance that requires judicial determination. There could be a finding that the basic responsibility is not transferable, since none of those who follow, during the construction of the project, carry the same liability and accountability. (See Appendix B for code-related specifications.)

Obviously, if the project is properly designed in regard to code compliance and is so described on the documents, the basic responsibility of the professional seems to have been achieved. If the construction faithfully follows the documents, there also would seem to be no problem with code compliance. However, if any changes are made that create noncompliance, there will be, at a minimum, an elongated process of establishing authority, responsibility, and so on. That could prove to be quite costly to the professional, although he/she is exonerated in the end.

It would be interesting to see how a provision in the various contracts, as follows, would hold up:

> The (owner) (construction manager) (contractor[s]) shall ensure that compliance with the applicable construction codes and regulations, as established by the design professional, and as described and depicted on the contract documents, is maintained and properly built into the finished work. Where revisions to the project are made, the (owner) (construction manager) (contractor[s]) shall ensure that necessary code compliance is maintained, reestablished, or achieved.

In a like manner, the following might be a good addition to a design contract for plans and specs only:

> The final documents produced under this contract shall depict a design and necessary construction details and information, which produce compliance with all programming parameters, including applicable codes and regulations, as listed in the documents, for the project intent, usage, and occupancy.
>
> Changes to the depicted work, during construction, may create noncompliance with such codes and regulations. The same shall not be the responsibility of the design professional who is not party to the construction contract or process.

Granted, code compliance is a relatively small issue overall, but it is a very important issue to the professional's liability exposure and to successful completion of the project. Also, it illustrates how other "small" issues can become pervasive if left with loose ends or, worst, not done at all.

ASSIGNING LIABILITY

The legal entanglements of the design professional engaged in contract administration are both pervasive and unavoidable. For the construction phase of the contract is where the anticipated execution of the contract becomes an active process; it is where design and construction details become time, money, and tangible evidence of compliance (or noncompliance) with all of the demands of the contract. It is a time when every member of the design and construction team must ensure that his/her action is commensurate, true to obligation, and within the parameters and requirements set out in the contract documents.

It is precisely here where the professional frequently finds himself/herself between a rock and a hard place, i.e., the owner and the contractor. Here, while practice, experience, and budgeting show that about 20 percent of the fee is expended, there is little doubt that 80 percent or more of the headaches occur. But this process is attempting to produce a "perfect" project through the use of personal, human-oriented procedures (design) and the imperfect science of construction. At best, it can be said that problems rooted in the other phases of the project will surely come to light and demand resolution in full here.

With this comes the intervention of fate. Usually describing unexpected adverse events, it must be forthrightly viewed as the description of faulty planning, analysis, design, concept, documentation, supervision, or execution. Specifically, in the construction scenario, fate seems to intervene when haste, avarice, ignorance, incompetence, carelessness, and/or confusion are present. Statistically, this produces some 17 percent of the claims against professionals (see Figure 6.4). In the norm, three types of errors are involved: design mistakes/incorrect information; changes from the designer's concept and intent; and unsatisfactory workmanship, materials, or both. The root causes of these errors are:

60%	lack of experience	2%	lack of ability to communicate
15%	negligence	1%	sharp practice
11%	lack of education	1%	lack of authority
9%	incompetence	1%	lack of formal qualifications

Certainly it appears easy to assign these faults to another party, but professionals can take little solace in the fact that studies show that the faults show up with:

---

1. Allowing non-code-approved material to be installed.
2. Ordering excess fill to be placed without proper soil testing/verification.
3. Failing to install or change to code-complying work.
4. Failing to condemn defective work.
5. Incompetent scheduling and coordination.
6. Failing to properly supervise and exercise that authority.
7. Not warning contractors of general precautions not known in the industry.
8. Failing to use or check consultants.
9. Failing to recommend work stoppage where appropriate.
10. Certifying improper pay requests.
11. Failing to detect deviations from approved design when reviewing shop drawings.

FIGURE 6.4 Examples of typical claims relating to the construction phase fall into these categories.

contractors   47% of the time
professionals   36% of the time
owners   15% of the time
building departments 2% of the time

However, in the case of structural failure, the fault lies with:

the designer   58% of the time
the contractor   53% of the time
the owner   8% of the time
(multiblame accounts result in total over 100)

Table 6.1 shows the results of a study of claims against design professionals.

It is easy to understand that the complex nature of today's construction and the presence of normal human imperfections in the related work will lead to disputes and claims. These are commonplace and should be expected as part of the process. The best effort should be directed toward minimizing such discord and preparing effective responses when they do arise. Every project participant needs to take precautions in every stage of the project. This can be accomplished by paying attention to a few crucial items:

1. **Competency:** Deal with reputable, experienced, technically expert, financially sound persons, who have a reputation for fair dealings.
2. **Document review:** Eliminate exploitation of any party and ensure a fair contractual arrangement. Define, clarify, modify (with extreme care), and make

TABLE 6.1 Study of 8,600 claims against design professionals, 1989–1995; $269 million was paid on claims.

Design Professional	% of Claims	% of Claim $	% of Fees	Claims as % of Fees
Structural Engineers	14	20%		7
condos	12	19	1	
commercial	15	17	25	
Mechanical Engineers				
HVAC	50 (61% of $ paid)			
design	9			
equipment	6			
Plumbing	5			
Electrical Engineers		(21% paid for bodily injury)		
schools	14		5	
Architects				
condos	7	14	1	
walls	9			
roofs	10			
floors	6			
code compliance	5			

Of the claims paid,
  50% brought by owner/developer
  30% brought by third parties
  14% brought by contractors

DPIC Companies.

appropriate all rights, obligations, requirements, procedures, and penalties through an in-depth, objective, and complete document review.

3. **Documentation:** Prepare complete, timely, coordinated, and accurate documentation of project events to dispose of claims and to aid communication, job monitoring, and informed decision-making.
4. **Communications:** The key to teamwork and, in turn, to cooperative working relationships, which are the key to successful projects. Establish an internal and external network of communications, with ease of transmission and consistent and complete filing systems.
5. **Accounting:** "Money talks" not only to the project as a whole, but in each dispute and claim. Establish good accounting measures to monitor progress, deal with claims properly, and give early indication of schedule problems or other job troubles.
6. **Scheduling:** Create a timely, reliable, well-conceived schedule for every job. It reveals the impact of contractor work segments and aids in the cooperative spirit of a successful project.

To anticipate disputes and claims is to moderate their impact and to resolve or dismiss them promptly.

The General Conditions of the contract (AIA document A201-1987 or a similar document) become the ground rules. They are operable, imposed, enforced, and argued on the job site. One is well advised to know every detail of these documents and how each party to the contract functions. In reality, the contract becomes self-enforcing to some extent, for if one member or party to the contract is willing to allow noncompliance or chooses to ignore violations of the requirements, the entire process begins to erode. Violations tolerated by one will create an opportunity for violations by others, until the contract becomes meaningless. Everyone must "buy into" the contract; everyone must pay attention to, fully support, and observe all of the requirements set forth.

One can easily understand the fruitless effort of a design professional, engaged in contract administration by proper means, trying to protect the project, when the other parties are not contributing to that effort by acting in a similar manner. If there is continual conflict, vying for position, petty disputes, and other irritations regarding issues covered in the General Conditions, the project will be impaired. Everyone will become immersed in the frivolous, nonproductive, and childish "push and shove" over administrative and nonconstruction items, to the detriment of the project and their reputations. Eventually they fail to please the client by being late with completion or over budget.

As seen in Chapters 2 and 10, the process of definition and explanation is really part of the necessary educational process required on each project. Even though the same client may be involved in several projects, the process is still required, perhaps in various forms. It is critical that the owner's demeanor be assessed and addressed (see Chapter 10 for directions and misconceptions).

Since about 60% of all claims against professionals are filed by owners, the risk management program must start early. Professionals have recognized this. Claims are down from 42 per 100 insured firms in the early 1980s to about 22 per 100 in 1991. Recognition of the need to manage their risks is paramount to the professionals' profitability.

Recent surveys give professionals some added insight into the source of some of their claims. Of contractors surveyed, some 84 percent found that specifications contain "major omissions" at least some of the time (the survey did not even speak to the continual and profuse nuisance of minor errors, gaps, redundancies, and other glitches). Complexity of construction, projects, and documentation, coupled with the need for speedy production of documentation (with inadequate time for objective checking), produce conditions that foster these problems. Some 50 percent of these contractors also found that specifications often require numerous other changes and modifications, which disrupt the smooth progress of construction. Their suggestion is that the "red tape/boilerplate" language be removed from the front of the specifications and that specific and appropriate specifications be carefully written for each project.

Another survey focused on the direct causes of professional liability. The leaders in this dubious category are (1) quality control problems, (2) failure to recognize risks, and (3) inadequate response to problems when they arise. This is puzzling, since these causes strike at the core of proper management. Surely, even the busiest professional knows of these threats. At the current premium rates, insurance simply cannot and should not be relied upon as the sole remedy. Active risk control, if not prevention, is mandatory.

Still, from time to time, successful claims for large amounts are made against design professionals. The average claim settlement paid by architects' and consulting engineers' errors and omissions insurance carriers was about $261,000 in 1993, compared with about $155,000 in 1985. Although the figures rise, this is not far outside mere inflationary factors; this shows a fairly stable claims atmosphere. However, these figures represent just the money paid by insurers. They do not reflect the added costs to the professional firms, i.e., payments made up to the deductible limit, nor the money lost to the offices for time spent in archived case research, deliberations with attorneys, case preparation, and other aspects of claims satisfaction. The more stable situation also indicates that design professionals seem to have become more sensitive to legal pitfalls. They have learned well from the crisis of the early 1980s. This, combined with greater stability in the insurance market, has resulted in decreasing premium costs, with no dramatic increase in the number of claims filed nationally against architects and engineers in the last few years.

Some of the reasons for the less than dramatic (and predicted) increase in claims and liability insurance costs is due to greater attention by the professionals to common acts, errors, and omissions that lead easily to claims; better contract language; avoidance of "bad" contracts; better client selection; and loss prevention programs in all of the firm's activities. In addition, it appears that design professionals may be making stronger early efforts to settle claims even though they may be paying a significant aggregate amount to settle. This seems to occur more frequently when the settlement cost is less than the deductible amount under their errors and omissions coverage, or when settlements cost less than the anticipated legal defense costs, if litigation results. As a result, the premiums paid to insurance carriers have declined, but the overall "payout" by some insureds to extinguish claims may have increased.

No doubt, too, the more active use of alternative resolution programs has greatly reduced the number of claims that are litigated. This is most important since, in almost all cases, it is only the attorneys who truly profit from litigation.

ALTERNATIVE DISPUTE RESOLUTION

Basically, all parties should seek the fastest and most positive manner to solve their problems. With proper groundwork laid with the owner early in the contract and education cycle, the professional can establish this philosophy, which is really to the owner's benefit. Several different programs of alternative dispute resolution (ADR) are used. These are negotiation, mediation, mediation-arbitration, arbitration, and in some states mandated pretrial panel requirements (reviews of claims prior to court action). All are aimed at defusing a situation prior to litigation. This list, in order of use (from the least formal, negotiation, which is widely and commonly used), should be addressed in the contract provisions. The primary goal in all of these programs is to resolve disagreements, claims, and disputes in a less than excessively costly process, promptly, and with some mutually suitable finality.

Obviously, negotiations go on almost daily in the common "give and take" on the construction site. This is easily done through plain old common sense. It is best resolved in writing, so there is full closure, understanding, and a track of the final agreement. All parties should sign the document of acceptance/approval. Of course, as other business demands may appear the process of negotiation can become more formal.

In the past, once a dispute became more involved than negotiations could contend with, the disputants would go to litigation. The process of suing, however, came to be one that served no particularly good purpose, was excessively expensive, and in many cases was massive overkill for the problem involved. Cases jammed court dockets to the point that courts came to require pretrial efforts at settlement. Often this was successful. This scenario led to establishing alternative methods. Primary among these methods was arbitration.

In 1993, the American Arbitration Association reported, some two-thirds of the claims filed were for amounts under $50,000. Almost 200 cases, though, were in amounts over a half million dollars. Arbitration has successfully resolved over $10 billion worth of claims in the last decade alone. It is realistic to think that this total is a fraction of what may have resulted from litigation.

But things are rapidly changing now. Arbitration, once seen as the most rigorous and definitive alternative to litigation, is waning dramatically in popularity. In both 1994 and 1995, less than 250 cases between design professionals and owners were heard each year, and 1,500 cases involving contractors (statistics of the American Arbitration Association). Most of the initial impetus behind its use was due to lower cost (relative to litigation), binding results, expedited time-frames for settlement, and the ability to use a panel of arbitrators with some construction experience. Time and costs, while significantly less than for litigation, were still formidable. Arbitration came to be seen as a giant step short of litigation, and a tremendous relief. Why there was no major effort toward elevating the level of negotiation or wider use of mediation is not known. The move was simply to the next less imposing system, arbitration.

Often arbitration results were distasteful to at least one party and even errant. The courts, however, encouraged and often compelled this method of claims resolution, since in general the good features far outweigh the bad. It is generally agreed that there is no perfect dispute resolution program, but then litigation is not always a perfect solution, either.

To alleviate some of the more prevalent problems with arbitration, the American Arbitration Association instituted several new procedures:

- A three-track system: "fast track" for standard 60-day resolution for cases involving less than $50,000 and heard by a three-arbitrator panel; "standard track" for claims from $50,000 to $1 million with added discovery roles for evidence gathering; and "large-case track" for complex cases worth over $1 million.
- Requiring arbitrators to have 10 years of experience in the construction industry. Attorney/arbitrators must have practices that service a good number of cases involving construction matters.
- Written opinions and explanation of awards to inform all parties of the basis of the decision.
- Defusing the common notion that arbitrators all too often "split the difference" in the cases (new studies show this to be false).

There was a need to address the downside to arbitration, mainly unjust and unexpected results, even though arbitration panels often include persons having expertise in the construction industry. It seems reasonable to assume that arbitration panels will produce better decisions overall. This is not always the case, however. Patently wrong, and irrational decisions occurred all too frequently. Unfortunately courts, almost without exception, tend to support even the most faulty arbitration decisions. Usually the only route to overturning a poor decision is to show "gross error," but many times even this is not viewed as sufficient grounds for reversal. Here is the main flaw in the arbitration process.

An arbitration award will not be overturned even in the case of mistake, improper interpretation, misapplication of law, or error in logic (see *Perini Corporation v. Greater Bay Hotel & Casino, Inc.*, 610 A.2d 364 [1994]). Hence, the best advice is to ensure that the arbitration agreement also has a clause stating that the rules of evidence of the particular state shall govern, and that either party to the arbitration decision may appeal the findings to the proper court.

Some other flaws exist, but can also be properly dealt with via good contract provisions. First, the award of punitive damages (penalties for wrongdoing) in addition to compensatory damages (payment to right the wrong and "heal" the plaintiff) is uneven. Many states prohibit the award of punitive damages by arbitration panels. Often such awards are unrelated to the main issue under dispute. Legal advisors note it is best to include a specific provision in the arbitration agreement to flatly exclude punitive damages.

A construction project is best governed by the laws of the state in which it is located. With contractors from various states and even extraneous contract language that survives from previous projects and yet other states, it is advisable to ensure that the contract documents and the arbitration agreement (where necessary) explicitly note which state's laws govern. Usually everything to do with a project should be controlled by the same state's laws (see *Mastrobuono v. Shearson Lehman Hutton, Inc.* 20F2d. 713 [1994]).

Some have suggested that there is a need to make some provisions for meritless arbitration cases (as done by Federal Rule 11 for court cases). In reality two fundamental changes would greatly aid the arbitration process. First, parties should be allowed to present evidence about the costs of the process and to argue that costs should be shifted where spurious actions were taken. Second, since arbitration awards are not automatically imposed or self-executing, the final "payment" can be forestalled. This can become so convoluted that additional court action is required to recover the award. Obviously this calls for imposition of the costs of collecting and enforcing the award as well as possible penalties for delaying tactics. Both of these changes are consistent with the intent of arbitration—a final and binding resolution without the need for further court action.

Because of the inherent problems with arbitration over the years, mediation has been brought to the forefront as the ADR program of choice. More than likely, since no formal statistics or organizations are involved, the use of mediation can be traced to its lack of structured formality, quicker turnaround time, and a more conciliatory approach to resolution. While the format can take many forms, the fundamental concept is the introduction of an unbiased third party whose primary goal is to induce the disputing parties to find their own settlement of the problem on a voluntary basis. It would appear to proceed very much on the "win-win" principle of resolution.

The crux of mediation lies in the attitude of the parties. If they approach resolution with a readiness to compromise, mediation will succeed. If this system fails, the parties are still free to pursue other ADR methods, or litigation if things really get out of hand. Still, in the courts the parties may find a judicial indication that they remediate or otherwise reach agreement.

Other efforts also lean toward a mediated solution. The new philosophical program of "partnering" brings parties into closer work relationships from the start. This obviously positions them to settle disputes in a much more cooperative manner, since they already agreed to work more closely, to maintain continuous communications, and to solve problems as they arise. This basic concept has also been put forth by the Construction Industry Dispute Avoidance and Resolution Task Force (DART). This organization has issued a "Declaration of Principles for the Prevention and Resolution of Disputes." Both the concept and this document has the endorsement of most of the professional organizations and many federal agencies. Interest is high, no doubt due to lower costs and better overall results.

A mutually agreed-to party, outside the contract, should be called in to mediate the situation. The contract administrator could be involved in these programs, where appropriate.

Regardless of brighter statistics, most professionals would agree that even the mere possibility of a claim is reason for concern and preventive measures, where possible. The loss of time and productivity along with the stress and anguish associated with mounting a defense in expectation of litigation, not to mention the results of an adverse judgment, take a toll in every case. For these reasons alone the proactive steps and preventive, if not defensive, practices that can result from a basic familiarity with procedures for avoiding errors, omissions, and other gaffes in practice are well worth utilizing.

Professionals need to continue and refine their risk management endeavors since risk is inherent and global, occurring throughout the process of each project and over a broad spectrum of activities, including the construction phase. Possibly the first step for an architect in developing a claims avoidance strategy is to examine the latest edition of AIA document B-141, Standard Form of Agreement between Owner and Architect.

Case law since 1986 seems to indicate that B-141 provides adequate protection for architects who provide contract administration. Where professionals carefully apply the basic definition of *contract administration* (the term used in B-141 to describe architects' services during construction), allegations of negligence for failure to exercise broad control over the construction process (as is implied with the term *supervision*) can often be refuted. This presumes that the architect did not supervise construction. If proper contract administration did, in fact, take place, not supervision, the language of the contract, together with the actions of the architect, will offer a strong legal defense.

All of this may seem foreign to the contract administrator. However, it is most important that administrators understand and respect these systems and their consequences. The on-site administrator will often have to deal with the results of some of these actions when they are taken during the conduct of the project. Where continuing problems and perhaps animosity sets in, the administrator will see an increase in the need for his/her most diplomatic, but pertinent, services.

By way of illustration, see the following discussion of a legal paper written about the overall liability of design professionals and excerpts from the paper itself. Particular emphasis is placed on liability and the nature of the services offered during construction.

Although the full text of this paper should be read by every architect, engineer, and design professional student, the overall situation cited in the paper has changed since its publication (1984). At the time of publication, many practicing professionals felt extremely threatened by the "eating away" of fundamental legal premises and protection. New styles, materials, and methods of construction, and larger amounts of money "demanded" unachievable "perfection" in the projects. What had been fairly minor instances became voluminous lawsuits. Professionals tried to run for cover, not to hide from their obligations, but to find some method to cope.

Portions of this paper are more apropos to contract administration than others. We have excerpt those that address the primary issue of this book. But the entire paper should be sought out and read. Note the dates of many of the court decisions (some pre-1900 and some in the early 1900s) and the elapsed time since publication. These provisions need to be compared with post-1984 court decisions and other changes to note how circumstances and precedents have changed, along with philosophical and physical changes in concepts, contracts, and professional methods used in the mid-1990s. Of course, we stress the need to seek legal advice in this venture.

ABSTRACT OF LEGAL LIABILITY PAPER

"The Crumbling Tower of Architectural Immunity: Evolution and Expansion of the Liability to Third Parties," by Jeffrey L. Nischwitz, provides an excellent orientation for design professionals regarding their professional liability.

It carries an ominous message in its title. At the time of publication (1984) a variety of circumstances in the insurance industry, along with an increase in claims, portended trouble for design professionals. After rising in the mid- to late 1980s, premium costs for professional liability insurance coverage, as a percentage of gross revenue in 1993, stood at one-half the rate encountered in the latter years of the 1980s.

For some thirty years, beginning in the mid-1950s, architects (and other design professionals) had been slowly exposed to an expansion of their liability from a variety of sources. Key issues were (1) elimination of the privity (contract) requirement, (2) abolition of the owner-acceptance rule, (3) application of the time of discovery rule to tort statutes of limitation, and (4) inadequacy in the worker's compensation system.

These issues acted upon by third parties affected two primary areas of professional practice and resulted in the expansion of liability. The two areas of expansion were (1) the preparation of contract documents (drawings and specifications), and (2) construction "supervision," and the role of the architect in each area.

Also, during this period the American people became a very litigious society. This concept is seated in the consumer movement of the time, the favoring of strict liability, liberal courts, more zealous plaintiffs' attorneys, and the search for a "deep pocket" (a substantial source of money), which found some satisfaction in the availability of professional liability insurance. The prognosis, as noted in the paper, was intimidating, if not outright threatening to the professional.

In this movement, the people were trying to turn the liability of professionals full circle. It was an effort to restore the "strict liability" theory prevalent in the ancient days of Babylon and epitomized by the "eye for an eye, tooth for a tooth" penalties of the code of King Hammurabi.

In general, even in 1984, strict liability was not applied to professional services. In fact, until the 1950s the professional was immune from suits filed by third parties and saw no threat of imposition of strict liability. Usually, strict liability was held for the mass production of faulty products. Most courts felt that professional responsibility was much more evident and that negligence could be more easily exposed, where present. In 1965, however, a court found that strict liability was appropriate in a case dealing with the design of mass-produced housing. While still an aberration (as far as professional services is concerned), this case did open the issue for further expansion. Still, three aspects must be satisfied for strict liability to be imposed: the difficulty in proving negligence, the economic burden, and deterrence.

The first, difficulty in proving negligence, seems to disappear when one considers that a project which has resulted in the litigation is a one-of-a-kind endeavor. Obviously the party injured knows exactly where to find any defect and where to find the party responsible. Thus, this aspect of strict liability is inapplicable.

The second aspect, economic burden, is the one open to most argument. In most cases, the client is owner of a business, as is the architect. Who can best stand the recovery of economic loss, since both have the ability to pass along any costs to their "customers"? The owner has the broader base of clientele, and therefore is able to spread any loss over a larger class and at a more moderate rate. The architect can only increase fees, even if insurance coverage pays the award (and increases the premium). In so doing, there may be a loss of clientele, and the loss of even one client is a much larger percentage of the architect's business. Obviously this is an argument open to review on a case-by-case basis.

The third issue of strict liability, deterrence, means simply that precautions need to be taken to prevent or minimize injuries. The manufacturer, who controls both a repetitive production process and the human element associated with it, has closer control over the process and, hence, better methods of minimizing injuries. The architect, on the other hand, faces a much more difficult quality control problem on the construction site. First, each project is a unique situation where the process cannot be tested or tightly controlled, adjusted as needed, and so forth. In addition, the architect has no control over the human element involved with the construction. Simply, the architect's control during the construction phase is restricted by contract, and certainly is not sufficient for deterring construction defects. Hence, it should not be the basis for targeting the professional for strict liability.

But strict liability was but one of the "new" threats to the professional. As noted, one of the first drastic changes came when privity of contract was discarded. Under this premise, only parties in a contract had standing to sue for breach of contract; third parties, with no such standing, could not sue. This restricted liability for negligence, to some degree, to contractual parties. Of course, the number of such parties for an architect is limited, since only two contracts (owner-architect and owner-contractor) usually are involved on a project. Obviously only the owner is a contractual party with the architect. However, two courts, some 40 years apart, decided that "foreseeability" was a better test than privity of contract.

Basically, these courts substituted the premise that anyone could foreseeably use the "product." This was extended to architects and their projects. Most courts now hold that even where the danger is patent (meaning evident, obvious, open, manifest) liability can be assessed. Later court decisions have avoided the "architect's liability to the entire world" by installing a flexible balancing approach, taking other considerations into account. This has provided a more realistic view of the extent of the architect's liability. This approach is superior to the inflexibility of the privity issue, but no doubt has significantly increased professional liability and the "pool" of potential plaintiffs. Subsequent cases have allowed other considerations and have taken the matter to a case-by-case basis. This has proved to be more acceptable because it recognizes (properly so) that architects do benefit society and thereby should have their liability and scope of duty limited.

The matter of the owner acceptance rule (OAR) has been overshadowed by the foreseeability test. In essence, the OAR held that the architect was not liable to third parties once the structure was completed and accepted by the owner; therefore, negligence was not an issue. This followed the general premise that the architect was in no position to control the owner's use and/or abuse of the structure, the lack of maintenance, and so on, and hence owed no duty to third parties injured after acceptance. The rule, though, often was tied to the matter of duty owed, which in turn was tied to privity of contract. Since privity was abolished, the OAR was undercut in that duty is not now a contractual issue, but is a factor established by the "foreseeability" test.

Time of discovery rule issues centered around the point or time at which the prevailing statute of limitations would start running. In actuality, the entire concept and its revision were aimed at fairness—a "fair" time frame to allow for action by injured parties, as well as "fairness" to professionals so they were not exposed to liability "throughout [their entire] professional life and into retirement." Courts recognized that endless expansion of liability is inconsistent with the principle that there should be an end to liability at some point. The initial enactments of statutes of limitation generally called for their initiation when the negligent act was committed; later this was changed to when some damage or injury occurred. The former premise worked in the professional's favor in that liability was held to a fixed period of time. This led to the discovery rule. Here, the statute time starts running only when the alleged wrong is discovered (or should have been discovered). Obviously this is favorable to situations that occur long after the negligence is committed.

There is obvious extension ("into perpetuity") of the professional liability. Courts have now, in some cases, limited the discovery rule to cases where time has not compounded the problem of proof or increased the danger of false/fraudulent claims, on a case-by-case determination. Still, construction cases appear fitted for discovery rule claims, despite the fact that the defense often is required to rely on inadequate resources. To counter this adversity (to the professions) most states or jurisdictions have enacted "special statutes (of limitations)." Here, the time commences running at the time the professional services are performed, or the time when the project reaches "substantial completion." They limit the term of the architect's liability, but do not affect the duty to third parties. Deemed "realistic and necessary" due to the longevity of an architect's work, and a good balance of issues, these statutes should not be the professionals' only defense/protection; they are, however, most effective in limiting periods of professional liability.

Many suits against professionals involve injury or death of construction workers. The prime impetus behind these actions is the very restricted coverage of typical workers' compensation programs. Normally, two factors are at issue: the programs are the sole recourse for the worker against the employer, and coverage is limited to medical expenses only (some provide partial compensation for lack of income production). These inadequacies almost force the worker to seek another "deep pocket" to pay for damage, pain, and suffering. One of the "deep pockets" has been the professionals—third parties who have at least some resources (like liability insurance) for added monetary recovery. The position of the professionals has been weakened by the courts overturning the privity requirement, and the continuing diversity of opinions over the professionals' "supervisory" duty.

Typically, negligence is, at least, part of the basis of any suit against professionals. With privity of contract discarded, negligence stands are the ones most likely to succeed. However, four standard elements are involved, and their proof lies with the plaintiff:

1. **duty** (plaintiff is protected by law from architect's action)
2. **breach** (architect's conduct violated the duty)
3. **proximate cause** (injury resulted from architect's conduct)
4. **damage** (plaintiff suffered a loss)

The prime content of the duty issue is the standard of care that applies to the professional. This entails expression of what duty is owed, to whom (or how many), and for how long. Courts are nearly unanimous in using the foreseeability test for establishing to whom duty is owed. "How long" is contained in the special statutes discussion and law.

Explanation of "duty" itself is a more involved process. One case, decided in 1917, holds that to fulfill duty the architect "must possess and exercise the care and skill of those ordinarily skilled in the business," and "has done all the law requires ... [w]hen he possesses the requisite skill and knowledge, and in the exercise thereof has used his best judgement." In 1953, another court provided further clarification and held that if the architect worked within the standard, no negligence finding could result, even if mistakes were made. Basically, this says that the architect is *not* the warrantor of his work and need only meet a reasonable standard of care contained in the law. The court's definition of duty was as follows:

> By undertaking professional service to the client, an architect impliedly represents that he possesses, and it is his duty to possess, that degree of learning and skill ordinarily possessed by architects of good standing, practicing in the same locality. It is his further duty to use the care ordinarily exercised in like cases by reputable members of his profession practicing in the same locality; to use reasonable diligence and his best judgement in the

exercise of his skill and the application of his learning, in an effort to accomplish the purpose for which he is employed.... The standard is that set by the learning, skill and care ordinarily possessed and practiced by others of the same profession in the same locality, at the same time.

The matter of "same locality" has been expanded by other courts, and is not now a limiting factor.

Even in view of the established standards of care, professionals still retain wide discretion in furnishing their skills and in determining what practices, principles, and techniques best suit their work. Most architectural questions are not answerable by "right" or "wrong," but rather utilize subjective considerations. This makes establishing an "ordinary and reasonable" standard difficult. Usually both sides in a suit rely on expert witnesses for help in this area.

Further, even though professional standards of care exist, the context in which they are applied must be considered. The duty of the architect must be examined through the professional services and functions; basically, preparation of documents and construction oversight.

To briefly speak to the issue of document preparation, one must understand that the design sequence is not a simple process. One court described the architect's duties in design as follows:

> Architects must have as a part of their competency a keen aesthetic sense to enable them to design structures of beauty and dignity; they must have a technical knowledge of many structural factors which lend strength and stability to their designs. The materials they recommend for use are produced by agencies beyond the control and influence of the architect. His work is to a certain degree experimental or depends on the experiments and on production of materials by others. Then, too, the law of physics, gravity and the rotation of the earth, must, in many projects, be taken into account.

Very insightful!

The foregoing coupled with the discussion regarding the standard of care form a pervasive discussion about the practice of architecture and the actions of the professional. This abstract has related, in summary to this point, the accompanying legal paper by Nischwitz. Let the paper speak for itself regarding construction contract administration.

The Crumbling Tower of Architectural Immunity: Evolution and Expansion of the Liability to Third Parties

Jeffrey L. Nischwitz

... The duty of an architect to exercise ordinary and reasonable care in the preparation of plans and specifications is strict and is not easily dismissed. Although the architect's liability is tempered by a heavy burden of proof on the plaintiff, the architect is still subject to extensive liability relating to the preparation of plans and specifications. In most cases in which a third party is injured because of a defect in plans or specifications, "no haven where an architect ... may safely take cover" exists.[188]

Copyright © 1984 The Ohio State University. 45 *Ohio State Law Journal* 217 (1984), 234–246. Used with permission of Ohio State Law Journal.

188. *Id.* at General Information—9(1).

C. Supervision of Construction

The architect's duty in preparing plans and specifications is fairly well established, but the architect's duty of supervision is the subject of a great deal of controversy. Courts differ about what is or should be included in the architect's duty to supervise.[189] The architect's duty to supervise encompasses two distinct areas: (1) supervision to prevent deviations from the plans and specifications,[190] and (2) supervision of construction methods and techniques.[191] The duty of the architect in each of these areas will be discussed to determine when the duty may arise.

Determining an architect's duty to supervise is significantly more difficult than determining the duty to use reasonable care in preparing plans and specifications. The courts must closely examine the relationships and transactions between the architect, owner, and contractor to determine the allocation of the supervisory responsibilities. Absent a contractual duty to supervise, the architect is generally not required to be present during the construction phase.[192] Occasionally an architect will desire limited supervisory duties, but, as noted by one commentator, "an owner generally wants to be sure that his building is put up as designed and who [is] better to see that this is done than the designer, to wit, the architect."[193]

Architects' views of their role in the construction process vary. The architects who favor an active role contend that the most complete set of construction drawings "can never express the entire design concept."[194] These architects believe that without an active role in the construction process, "the design concepts will not be executed."[195] This activist school of thought is based on the traditional role of the architect as the provider of complete services to the client, including complete supervision during the construction phase.[196] The traditional role came into existence at a time when buildings were relatively uncomplicated and the architect was in effect both the designer and the superintendent of construction.[197] In some cases the architect was also the contractor.[198] As a result of this tradition, many owners assume that the architect will provide complete supervision.[199]

At the other extreme are architects who favor a passive role. As structures have become more complicated and designs more technically demanding, the design professional has become less concerned with the day-to-day operations of construction.[200] Therefore, the proponents of a passive role for architects claim that they "are not skilled at construction administration and supervision,"[201] and further, that if they become too heavily involved in the construction process they may be held "responsible for everything that goes wrong."[202] From the discussion to follow it appears that these fears are well founded.

The standard of care applicable to an architect's supervision is generally that care ordinarily required of "a professional skilled architect under the same or

---

189. *See infra* text accompanying notes 194–202, 263–89.
190. *See infra* subpart III(C)(1).
191. *See infra* subpart III(C)(2).
192. *See* Duggan v. Arnold N. May Builders, Inc., 33 Wis. 2d 49, 53, 146 N.W.2d 410, 412 (1966).
193. Goodin, *Architects and Malpractice,* 34 INS. COUNS. J. 290, 292 (1967).
194. J. SWEET, LEGAL ASPECTS OF ARCHITECTURE, ENGINEERING AND THE CONSTRUCTION PROCESS 122 (1970).
195. *Id.*
196. *See. e.g.,* Schreiner v. Miller, 67 Iowa 91, 24 N.W. 738 (1885); Louisiana Molasses Co. v. Le Sassier, 52 La. Ann. 2070, 28 So. 217 (1900). These cases indicate the extensive role that the architect historically played in the construction process.
197. *See* Schreiner v. Miller, 67 Iowa 91, 24 N.W. 738 (1885).
198. *See* Louisiana Molasses Co. v. Le Sassier, 52 La. Ann. 2070, 28 So. 217 (1900).
199. *See* J. SWEET, *supra* note 194, at 122.
200. Note, *supra* note 16, at 1237.
201. J. SWEET, *supra* note 194, at 122.
202. Note, *Supervisory Duties of an Architect.* 3 MEM. ST. U.L. REV. 139, 139 (1972); *see* J. SWEET, *supra* note 194, at 122.

similar circumstances in carrying out his technical duties in relation to the services undertaken by his agreement."²⁰³ This standard is accepted by most jurisdictions²⁰⁴ and applies whenever a duty of supervision exists.²⁰⁵ Establishing the existence of that duty is the vital inquiry.

The employment of an architect is generally a matter of contract and the terms of employment are governed by the terms of the contract. Therefore, the courts must closely examine the contractual relationship between the architect and the owner to determine what supervisory duties exist before they can determine if the architect has been negligent in discharging those duties.

1. Supervision to Assure Substantial Conformity with the Plans and Specifications

The least controversial supervisory duty is the duty to assure that the building or structure is constructed substantially according to the architectural plans. Whether an architect supports a passive or active role in supervision of the construction process, it is clear that in most jurisdictions "the architect does owe some duty . . . to see that the building ends up built substantially according to the plans and specifications."²⁰⁶ Any deviation from the plans by the contractor may be evidence that the architect was negligent in his duty to supervise for substantial conformity.²⁰⁷

The importance of this supervisory duty is clear from the cases that set forth this duty. One of the earliest cases dealing with the architect's duty to see that a building is constructed according to the plans was *Schreiner v. Miller*.²⁰⁸ In that case, the Supreme Court of Iowa found that an architect had a duty to assure substantial conformity.²⁰⁹ The court also held the architect to a duty to use reasonable care in construction methods, but this was at a time when the architect was regarded as the sole superintendent of the project.²¹⁰

The leading case on the nature of the architect's responsibility to supervise construction is *Clinton v. Boehm*.²¹¹ In *Clinton*, a New York Appellate Division Court held that the "very utmost obligation" assumed by the architect was "to see that the building was properly constructed" and, generally, to see that the owner received the building for which he contracted.²¹²

One unsettled issue is whether the architect must have notice of the contractor's deviation to be held liable for negligent supervision. In *Paxton v. Alameda County*²¹³ a California appellate court held that an architect was under a duty to supervise with reasonable care when he was put on notice that the contractor had deviated or was about to deviate from the architect's plans.²¹⁴ The court

---

203. Aetna Ins. Co. v. Hellmuth, Obata & Kassabaum, Inc., 392 F.2d 472, 477 (8th Cir. 1968).
204. *See, e.g.,* Aetna Ins. Co. v. Hellmuth, Obata & Kassabaum, Inc., 392 F.2d 472 (8th Cir. 1968); Peerless Ins. Co. v. Cerny & Assocs., 199 F. Supp. 951 (D. Minn. 1961); Covil v. Robert & Co. Assocs., 112 Ga. App. 163, 144 S.E.2d 450 (1965); Coombs v. Beede, 89 Me. 187, 36 A. 104 (1896); Chapel v. Clark, 117 Mich. 638, 76 N.W. 62 (1898); Cowles v. City of Minneapolis, 128 Minn. 452, 151 N.W. 184 (1915); Scott v. Potomac Ins. Co., 217 Or. 323, 341 P.2d 1083 (1959); Willner v. Woodward, 201 Va. 104, 109 S.E.2d 132 (1959).
205. *See. e.g.,* Aetna Ins. Co. v. Hellmuth, Obata & Kassabaum, Inc., 392 F.2d 472, 477 (8th Cir. 1968).
206. Goodin, *supra* note 193, at 292; *see* Aetna Ins. Co. v. Hellmuth, Obata & Kassabaum, Inc., 392 F.2d 472, 474 (8th Cir. 1968); United States v. Rogers & Rogers, 161 F. Supp. 132, 136 (S.D. Cal. 1958); Dysart-Cook Mule Co. v. Reed & Heckenlively, 114 Mo. App. 296, 299, 89 S.W. 591, 592–93 (1905); Clemens v. Benzinger, 211 A.D. 586, 590, 207 N.Y.S. 539, 543 (1925).
207. *See* Johnson v. O'Neill, 172 Mich. 334, 339, 137 N.W. 713, 715–16 (1912).
208. 67 Iowa 91, 24 N.W. 738 (1885).
209. *Id.*
210. *Id.* at 92–93, 24 N.W. at 738; *see supra* text accompanying notes 196–98.
211. 139 A.D. 73, 124 N.Y.S. 789 (1910).
212. *Id.* at 75, 124 N.Y.S. at 792; *see also* Olsen v. Chase Manhattan Bank. 10 A.D.2d 539, 205 N.Y.S.2d 60 (1960), *aff'd mem.,* 9 N.Y.2d 829, 175 N.E.2d 350, 215 N.Y.S.2d 773 (1961).
213. 119 Cal. App. 2d 393, 259 P.2d 934 (1953).
214. *Id.* at 410, 259 P.2d at 944–45.

found that the architect was under a duty in these circumstances to make certain that the situation was corrected.[215] In *Lotholz v. Fiedler*,[216] however, an Illinois appellate court held that an architect should have prevented a variance from the plans, even though he had no knowledge of the deviation, which was impossible for him to see.[217] In this case the lack of a notice requirement made the duty to supervise for conformity an extremely difficult standard to satisfy. Nevertheless, the tendency is apparently to find liability despite a lack of notice.[218]

Many cases have established that an architect has a duty to ensure a building's conformity with its design. Some jurisdictions have held that architects have too much authority over contractors and job progress to be immune from liability.[219] Others have based the duty on the novel features involved in a design.[220] In *Bayuk v. Edson*[221] a California appellate court found that the contract between the architect and owner did not include supervision.[222] The court held, however, that the "novel and untried features ... required close supervision."[223] The court apparently implied a duty based on the importance of supervision in assuring that an owner receives the building for which he contracts.

The recognition by the courts of a supervisory duty to prevent deviations from the plans and specifications is supported by the architectural profession itself. The terms in form contracts frequently used by architects, such as the American Institute of Architects' *Standard Form of Agreement Between Owner and Architect*,[224] show that architects recognize the duty to supervise for building conformity. According to these forms, "The Architect shall visit the site at intervals appropriate to the stage of construction ... to become generally familiar with the progress and quality of Work and to determine in general if Work is proceeding in accordance with the Contract Documents."[225] Clearly, then, the architectural profession recognizes a duty to supervise to prevent deviations from designs, but even this has not settled the issue.

Architects would like to limit liability for failure to conform to this supervisory duty to cases in which they are at least aware or on notice of the deviation and the dangers it presents.[226] As noted earlier, however, *Lotholz v. Fiedler*[227] held an architect liable for a variance from the plans even though it was impossible for the architect to see it.[228] Subsequently, another court held that an engineer should have been aware of a deviation from his design, even though he had no notice of the variance.[229] Architects have attempted through disclaimers[230] to limit liability for these unknown deviations, but the effectiveness of the disclaimers is left to the ultimate determination of the courts. Given the trend toward findings of negligence, the disclaimers may not be very effective.[231]

---

215. *Id.*
216. 59 Ill. App. 379 (1895).
217. *Id.* at 380–81.
218. *See infra* notes 226–29 and accompanying text.
219. Allen, *supra* note 1, at 460.
220. *See, e.g.,* Bayuk v. Edson, 236 Cal. App. 2d 309, 316, 46 Cal. Rptr. 49, 54 (1965).
221. 236 Cal. App. 2d 309, 46 Cal. Rptr. 49 (1965).
222. *Id.* at 312, 46 Cal. Rptr. at 51.
223. *Id.* at 316, 46 Cal. Rptr. at 54; *see also* Note, *supra* note 202, at 141.
224. AIA Document B141/CM, *supra* note 26.
225. *Id.* § 1.5.4.
226. Note, *supra* note 7, at 250.
227. 59 Ill. App. 379 (1895).
228. *Id.* at 380–81: *see supra* text accompanying notes 216–17.
229. Pastorelli v. Associated Eng'rs, Inc., 176 F. Supp. 159, 167 (D.R.I. 1959).
230. AIA Document B141/CM, *supra* note 26, § 1.5.5.
231. *See* Miller v. DeWitt, 37 Ill. 2d 273, 293, 226 N.E.2d 630, 642–43 (1967) (House, J., dissenting). The architect disclaimed any guarantees of performance by the contractor, and the contract explicitly provided that the contractor was responsible for the safety of employees. *Id.* at 280–81, 226 N.E.2d at 635–36. The court found a duty of supervision despite the disclaimers.

Although the details of the duty to supervise to assure substantial conformity with the plans are the subject of litigation, the existence of the duty is generally recognized,[232] and the negligence standard of care is applied to determine whether the architect was negligent in carrying out his supervisory duty.[233] The duty to supervise construction methods and techniques, however, presents a more difficult issue.

2. Supervision of Construction Methods and Techniques

The greatest growth in claims by third parties against architects has been based on contracts between the architect and the owner that require the architect to supervise construction.[234] This area is also subject to the most controversy. Controversial issues are whether an architect should have a duty to supervise construction methods and techniques and when the duty should apply.

The aspect of this duty applicable to third parties arises primarily from physical injury caused by faulty construction or improper construction methods. Construction workers comprise a large part of the claiming third parties, and the predominant issue is the duty to supervise for site safety. The two relevant inquiries are, first, whether the architect has undertaken to supervise the construction and, second, what the promise to supervise entails.

a. Early Cases

Lottman v. Barnett[235] is an early case that held an architect liable for improper supervision of construction methods. In that case, the architect approved the use of a jackscrew in an unsafe manner and was held liable for advising the use of an improper construction technique.[236] Liability was based on a theory of misfeasance[237] rather than breach of a duty to supervise construction methods. Thus, *Lottman* established the proposition that when an architect engages in positive acts of misfeasance that endanger a third party, a duty of care arises to the injured party.[238]

Another area in which a duty of care relating to construction methods arises to third parties is when an architect has prior knowledge of a hazardous condition that eventually causes personal injury to a third party.[239] In *Swarthout v. Beard*[240] an action was brought against an architect for the wrongful death of a contractor's employee resulting from a cave-in of an excavation.[241] The court held that since the architect had knowledge of the dangerous condition, his failure to act could constitute negligence.[242]

---

232. See *supra* notes 206–07 and accompanying text.
233. See *supra* text accompanying notes 128–37, 20–05 (discussion of the negligence standard of care).
234. Crisham, *supra* note 166, at 184.
235. 62 Mo. 159 (1876).
236. *Id.*
237. Misfeasance is defined as "[t]he improper performance of some act which a man may lawfully do." BLACK'S LAW DICTIONARY 902 (5th ed. 1979). In the architectural context, misfeasance denotes an architect's improper act (*e.g.,* directing the use of an unsafe construction method), rather than a negligent failure to prevent the use of the unsafe methods.
238. See Day v. National U.S. Radiator Corp., 241 La. 288, 128 So. 2d 660 (1961): Olsen v. Chase Manhattan Bank, 10 A.D.2d 539, 205 N.Y.S.2d 60 (1960), *aff'd mem.,* 9 N.Y.2d 829, 175 N.E.2d 350, 215 N.Y.S.2d 773 (1961); Clemens v. Benzinger, 211 A.D. 586, 207 N.Y.S. 539 (1925).
239. See Erhart v. Hummonds, 232 Ark. 133, 334 S.W.2d 869 (1960); Swarthout v. Beard, 33 Mich. App. 395, 190 N.W.2d 373 (1971); *see also* Potter v. Gilbert, 130 A.D. 632, 634, 115 N.Y.S. 425, 427, *aff'd mem.,* 196 N.Y. 576, 90 N.E. 1165 (1909).
240. 33 Mich. App. 395, 190 N.W.2d 373 (1971).
241. *Id.* at 398–99, 190 N.W.2d at 374.
242. *Id.* at 402–03, 190 N.W.2d at 376.

Lottman and *Swarthout* were not typical cases of negligent supervision of construction methods. These cases did not impose a duty to supervise construction methods, but were based on misfeasance (*Lottman*) and on a failure to act despite notice of a dangerous condition (*Swarthout*). Thus, these cases are distinguishable from cases in which the architect is held to have a duty to ensure that no dangerous condition arises.

The contract between architect and owner, which typically sets forth the responsibilities of each of the parties,[243] is vital in determining the architect's duty to supervise construction methods since it is generally recognized that the architect has no duty of supervision unless he assumes it in some manner.[244] Many cases in which third parties (including construction workers) have brought successful suits against architects have based liability on "provisions in the professional service contract with the client or the general conditions in the construction contract (or both)."[245]

The trend for courts to find that an architect has a duty to supervise construction methods began with the view that architects have too much authority to be immune from liability.[246] *Pancoast v. Russell*[247] is an early case that held an architect to a duty to supervise construction methods. The opinion, by a California appellate court, *reflects the activist school of thought.*[248] The court held that "the term 'general supervision,' as used in the instant agreement, must mean something other than mere superficial supervision. Obviously, there can be no real value in supervision unless the same be directed towards securing a workmanlike adherence to specifications and adequate performance on the part of the contractor."[249] In *Pancoast* the action was by a homeowner against the architect for negligent performance of the duty to inspect and approve the contractor's work.[250] The opinion clearly demonstrates the view that the architect's contractual obligation of "general supervision" means significantly more than supervision for conformity with the design. Thus, this case marks the beginning of the liberal expansion of the architect's duty.

Although the duty of supervision generally had been limited to assuring conformity with plans and specifications,[251] many jurisdictions began to adopt expansive views of the architect's duty.[252] In *Pastorelli v. Associated Engineers, Inc.*,[253] which dealt with the duty of supervising engineers, a federal district court held that an architect or engineer who has general supervision and con-

---

243. See generally the American Institute of Architects' Documents reprinted in J. LAMBERT & L. WHITE, HANDBOOK OF MODERN CONSTRUCTION LAW at 231–347 (1982).
244. *See, e.g.,* Duggan v. Arnold N. May Builders, Inc., 33 Wis. 2d 49, 53, 146 N.W.2d 410, 412 (1966).
245. *Types of Professional Liability Claims,* in GUIDELINES FOR IMPROVING PRACTICE at Special Studies—5.—5(1) (Victor O. Schinnerer & Co. ed. 1979); *see* Clinton v. Boehm, 139 A.D. 73, 124 N.Y.S. 789 (1910). *Clinton* was an early case setting forth the architect's duty to supervise construction methods and techniques. The court examined the relationship between the architect, owner, and contractor to determine whether or not the architect had assumed the duty to supervise construction methods. *Id.* The court held that the architect had not assumed a duty to supervise ithe methods of construction and, therefore, was not liable. *Id.* This case indicates the analysis of the contractual relationships that courts undertake when considering the duty of supervision.
246. Allen, *supra* note 1, at 460.
247. 148 Cal. App. 2d 909, 307 P.2d 719 (1957).
248. *See supra* text accompanying notes 194–99 (discussion of the "active" architectural role).
249. 148 Cal. App. 2d 909, 913, 307 P.2d 719, 722–23 (1957).
250. *Id.* at 910–11, 307 P.2d at 721.
251. *See* Day v. National U.S. Radiator Corp., 241 La. 288, 128 So. 2d 660 (1961); Wells v. Stanley J. Thill & Assocs., 153 Mont. 28, 452 P.2d 1015 (1969); Olsen v. Chase Manhattan Bank, 10 A.D.2d 539, 205 N.Y.S.2d 60 (1960), *aff'd mem.,* 9 N.Y.2d 829, 175 N.E.2d 350, 215 N.Y.S.2d 773 (1961).
252. *See* Associated Eng'rs, Inc. v. Job, 370 F.2d 633 (8th Cir.), *cert. denied,* 387 U.S. 907 (1966); Fidelity & Casualty Co. v. J. A. Jones Constr. Co., 325 F.2d 605 (8th Cir. 1963); Pastorelli v. Associated Eng'rs, Inc., 176 F. Supp. 159 (D.R.I. 1959) (supervising engineer); Erhart v. Hummonds, 232 Ark. 133, 334 S.W.2d 869 (1960); Larson v. Commonwealth Edison Co., 33 Ill. 2d 316, 211 N.E.2d 247 (1965); Nauman v. Harold K. Beecher & Assocs., 19 Utah 2d 101, 426 P.2d 621 (1967).
253. 176 F. Supp. 159 (D.R.I. 1959).

trol of construction must exercise reasonable care to see that the contractors do their work properly.[254] The application of a negligence standard indicates that the court implied a duty to supervise construction methods. This principle was made explicit in *Erhart v. Hummonds*,[255] an action against architects for the deaths of workmen who were killed when the wall of an excavation caved in.[256] The Supreme Court of Arkansas expressly held that the architects had a duty to supervise construction methods.[257] The duty, reasoned the court, arose from the general supervisory responsibilities of architects coupled with the architect's contractual authority to stop work to ensure " 'proper execution of the contract.' "[258] This theme is common among early cases finding a duty to supervise construction methods.[259]

While *Erhart* established the existence of a duty to supervise construction methods, *Day v. National U.S. Radiator Corp.*[260] limited the architect's duty to that of reasonably ensuring conformity with the plans and specifications. In *Erhart* the architect had an express right to stop work to assure proper execution of the contract.[261] In *Day,* however, the contract gave the architect no express authority to stop work, although this authority could have been implied from the architect's responsibility to assure conformity with the design.[262] Thus, the presence of a contractual right to stop work played an important role in the expansion of the supervisory duty.

b. The *Miller* Doctrine

The right to stop work also played a key role in *Miller v. DeWitt*.[263] *Miller* was a landmark decision and the culmination of the cases establishing the duty to supervise construction methods.[264] In *Miller* a contractor's employees, who had been injured when the roof on a building they were renovating collapsed,[265] alleged that the architect's failure to prevent the contractor from improperly shoring the roof constituted negligence.[266] The Supreme Court of Illinois held that the architect had a duty to supervise construction methods and techniques, basing the existence of this duty on the contract between the architect and the owner.[267]

The contract provided that the architect's duties included " 'general supervision and direction of the work,' " and the contract gave the architect " 'authority to stop the work whenever such stoppage may be necessary to insure the proper execution of the contract.' "[268] The contract provided that the contractor was

---

254. *Id.* at 166.
255. 232 Ark. 133, 334 S.W.2d 869 (1960). *Contra* Day v. National U.S. Radiator Corp., 241 La. 288, 128 So. 2d 660 (1961).
256. 232 Ark. 133, 135–36, 334 S.W.2d 869, 871 (1960).
257. *Id.* at 138, 334 S.W.2d at 872.
258. *Id.*
259. *See supra* note 252.
260. 241 La 288, 128 So. 2d 660 (1961). In *Day* a workman sued the architect for injuries sustained when a boiler exploded during installation and testing in the building designed by the architect. The court held that supervision of installation and testing methods was outside the architect's realm of authority. *See also* Olsen v. Chase Manhattan Bank, 10 A.D.2d 539, 205 N.Y.S.2d 60 (1960), *aff'd mem.*, 9 N.Y.2d 829, 175 N.E.2d 350, 215 N.Y.S.2d 773 (1961).
261. 232 Ark. 133, 138, 334 S.W.2d 869, 872 (1960).
262. *See* 241 La. 288, 303–05, 128 So. 2d 660, 666–67 (1961).
263. 37 Ill. 2d 273, 226 N.E.2d 630 (1967).
264. *See supra* text accompanying notes 252–62.
265. 37 Ill. 2d 273, 275, 226 N.E.2d 630, 633 (1967).
266. *Id.* at 285, 226 N.E.2d at 638.
267. *Id.* at 284–85, 226 N.E.2d at 638.
268. *Id.* at 284, 226 N.E.2d at 638.

responsible for site safety[269] and that the architect did not guarantee the work of the contractor.[270] The court noted that generally the duty of an architect to supervise merely creates a duty to see that the building meets the plans and specifications,[271] but imposed a greater duty, one that was not expressly part of the contract between the owner and architect. The court considered both the owner-architect and owner-contractor agreements, and interpreted the sum of the agreements to impose on the architect the duty to interfere or even stop work if the contractor began to act in an unsafe manner or to use a hazardous method in violation of the contractor's agreement with the owner.[272] Thus, the court imposed on the architect a duty of ensuring that the contractor did not violate the owner-contractor agreement and that safe and adequate construction methods were used[273]—an expansion of duty that neither party intended.

Justice House, dissenting in *Miller*, argued that the contract did not impose such a duty, but provided only for limited supervision,[274] noting that in the contract the architect agreed to attempt to prevent defects, but specifically disclaimed any guarantee of the contractor's performance.[275] Justice House acknowledged that the architect had a right to insist upon the safe and adequate use of construction methods, but argued that to transform the right into a duty was inconsistent with common usage and with the contract itself.[276] An architect, according to Justice House, does not normally contract for continuous supervision, but if the duty to supervise construction methods is expanded, the architect will reflect the increased responsibility through an increase in fees, a cost that will ultimately be borne by the public.[277] Finally, Justice House argued that an owner does not want such a duty imposed since it would result in chaos at the work site, with architects stopping work regularly—a result that is inefficient and costly to the owner.[278]

The effect of *Miller* is unclear. One writer considers the decision significant because it based the architect's duty to supervise construction methods solely upon the contract between the owner and the architect.[279] This view is not accurate. The court in *Miller* looked at all the agreements, including the owner-contractor agreement. More important, the case is significant because it extended the architect's liability beyond that contemplated in the owner-architect agreement and, apparently, beyond the intent of the parties.[280]

Miller is inconsistent with the view that methods of construction are within the realm of the contractor's authority and control rather than the architect's.[281] It is also inconsistent with the view that contractors may be better able to control construction methods than architects, whose primary responsibility is design.[282] These inconsistencies have led some to believe that the courts have

---

269. *Id.* at 281, 227 N.E.2d at 636.
270. *Id.* at 280, 226 N.E.2d at 635.
271. *Id.* at 284, 226 N.E.2d at 638.
272. *Id.*
273. *Id.*
274. *Id.* at 293, 226 N.E.2d at 642 (House, J., dissenting).
275. *Id.*
276. *Id.* at 293–94, 226 N.E.2d at 643.
277. *Id.* at 295, 226 N.E.2d at 643. While there will be an increased cost no matter who bears the burden, deciding which party will bear the burden is important and can have significant ramifications. See *infra* text accompanying notes 340–44, 444–48 for a discussion of the architect's ability to allocate these costs and an argument indicating that costs to the public need not necessarily increase.
278. 37 Ill. 2d 273, 294, 226 N.E.2d 630, 643 (1967) (House, J., dissenting).
279. Note, *supra* note 7, at 252.
280. Carey, *Assessing Liability of Architects and Engineers for Construction Supervision*, 672 INS. L.J. 147, 154 (1979).
281. *See* Note, *supra* note 7, at 250.
282. Interview with practicing engineer, Dayton, Ohio (December 28, 1982).

created a duty that "[requires] more than conduct reasonably to be expected of a prudent design professional and amounts, in effect, to liability without proof of negligence—strict liability."[283]

The *Miller* decision has not been overruled, but a later Illinois case has distinguished *Miller*, finding that an architect did not have a duty to supervise construction methods.[284] In *McGovern v. Standish*[285] an injured employee of the contractor brought an action against the architect. The Illinois Supreme Court held that the architect was not liable to the worker because the architect did not have the right to control or direct methods of construction.[286] The contract terms were similar to those in *Miller*, except that the architect did not have an express right to stop work.[287] In *McGovern*, the court found that the architect did have the right to reject defective materials and require their correction, but did not have the right to stop work because it was being done in a dangerous manner.[288]

Although *McGovern* apparently held that no duty to supervise construction methods exists without a contractual right to stop work, the court did indicate some exceptions. The opinion implies that the existence of expansive authority vested with the architect may create a duty to workers and, further, that attempts to exercise control over the work by issuing orders or directions may create such a duty.[289] Therefore, this decision does not establish nonliability based on carefully worded contracts. Instead, the opinion reflects the court's determination that in the circumstances of the *McGovern* case the architect did not have sufficient control to warrant imposing a duty, but that an architect could have such control even absent a right to stop work. Thus, strategic use of contract language may not protect the architect if the judiciary determines that liability should be extended.

The *Miller* decision is not an isolated case. In *Geer v. Bennett*[290] a Florida appellate court held that an architect may be liable for his failure to direct the contractor to install a guardrail to prevent persons from falling from a twelve-foot-high construction area. The plaintiff was injured in a fall from the construction area.[291] The court based the duty to supervise construction methods on the duty of "supervision" specified in the contract, although the contract did not give the architect the right to stop work.[292] The decision is a simple expansion of the duty established by the word "supervision," an analysis that holds an architect to a duty that neither party intended.

The expansion of duty has not been accepted by all courts. In *Réber v. Chandler High School District No. 202*[293] an Arizona appellate court expressly rejected the *Miller* doctrine.[294] *Reber* was an action by the contractor's employees against an owner and his architect for injuries sustained when a gymnasium under construction collapsed.[295] The court held that the owner-contractor agreement can be used to settle ambiguities in the owner-architect agreement if the architect provided

---

283. Note, *supra* note 16, at 1243; *see also infra* text subpart IV(A) (discussion of the strict liability theory).
284. McGovern v. Standish, 65 Ill. 2d 54, 357 N.E.2d 1134 (1976).
285. *Id.*
286. *Id.* at 69, 357 N.E.2d at 1142.
287. *Id.* at 63–65, 357 N.E.2d at 1139–40; *see supra* text accompanying notes 268–70.
288. 65 Ill. 2d 54, 68, 357 N.E.2d 1134, 1141 (1976).
289. *Id.* at 68–69, 357 N.E.2d at 1142.
290. 237 So. 2d 311 (Fla. Dist. Ct. App. 1970).
291. *Id.* at 313.
292. *Id.* at 313–14.
293. 13 Ariz. App. 133, 474 P.2d 852 (1970).
294. *Id.* at 135–36, 474 P.2d at 854–55; *see supra* text accompanying notes 263–73.
295. 13 Ariz. App. 133, 134, 474 P.2d 852, 853–54 (1970).

both agreements,[296] but found that the contracts did not provide for the architect to exercise control over the method and manner of performing the details of the work.[297] While the contract in *Reber* did not include the right to stop work, the court rejected the argument from *Miller* that such a right gives rise to a corresponding duty.[298] The court held that "liability for negligent exercise of retained supervisory powers can attach only when there is a showing that a *duty* has been created by the [architect's] reservation of . . . 'the right to exercise day-by-day control over the manner in which the details of the work are performed.' "[299]

The absence of a right to stop work is apparently not the basis of the decision since the court expressly rejected *Miller* and adopted the view of the *Miller* dissent.[300] Neither the contract in *Reber* nor that found in *Geer*[301] contained a right to stop work, yet the results of the two cases were different, presumably because, in *Reber,* the court determined that the duty should not be extended unless expressly assumed and undertaken by the architect.[302]

The doctrine established by *Miller*[303] still has viability, and architects have been held to a duty of supervision of construction methods in cases subsequent to *Miller.*[304] The trend established in *Miller* still appears predominant, but a movement has occurred away from the extensive duty of supervision set forth in *Miller* back to a consideration of the agreements on a more objective level.[305] Recent decisions in some jurisdictions have adopted a less expansive view of the architect's duty to supervise. It is well recognized that the contract relationships between architect, owner, and contractor play an important role in determining whether an architect has a duty to supervise construction methods for the protection of third parties. *The intent of the parties is also recognized as an important consideration.*[306] Some recent cases have taken a narrow view of the contract so that it is interpreted to strictly conform to the intent of the parties.[307]

---

296. *Id.* at 136, 474 P.2d at 855.
297. *Id.* at 137, 474 P.2d at 856.
298. *Id.* at 135–36, 474 P.2d at 854–55.
299. *Id.* at 135, 474 P.2d at 854 (emphasis in original).
300. *Id.* at 135–36, 474 P.2d at 854–55 (citing Miller v. DeWitt, 37 Ill. 2d 273, 226 N.E.2d 630 (1967)). The dissenters in *Miller* believed that the contract required only limited supervision and that the contract should be interpreted to fulfill the intent of the parties. Justice House indicated that extensive supervisory powers should be based on clear enumeration of them in the contract and that in the typical contract, supervision only includes ensuring substantial conformity of the structure to the owner's requirements. 37 Ill. 2d 273, 293–95, 226 N.E.2d 630, 642–43 (1967) (House, J., dissenting).
301. *See supra* notes 292–98 and accompanying text.
302. 13 Ariz. App. 133, 135, 474 P.2d 852, 854 (1970).
303. *See supra* text accompanying notes 263–73.
304. *See e.g.,* Duncan v. Pennington County Hous. Auth., 283 N.W.2d 546, 548–49 (S.D. 1979) (architect liable for site safety based on architect's contractual requirement of "obtain[ing] compliance with the contract documents" through on-site inspections).
305. *See* McGovern v. Standish, 65 Ill. 2d 54, 357 N.E.2d 1134 (1976). However, the court still relied on the overall circumstances and a review of the architect's role and authority rather than relying strictly on the agreement.
306. *See* Wheeler & Lewis v. Slifer, 195 Colo. 291, 577 P.2d 1092 (1978) (intent is critical determination); Porter v. Iowa Power & Light Co., 217 N.W.2d 221 (Iowa 1974) (right to inspect work for conformity with plans and right to stop work if provisions of contract are not carried out does not constitute retention of control sufficient to hold inspection engineer liable to contractor's employees); Duggan v. Arnold N. May Builders, Inc., 33 Wis. 2d 49, 146 N.W.2d 410 (1946) (architect held to no supervisory duty at all).
307. Wheeler & Lewis v. Slifer. 195 Colo. 291, 577 P.2d 1092 (1978); Fruzyna v. Walter C. Carlson Assocs., Inc., 78 Ill. App. 3d 1050, 398 N.E.2d 60 (1979); Krieger v. J.E. Greiner Co., 282 Md. 50, 382 A.2d 1069 (1978); Moundsview Indep. School Dist. No. 621 v. Buetow & Assocs., Inc., 253 N.W.2d 836 (Minn. 1977); Brown v. Gamble Constr. Co., 537 S.W.2d 685 (Mo. Ct. App. 1946); Porter v. Stevens, Thompson & Runyan, Inc., 24 Wash. App. 624, 602 P.2d 1192 (1949); Luterbach v. Mochon, Schutte, Hackworthy, Juerison, Inc., 84 Wis. 2d 1, 267 N.W.2d 13 (1978); Vonasek v. Hirsch & Stevens, Inc., 65 Wis. 2d 1, 221 N.W.2d 815 (1974).

c. Architects' Attempts to Limit the Liability Arising from Supervision of Construction Methods

Architects have responded to decisions such as *Miller* and *Geer* by attempting to limit their liability through changes in contract language. The current owner-architect contract form of the American Institute of Architects does not use the word "supervision."[308] The form provides only for visits to the site, at intervals to be determined by the architect, to verify that the work is proceeding according to the contract.[309] It also contains an express disclaimer providing that the architect is not responsible for construction methods, techniques, or safety precautions.[310] The contract also has omitted the "right to stop work," and retained only a right to reject work that does not conform to the contract documents.[311]

The effectiveness of these changes is uncertain. Architects have eliminated the word "supervision" from the contracts because they believe that it is too broad to describe architects' duties and it allows courts to hold architects responsible for many aspects of the construction process for which the architects did not intend to assume responsibility.[312] The deletion of the word "supervision," however, may not make a substantial difference. Since architects still perform the same functions despite the reworded definition, a court, consistent with *Miller*,[313] may look beyond the language and impose a duty of supervision of construction methods on the architect.[314] Similarly, the deletion of the "right to stop work" may not have the desired effect since the architect still retains the same functions at the work site, and the power to "reject work"[315] may be interpreted to be substantially equivalent to the power to stop work. Last, the express disclaimers in the current form contracts[316] are similar to provisions in earlier contracts, provisions that the courts have given little or no effect.[317]

If, as the cases indicate, the courts base the duty of supervision on the role of the architect and the control that he has,[318] the change in contract language will have little effect. The outcome will instead depend on the architect's control and the willingness of a court to expand the architect's liability. Given the traditional role of the architect in the construction process, it may be very easy for a court to determine that an architect had sufficient authority and control to justify imposing a duty to supervise construction methods, techniques, and site safety. Architects must be aware of this expansive view of their roles and responsibilities. Architects must also be aware of other theories that impose liability to third parties on them.

---

308. AIA Document B141/CM, *supra* note 26.
309. *Id.* § 1.5.4.
310. *Id.* § 1.5.5.
311. *Id.* § 1.5.12.
312. *Types of Professional Liability Claims, supra* note 245, at Special Studies—5(3).
313. 37 Ill. 2d 273, 226 N.E.2d 630 (1967); *see supra* text accompanying notes 263–73.
314. *See supra* text accompanying notes 279–80, 288–89; *see also* Miller v. DeWitt, 37 Ill. 2d 273, 226 N.E.2d 630 (1967); McGovern v. Standish, 65 Ill. 2d 54, 357 N.E.2d 1134 (1976). *But see* Reber v. Chandler High School Dist. No. 202, 13 Ariz. App. 133, 474 P.2d 852 (1970).
315. AIA Document B141/CM, *supra* note 26. § 1.5.12.
316. *Id.* § 1.5.5.
317. *See, e.g.,* Miller v. DeWitt, 37 Ill. 2d 273, 226 N.E.2d 630 (1967). *Miller* dealt with a contract with a disclaimer similar to those currently used, but the court apparently refused to give the disclaimer any effect. *See supra* text accompanying notes 269–75.
318. *See, e.g.,* Geer v. Bennett, 237 So. 2d 311 (Fla. Dist. Ct. App. 1970); Miller v. DeWitt, 37 Ill. 2d 273, 226 N.E.2d 630 (1967); McGovern v. Standish, 65 Ill. 2d 54, 357 N.E.2d 1134 (1976); *see also supra* text accompanying notes 263–92.

Limiting Professional Liability

It is all too clear that the evolutionary expansion of the professional's liability has taken a costly toll, the least of which is not the diminution of the architect/engineer's role in the construction process. Opinions vary, but one source noted, "Once towering over the chain of command, demanding authority to control the work of virtually all who breathed life into their designs, architects now have withered into the shadows.... leaving owners the task of overseeing the builders." The decisions that expanded liability supposedly were based on contractual responsibilities, but in reality they were based on actual control exercised by the architect. Less control obviously leads to disastrous results. Stung by the various legal intrusions, professionals naturally narrowed their responsibilities, spread risks, and retrenched to be part of a balanced array of interests.

Oddly enough, in a profession where innovation is celebrated, it is within that very creative venture where the professional is most legally vulnerable. Properly limiting professional liability, in a reasonable manner, is the best posture for producing quality, attractive, and safe architecture. Until that day, professionals are best advised to protect their projects, concepts, and interests, on-site, through a comprehensive, skilled, and ongoing construction contract administration program.

THE PROFESSIONAL'S ROLE IN DEALING WITH PROBLEMS OF LIABILITY

Some clarification of the professional's situation has occurred with newly revised legal and professional determinations, revision of standard contract text provisions, more education, and different modes of practice adopted by professional firms. Almost all of this was driven by court decisions that imposed new—and expanded existing—liability exposure. From the mid-1960s until 1984, things were becoming more and more threatening. Added to this was the crucial "reassessment" of professional liability insurance; the professional support and safety net. The two issues were divergent, causing further concern to professionals. Court decisions were of all kinds, but did not address the changed professional conditions.

There are several causes for the professional liability insurance malaise of the 1980s. None of these are directly attributable to design professionals, except for the fact that they are engaged in a high-risk, imperfect industry—construction. Juries composed of the general public have seen fit to impose huge monetary awards (settlements) in liability cases. This reflects a general attitude that there must be a risk-free society (in spite of the fact that humans cannot yet produce that scenario), penalty for all wrongs, no matter how insignificant the "crime," and a "deep pocket" to support massive awards without real regard to fault. But in this prevailing attitude lies the fact that all of the imposed costs are eventually recouped through the price structure imposed on the general public in the cost of construction materials and services.

Without doubt there is some malpractice in the construction design professions (as in all professions) that justifies proper imposition of liability. However, it is clear that the expectations of the plaintiffs are more often than not ill-informed and unrealistic. There are limits in every field of technology and science; no profession or industry is "perfect." Everyone involved in construction litigation needs to understand this; they must also try for a more modest approach to resolutions.

The issue, although calmer now than in the 1980s, continues. There is still an effort at the federal level to enact tort (damages) reform, to restrict the amount of punitive damages that can be assessed, and to deal with frivolous suits. Arbitration continues to be one of the prime vehicles for avoiding needless suits and for achieving reasonable and prompt resolutions.

How can the contract administrator contribute to the firm's liability situation? The following thoughts are reasonable:

- Look for and anticipate problems in the contract documents and in the actual construction. Give proper and prompt notice. Actively seek prompt responses and/or solutions.
- Maintain good records and documentation of all kinds, from established forms to personal "Memos to File," which may prove extremely helpful if litigation occurs. Litigation is long-term, tedious, arduous, and often nasty. It could demand your recall of seemingly inconsequential incidents that took place years earlier.
- Thoroughly know your authority and its limits. Exercise it continually, fairly, promptly, and properly.

Since the 1984 publication of Nischwitz's paper, professional affairs have quieted down. Many made the necessary adjustments in practice procedures, insurance coverage, greater efforts at claims and risk control, and so forth. Of course, suits, claims, and disputes still exist. Some situations still are litigated. Each court case must be closely examined for nuances and even slight deviations in circumstances from those in previously decided cases. Even minor (seemingly) changes have led courts to different and even "odd" decisions. All courts, though, can still rely on any case (precedent) that reasonably supports the circumstances at hand.

From this the professional, the project representative, and the contract administrator are well advised that the courts are sending a direct message:

1. Carefully write the full content, intent, and impact of your services contract, how you are directed or inhibited, and what ramifications may befall you (avoid them with legal assistance).
2. Understand and work every aspect of your project carefully within the prevailing standard of care. Make reasonable and rational decisions. You are not bound to be "perfect" unless you allow your contract to require that.
3. Better to err on the side of too much pertinent correspondence or too much verbal information.
4. Be aware of problem areas or work. Anticipate difficulties in the work. Prepare yourself and others for some adversity as the work progresses, understanding that late/slow/long deliveries, strikes, unavailability of workers, weather, unseen site conditions, need for remedial work, and similar circumstances can play havoc with schedules, progress, and completion.
5. Stay ahead of the project, looking to new work, work areas, and needs prior to their actual implementation.

Such efforts can greatly reduce, if not eliminate, many root causes of claims.

The overwhelming legal message to the design professional, and down to the on-site contract administrator, is that they must take advantage of the very limited number of opportunities afforded them to bring the project on line, in full accord with all contractual, regulatory, and legal parameters. No matter what decisions are made as to the limits of duties and responsibilities, these decisions need to be exercised to those limits. Simply, projects can pose threats both to the general public (physically) and to professionals (legally). Surely every professional wants to meet the legal and necessary regulations imposed on the project, so why not also meet the more personal implications around the professional, the practice, and the staff.

Many people maintain that architecture is a "dreamer's" profession and that most designs are unrealistic. Others, with more understanding, see it as a combination of art and engineering, design and construction. Certainly the latter view is much closer to reality. But brutal reality does play a major part in architecture. Everything is not as one envisions it or desires it. Most design concepts anticipate the "perfect" solution. But the process of bringing this concept to useable reality necessitates the use of an imperfect science called construction. To the understandable limits of human production in the actual construction must be added the human realities of designing, detailing, specifying, and envisioning. The absolutely "perfect in every way" project has yet to be built. This, too, needs to be explained and/or demonstrated to the client, so both client and professional proceed on the same track, with parallel expectations that are reasonable, reachable, and realistic.

Professionals, then, are well advised to continue to upgrade and refine their risk management endeavors. Professionals need to produce designs that are both prudent and better conceived, produce better quality and better coordinated contract documents, improve relationships and communications, and enhance construction contract administration practices. In embracing all of the above and providing their clients with a more informed view, professionals can directly contribute to claims reduction.

Every participant on the project, professional, owner, and contractors, take risks. Risks and their companion claims can be neither completely predicted nor totally avoided. This is fairly obvious since construction contract risks are of two types:

1. contractual: those produced by reduced clarity in the contract, inadequate/imperfect communications, and improper, inadequate, or less than timely contract administration
2. construction: those produced by factors such as site conditions, resource availability, weather, and so on

To manage these risks, there is a need to avoid them, abate them, retain them, or transfer them. While it is obviously better to share the liability involved and the unavoidable associated cost, there is a strong inclination to assign any risk to that party best able to evaluate, control, pay for, and benefit from the assumption. Design professionals often are caught in the middle of this. However, as a major contributor to the contract documents as well as the technical adviser to the owner, the professional is uniquely qualified to aid both him/herself and the client in these matters.

It is well to follow the advice of practitioners of construction law in this regard. They are well aware that problems occur when the various parties fail to clearly define the concepts and details that are embodied in the construction process. This is further highlighted by the recent and drastic changes in the traditional roles, contracts, processes, and procedures. What with the traditional methods still in place and working, and the addition of such programs as design/build (turnkey), phased design and construction, and fast-track construction methods, clarification and definition (as well as "education") have become much more important. Numerous variations of these systems have been devised, so one cannot rely on the "last time," "that other job," standardized documents, or established procedures that have not been recently reviewed. Almost all of the changes in the industry have been aimed at (1) reducing unrealized expectations or unexpected results for the parties involved, (2) meeting the evolving basic legal concepts, which have shifted the rights and responsibilities of the principals, and (3) economic aspects and technological improvements, which motivated experimentation with new concepts to procure and manage construction projects.

Unrealized expectations are fostered by misunderstandings rooted in a basic lack of communication and mutual understanding at the very outset of the contract, and ambiguous text. These are recognized as the primary causes of disputes and litigation. Some are as simple as expecting a "two by four" to be exactly that and not a nominal size, or that a specification calling for concrete form work to be "true to line" does not mean no tolerance is permitted. To overcome these communication gaffes, more is being reduced to paper, both written and graphically.

Chapter 6 Obligation, Education, and Liability

Simply, and particularly in the case of an inexperienced (as to construction) owner, it is necessary to provide a format for the discussion and understanding of construction legal formats, procedures, jargon (including misnomers), nomenclature, symbols, tolerances, standards, practices/customs, relationships, jurisdictions, manufacturers, contractual obligations, and other nuances. Obviously, the owner cannot be brought to the same level of expertise as the other project participants, but familiarization and orientation of the owner in these terms is necessary.

A "meeting of the minds" is essential for a successful contract. This entails a common understanding of what is to be achieved, how, and what each party is to contribute. The perspective of each member of the team (or party to the contract) must become part of the greater perspective of the project itself. More than likely, each party will be required to give a little for the sake of the project as a whole; no one party can take the position that nothing within their purview can be changed. This is not to say that subservience is necessary, but flexibility is required. The true test of this configuration comes with the first problem of any consequence. Addressing and resolving this problem will tell, to a large degree, how the project will proceed.

Early on, then, it is essential to recognize that contract administration produces a sizable amount of litigation, and usually that litigation involves massive sums of money. It is vitally important that design professionals understand the concept of litigation; such understanding will aid in minimizing or eliminating at least part of it. Any time, though that there is a perceived wrong or damage, or real personal injury or death, there will be litigation. Obviously, the professional must always be in a position to explain, describe, verify, and defend the project, the design, and the construction; along with decision making, procedures, approvals, communications, and so forth.

With the addition of numerous people on the site and the innumerable relationships, obligations, and responsibilities, the path to a fully successful project is often convoluted. It is no place for unilateral, "off the wall" decisions that in any way deviate from the pronouncements of the contact documents and/or the conditions of the contracts. Even situations that cry out for a quick and decisive answer should be pursued very carefully. To do otherwise is to flirt with the legal system. Contract administration is not an informal, "gentlemen's agreement" operation. It must be pragmatic to a fault! Communication and documentation are more than essential—they are imperative.

No matter how professionally or diligently you may practice architecture (or any other design profession), anyone involved in a project and some third parties can sue you, at any time, on almost any grounds (no matter how tenuous), and there is nothing you can do about it. Diligence, however, is an important aid in reducing the success of claims of all types. Claims reduction and litigation are time-consuming and excessively costly. Their cost, however, pales compared to a successful award against the professional. In addition, the professional is open to other civil actions: loss of license, loss of practice, and so on.

In this entire scenario, as in the complete issue of contract administration, current professional education, i.e., professional practice instruction in professional degree programs, is woefully lacking in alerting prospective professionals of their legal status and the varied problems they will encounter in practice. Of course, this is also true for other phases of the professional's practice and work. It is necessary that each practitioner be fully aware of the legal network in which practice is conducted.

The saga of legal threats to the design professions took yet another aberrant turn in early 1997, when an Ohio appeals court ruled that design professionals *must proceed to full trial* in *every* instance when made a defendant to *any* allegation of construction liability, even though no allegation is made that the professionals were part of the factual sequence that created liability (Franklin County, Ohio, Court of Appeals, case nos. 96APE03-291 and 96APE03-205). In essence, no summary judgment (declaration of dismissal by the judge without a major hearing) would be available to the professionals, and they would be required to stand the cost (in both time and money) of a full trial. Of course, all design professionals and their associations in the state are active in attempting to overturn this decision. A statute in a similar direction in the state of Illinois was overturned just a few years ago.

This situation drastically illustrates the hazard of the design professional (and the contract administrator) engaging in any safety program, instructions, specifications, directions, advice, etc. (including mere comments). Other courts have noted that where a professional observes an unsafe condition, it is proper to pass this observation (but with no comment as to its resolution) on to the proper persons who hold responsibility for safety on the job site.

Additionally, this is an excellent example of the ongoing need to stay in touch with your profession and of being fully aware of the legal as well as the technical aspects of your work as they continue to evolve. To do less can lead to adverse situations involving all personnel, even the contract administrator.

In attempting to understand the law and one's legal position, it is advisable to understand that the law is dynamic and continually open to new interpretation. At a basic level, one is well advised to meticulously read all contracts in order to ascertain what, specifically, is required of all parties and what is not required. The more precise the contract language, the better. One is usually doomed, though, in any attempt to write the "perfect" contract. Conversely, one should never agree to a contract that assigns, directly or by reasonable implication, duties, obligations, and responsibilities that are extraordinary or for which one possesses no expertise (construction job safety is a good example).

It goes without saying that legal advice should be sought in many, if not all, contract situations. Often smaller projects seem "ripe" for merely filling in the blanks of a standard contract form, but caution should prevail. Besides seeking the best legal position, it is also obvious that the professional must endeavor to take every action possible to ensure that the project is built *exactly* as designed, detailed, and contracted.

In maintaining the "professional edge" or in developing stronger legal background, a number of good resources can be utilized. Many refer to narrow aspects of contract administration, while others take a broad approach. The resources below can be useful:

Books

Justin Sweet, *Legal Aspects of Architecture, Engineering and the Construction Process*, 5th ed. St. Paul, MN: West Publishing Co. (1994).

Keith Collier, *Managing Construction: The Contractual Viewpoint.* Albany, NY: Delmar Publishers (1994).

Book Insert

J. Russell Groves, Jr., AIA, "Contract Administration", pp. 131–142 of *Encyclopedia of Architecture: Design, Engineering & Construction*, Vol. 2, ed. Joseph A. Wilkes. New York: John Wiley and Sons (1988). Excellent summary of contract administration; extensive bibliography.

Seminars

Annual presentations sponsored by the University of Wisconsin Extension, 432 North Lake Street, Madison, WI 53706. Held for various terms through the year. All presentations utilize expert speakers with varied professional expertise and backgrounds; numerous handouts are distributed.

Manual

Published by Office of Professional Liability Research, Inc., Schinnerer Management Services, Inc., Two Wisconsin Circle, Chevy Chase, MD 20815-7003. Contains an on-going series of narrow-scope pamphlets dealing with all phases of professional practice including contract administration; back issues available, as well as an annual subscription service for new and updated material.

Legal

The construction department of Smith, Currie & Hancock, Attorneys, 2600 Harris Tower, Peachtree Center, 223 Peachtree Street, N.E., Atlanta, GA 30043, provides seminars, handouts, booklets, and newsletters dealing with construction law, dispute resolution, documentation, and the like.

Newsletters

"Common Sense Contracting" by Smith, Currie & Hancock (address above).

"Construction Contracting Alert" by Wiley Law Publications, John Wiley & Sons, P. O. Box 2575, Secaucus, NJ 07096-2575.

"The Guidelines Letter," ed. Fred A. Stitt, publisher. A monthly with excellent coverage on all aspects of practice. P.O. Box 456, Orinda, CA 94563.

Booklet

"Chapter 11—Contract Documents and Their Interpretation and Construction" (1984), National Institute of Construction, Inc., The Michie Co., P.O. Box 7585, Charlottesville, VA 22906-7585. Part of a two-volume set of booklets on various related topics.

Handbook Module

"Construction Contract Administration." New module of the *Manual of Practice,* available separately. Construction Specifications Institute (CSI), 601 Madison Street, Alexandria, VA 22314. It has been the basis for a two-day seminar and is being developed into a certification program by an ad hoc committee of CSI.

Dictionary

Construction Dictionary, 9th ed., including 17,000 definitions, 16,000 terms, symbols, associations, agencies, formulas, conversion factors, and abbreviations; by and available from the Greater Phoenix chapter of the National Association of Women in Construction, P.O. Box 6142, Phoenix, AZ 85005, (602) 263-7680.

Glossary

"A Glossary of Construction Specifications Terminology," District of Columbia Metropolitan Chapter, Construction Specifications Institute, 1777 Church Street, N.W., Washington, D.C. 20036. Terms included taken from the CSI specifications format, Division 1–16. Excellent basic information.

Resource Catalog

"ARCAT," listing of 7,100 construction product manufacturers and some 1,600 trade associations, including addresses, telephone and fax numbers, and contact names; published annually by The Architect's Catalog, Inc., 1305 Post Road, Fairfield, CT 06430-9709, telephone (203) 256-1600, fax (888) 329-2722.

In addition, many local AIA, CSI, and other construction-related organizations and local colleges and technical schools present short courses, seminars, and programs pertinent to contract administration, such as the AIA Intern Development Program.

EXERCISES

The following tasks are based on the concepts and information in the chapter. All of the material or information required to complete the tasks may not be contained in the chapter text. Independent research is required. Consult your instructor for guidance.

1. Sometimes the contract administrator will be asked to "promise," "attest," or "certify" work, work processes, or situations. Should this be done? Defend your position.
2. List and explain four valid bases for a claim by a contractor.
3. According to AIA A201, on what contractual grounds may work be rejected?
4. Argue for or against the proposition that deficiencies in documentation and communication between designer and contractor contribute to deficiencies in the work.
5. In which contract document is quality addressed? Regarding quality, what should be the relationship between contract documents, and what about quality of work not mentioned in the contract?
6. Discuss the owner's right to uncover work and inspect it when the contract has no specific provisions for such inspection.
7. Argue for or against the proposition that the basic responsibility for code compliance lies with the design professional.
8. Upon inspection you find work with a serious deficiency. In checking the specs you find the particular point is not covered. What other contractual authority do you have or could you turn to in support of your rejection of the work? Cite an example and give details.
9. The work you inspect has not been done, but other work of equal quality has been substituted. What do you do? Answer in detail.
10. What is the difference between General Conditions and General Requirements (Division 1 of the specifications)?
11. Are standard form General Conditions amendable? If so, how? How are these changes made part of the contract? Which ones should the contract administrator enforce?
12. You find a marked difference between the General Conditions and the General Requirements. What action do you take? What appears to be the resolution?
13. In a meeting your boss, without talking to you, is ready to sign off on a questionable item. You disagree with the solution. What do you do? When? How?

7

Forms for the Process

The process itself is the actuality.
—Alfred North Whitehead, English mathematician and philosopher

IMPORTANCE OF DOCUMENTATION

Perhaps the most widely used program of contract administration is delineated in AIA document B141, the Standard Owner-Architect Agreement. Of course, other varied scenarios are in use in *direct reaction to the conditions set forth for the projects at hand:* construction management, owner-management, design/build, and others.

Study the details of Section 2.6 of document B141 to capture the flavor of the program, as well as its detailed requirements. The document lists not only what is to be done, but by whom and, in some cases, when. The document does, though, suffer from various instances of ambiguous language. More than likely, this was a planned maneuver by those drafting the document, who sought to both explain the program and detail it, without inhibiting its use. While this provides a good complete plan, it is necessary that the ambiguity be resolved, as well as the specific desires and capabilities of the owner. The mere execution of B141, even where it retains the provisions for contract administration, is not going to get the job done properly. In this, the document is a little deceiving, if one is not familiar with what is included and what is not.

The discussion of construction contract administration in Chapter 2.8 of the AIA *Manual of Professional Practice,* Appendix B, suggests that AIA document D200, Project Checklist, is also available. While one may choose to prepare a personal checklist, D200 provides good insight into the detail work that must be done in this phase. (Document D200 is shown in Figure 7.1.) Of course, each of the items listed requires further refinement, detailed information, additional forms, and the other procedural and administrative functions to permit proper application of the listed items to the work involved. In essence, there is a need for development of a set of documents that can be used on a repetitive basis on the project. These may vary slightly from one project to another, but a basic set of such documents aids the firm or individual in the actual work of contract administration. Further, it provides a consistent and uniform system for the processing of information.

Participation in contract administration by the design professional should never be considered as anything less than a vigorous, expert service. It is a matter of control. This critical concept should be instilled in staffers, and is something to keep in mind throughout the project. The professional is in the position of doing something for the owner, which he/she may not be able to do or may have limited resources to do. In addition, the professional is the sole link, through this program, between a highly technical orientation (the contractors) and the lay person (the owners).

To be effective, the professional must be able to function in both arenas with equal skill, and must be able to bring these factions together as the project may require. Obviously a support system is necessary, from procedures to forms, meetings, specifications and other contract provisions and authority, and so forth.

Personnel selection is also important, because the chore of administration must be fitted to the skills of the personnel. A pure technician, while highly expert as to the nuts and bolts of the project, may fail as a contract administrator for lack of good management, communications, and administrative skills; of course, the reverse is true of a pure manager. There is a need for a well-rounded person, perhaps tilted a little more to "paper skills" and communications than to technical matters. The selection of the right person is the beginning of the needed support system that leads to the success of the task and, directly, to the success of the project. On this subject Gerald Hammond, AIA (Hamilton, OH) commented at an AIA roundtable discussion on practice, in October, 1986:

> Construction administrators need a wealth of experience, a sense of fair play, and a sufficient understanding of design to be able to make on-the-spot decisions in the field. . . . The construction contract administrator is at the front line of client contact, and his/her performance is probably the last impression the owner has of an architectural firm's project performance.

Chapter 7 Forms for the Process

In any attempt to develop an all-inclusive program, the design professional should utilize every opportunity to undergird the contract administration effort in all of the contract documents. This effort will arm the on-site personnel with the specifics of the authority they should possess, and with exact explanations and procedures to be used. Usually, this can be accomplished within the purview of three catchwords: *clarity, detail,* and *explanation.* For instance, the inclusion of administrative definitions can go a long

AIA Document D200

PROJECT:

NUMBER:

Reproduced with permission of The American Institute of Architects under license number #97016. This license expires April 30, 1998. FURTHER REPRODUCTION IS PROHIBITED. Because AIA Documents are revised from time to time, users should ascertain from the AIA the current edition of this document. Copies of the current edition of this AIA document may be purchased from The American Institute of Architects or its local distributors. The text of this document is not "model language" and is not intended for use in other documents without permission of the AIA.

 CAUTION: You should use an original AIA document which has this caution printed in red. An original assures that changes will not be obscured as may occur when documents are reproduced.

FIGURE 7.1   AIA document D200 referring to construction contract administration.

Dates	Initials	Item	Remarks
10.		Determine Owner's time schedule for bidding and occupancy.	
11.		Determine Owner's budget and determine its basis (e.g., cost estimate, available funds, etc.).	
12.		Determine whether the budget includes:	
		A. *Site acquisition*	
		1. Land costs	
		2. Demolition	
		3. Title insurance	
		4. Real estate fees	
		5. Rezoning	
		6. Legal fees	
		7. Others	
		B. *Compensation*	
		1. Civil engineering	
		a. Off-site utilities/street work	
		(1) Basic compensation	
		(2) Reimbursable expenses	
		(3) Contingency	
		b. On-site utilities/grading and paving	
		(1) Basic compensation	
		(2) Reimbursable expenses	
		(3) Contingency	
		2. Architecture/engineering	
		a. Basic compensation	
		b. Reimbursable expenses	
		c. Contingency	
		3. Interior design	
		a. Basic compensation	
		b. Reimbursable expenses	
		c. Contingency	
		4. Procurement and purchasing furniture, furnishings and equipment	
		a. Basic compensation	
		b. Reimbursable expenses	
		c. Contingency	

4   AIA DOCUMENT D200 • PROJECT CHECKLIST • AUGUST 1982 EDITION • AIA® • ©1982 • THE AMERICAN INSTITUTE OF ARCHITECTS, 1735 NEW YORK AVE., N.W., WASHINGTON, D.C. 20006

WARNING: Unlicensed photocopying violates U.S. copyright laws and is subject to legal prosecution.

FIGURE 7.1   *(Continued).*

way in explaining the true intent of the design professional. To prevent verbosity, one can "explain, in clear detail." If there is no explanation, there is no information—no communication. This will produce a scenario in which everyone is left to guess at what is intended, and usually these guesses will not all be the same. Hence, conflict, and all its bad ramifications, will be created needlessly by the professional organization itself.

Another aspect is that the person performing the contract administration is, by that obligation, in control of the project. Since this service is provided through the

Chapter 7 Forms for the Process 203

Dates	Initials	Item	Remarks
		5. Furniture, furnishings and equipment	
		a. Basic estimate	
		b. Contingency	
		6. Landscaping	
		a. Basic estimate	
		b. Contingency	
		7. Owner-furnished equipment —procurement	
		a. Basic estimate	
		b. Contingency	
		8. Owner-furnished equipment —installation	
		a. Basic estimate	
		b. Contingency	
		9. Allowances	
		a. Artwork	
		b. Hardware	
		c. Carpet	
		d. Brick	
		e. Landscaping	
		f.	
		g.	
		E. *Miscellaneous costs*	
		1. Survey—topographical, boundary and utilities	
		2. Adjacent building damage survey	
		3. Soils report	
		4. Full-size mock-ups	
		5. Construction testing laboratory inspection	
		6. Construction soils inspection	
		7. Full-time project representatives	
		8. Performance bond, labor and material payment bond	

AIA DOCUMENT D200 • PROJECT CHECKLIST • AUGUST 1982 EDITION • AIA® • ©1982 • THE AMERICAN INSTITUTE OF ARCHITECTS, 1735 NEW YORK AVE., N.W., WASHINGTON, D.C. 20006

WARNING: Unlicensed photocopying violates U.S. copyright laws and is subject to legal prosecution.

FIGURE 7.1 *(Continued).*

owner-architect (engineer) contract, the design professional becomes the key performer in the administration process. Everything from the ground rules to the forms should be under the purview of the professional. His/her client is the beneficiary of the project and has deemed it necessary that someone administer ("control") the project through the construction contract phase. The charge, then, is obvious.

The professional must be flexible enough to vary the procedures, sequences, forms, lines of communications, and all of the other administration elements to the

	Dates	Initials	Item	Remarks
18.			Prepare Owner/Architect Agreement form and submit to Owner. (AIA Doc. B141, B141/CM, B151, B161/162, B171, B181 or B727) Have legal counsel review any modifications or specific provisions required by Owner.	
19.			Review final agreement with your legal counsel. Approve any modifications made by Owner, or renegotiate.	
20.			Verify authorization of the party signing for Owner (required for public agency, institutional and corporate clients).	
21.			Complete execution of the Agreement.	
22.			Submit initial payment statement to Owner. (AIA Doc. F5002)	
23.			Submit information required by the Agreement.	
24.			Assign project number.	

FIGURE 7.1 *(Continued).*

nuances of the project. These will vary with the other contractual arrangements, with the size of the project, and often with the schedule of the project's work. Thus, no "stock" plan of attack or array of documents will meet every need. No such instruments should be forced upon a project; remedy must match need.

It may seem that an active, aggressive program of documentation on a construction site, or within the project process overall, is a defensive operation. However, in the context of modern-day construction, it is just the opposite. A properly conducted program

Chapter 7 Forms for the Process

	Dates	Initials	Item	Remarks
39.			Request and receive from each consultant proof of professional liability insurance coverage.	
40.			Negotiate, prepare and execute consultants' agreements. When required, obtain Owner's approval of consultants. (AIA C-Series Documents)	
41.			Distribute copies of pertinent portions of Owner/Architect Agreement to staff and consultants who require copies.	
42.			Distribute copies of pertinent portions of consultants' agreements to staff members who require copies.	
43.			Obtain land survey from Owner. When necessary, assist Owner in securing survey. Request from owner any information required from surveyor. (AIA Doc. G601)	
44.			Obtain from appropriate consultants requirements for investigations and tests including soil borings, test pits, percolation tests, soil boring values, etc. necessary for proper execution of their work and request such information from Owner. Assist Owner in securing proposals for this work. (AIA Doc. G602)	
45.			Have mechanical, electrical, structural and other consultants review site information.	
46.			Have appropriate staff members and consultants examine the site.	
47.			Using Owner's statement of space needs and program requirements of the project, determine tentative space and volume requirements and obtain Owner's written approval of them and written authority to proceed. (AIA Doc. D101)	

AIA DOCUMENT D200 • PROJECT CHECKLIST • AUGUST 1982 EDITION • AIA® • ©1982 • THE AMERICAN INSTITUTE OF ARCHITECTS, 1735 NEW YORK AVE., N.W., WASHINGTON, D.C. 20006

WARNING: Unlicensed photocopying violates U.S. copyright laws and is subject to legal prosecution.

FIGURE 7.1   *(Continued).*

is prudent and businesslike. Complete and accurate record-keeping and documentation is wise on the part of all concerned. Memories can fail or be distorted and, too often, verbal instructions or agreements are misunderstood or lost through repeated recitation or interpretation. The written word usually survives, if properly handled and stored (including, of course, computer programs carefully protected from erasure).

It is somewhat sad that verbal instructions and agreements (including "gentlemen's agreements") are no longer viable. In bygone days, professionals and the contractors

Dates	Initials	Item	Remarks
48.		If directed, as an Additional Service or provided as a Designated Service, prepare facility program and reconcile it with owner's budget. Advise Owner if budget and program are not compatible. Submit facility program to Owner, review it with the Owner and obtain written authority to proceed.	

Part C—Tasks to be started after completion of Part B.

Dates	Initials	Item	Remarks
49.		Review all data furnished, including program, budget, legal, site, code, space and special owner requirements. Record the source of all data. Be sure the design complies with all requirements.	
50.		Prepare functional space diagrams.	
51.		Provide engineers and consultants with pertinent program data and functional space diagrams.	
52.		Confer with engineers and consultants to determine systems to be used in the project and obtain from them analyses of comparative systems, with recommendations. Upon selection of a system, obtain their space and location requirements. ☐ Structural ☐ Mechanical ☐ Electrical ☐ Other consultants.	
53.		Prepare basic design documents to include: ☐ Site plan with diagrammatic indications showing relationships ☐ Vertical sections through the site, if required ☐ Principal floor plans ☐ General descriptive views ☐ Illustrative sketches, models or renderings, if required.	
54.		Calculate areas and volumes, and analyze plan efficiency of the design by usable area, area per person or other method. (AIA Doc. D101)	
55.		Prepare general description of the project, and construction and equipment outlines.	

AIA DOCUMENT D200 • PROJECT CHECKLIST • AUGUST 1982 EDITION • AIA® • ©1982 • THE AMERICAN INSTITUTE OF ARCHITECTS, 1735 NEW YORK AVE., N.W., WASHINGTON, D.C. 20006

WARNING: Unlicensed photocopying violates U.S. copyright laws and is subject to legal prosecution.

FIGURE 7.1 *(Continued).*

operated in a closer and far less complex atmosphere, where personal relationships, pride, and care prevailed. Today, projects are much more complex and so too the responsibilities and legal obligations of the parties. This has caused a regrettable separation that requires all parties to interface in a more "mechanical" manner.

Of course, the growing use of computers on the job site, particularly by construction managers and contract administrators, is a great aid to documentation. They use either more permanent PCs or laptop versions. Professionals, more than likely, will opt for

Dates	Initials	Item	Remarks
56.		From each consultant obtain and review statements of probable construction cost.	
57.		Prepare written statement of probable construction cost based on all available data. Include appropriate contingency to cover future development of project.	
58.		Submit schematic design documents (drawings, descriptions, calculations, outlines and statements of probable construction cost) to Owner.	
59.		Obtain Owner's written approval of schematic design documents.	
60.		Obtain Owner's written authority to proceed to the design development phase. (HBC D-1)	
61.		Submit end of phase statement to Owner for payment. (AIA Doc. F5002)	

FIGURE 7.1 *(Continued).*

laptops when they are rotating among several projects at a time. A PC type of setup into which the laptop can be docked (connected) combines the best of both systems.

Contract administrators can easily find various software packages, which can be adapted to the work required. Many programs can also be adjusted and new, specific, individualized forms, spreadsheets, and programs can be installed. These programs should be coordinated with the same items utilized in the office, so interconnection and communication is easier.

DESIGN DEVELOPMENT
Refer to AIA Handbook Chapter B-4 and B-Series Documents

Part A—Ongoing or periodic tasks accomplished throughout this phase.

	Dates	Initials	Item	Remarks
62.			Have Owner review the documents and render prompt decisions as the need arises. (HBC D-1)	
63.			Periodically review internal office budgets and production schedule and check against actual progress.	
64.			Review and update schedule of completion dates for this and all subsequent phases.	
65.			Review and update personnel time and production cost projections.	
66.			Adjust number and type of personnel as required.	
67.			Update project record book.	
68.			Update project directory. (AIA Doc. G807)	
69.			Update project data. (AIA Doc. G809)	
70.			Review and update project flow chart.	
71.			As documents develop, confer with and obtain preliminary review from regulatory agencies such as: ☐ Building department ☐ Fire marshal (state & local) ☐ Dept. of Health ☐ Dept. of Education ☐ Zoning commission ☐ Planning commission ☐ Design review board ☐	
72.			Maintain expense accounting records.	
73.			Submit monthly or periodic statements to Owner for payment. (AIA Doc. F5002)	
74.			Submit monthly or periodic reimbursable expense statements to Owner for payment. (AIA Doc. F5002)	

AIA DOCUMENT D200 • PROJECT CHECKLIST • AUGUST 1982 EDITION • AIA® • ©1982 • THE AMERICAN INSTITUTE OF ARCHITECTS, 1735 NEW YORK AVE., N.W., WASHINGTON, D.C. 20006

WARNING: Unlicensed photocopying violates U.S. copyright laws and is subject to legal prosecution.

FIGURE 7.1 *(Continued).*

Beyond the obvious administrative functions of computer use—record keeping, making notes and memos, writing agendas and minutes, preparing forms, sending e-mail messages—a new aspect is coming into its own. This is the use of digital photography.

Many larger projects require in-progress photos. In the past, a professional photographer would make periodic visits to the site and take still photos, normally 8″ × 10″; usually this coincided with the monthly payment request (to document progress). Now continuous video monitoring is often done. Photos are taken on a

Chapter 7 Forms for the Process

Dates	Initials	Item	Remarks
		Part B—Tasks prior to starting design development.	
75.		Review the schematic design checklist to assure yourself that it has been completed and that all required data has been obtained.	
76.		Have Owner delineate any additional or special requirements.	
77.		If models, perspectives or renderings will be required, obtain Owner's approval of expenditure for them. Take appropriate steps to expedite completion at the proper time.	
78.		Review the program and verify compliance.	
79.		Re-check schematic documents against all codes and regulations.	
80.		Select additional consultants, if required, and establish contractual relationships. Obtain Owner's approval when required. (HBC D-2)	
81.		Receive results of all investigations and tests, including soil borings and analysis. Request additional information, if necessary. Forward final information to appropriate consultants. (AIA Doc. G602)	
82.		Review all other data received from the Owner, consultants, etc. Request additional data if necessary.	
83.		Obtain Owner's standards and requirements, if any, for drawings and for other materials.	
84.		Develop and forward to the consultants a checklist of systems required such as: ☐ clock ☐ paging ☐ intercom ☐ building protection ☐ lighting ☐ telephone ☐ cable TV ☐ closed circuit TV ☐ oxygen ☐ vacuum ☐ compressed air ☐ distilled water	

AIA DOCUMENT D200 • PROJECT CHECKLIST • AUGUST 1982 EDITION • AIA® • ©1982 • THE AMERICAN INSTITUTE OF ARCHITECTS, 1735 NEW YORK AVE., N.W., WASHINGTON, D.C. 20006
WARNING: Unlicensed photocopying violates U.S. copyright laws and is subject to legal prosecution.

FIGURE 7.1 *(Continued).*

timed basis (every fifteen minutes, for example) to note progress and procedures. These photos prove quite handy where there is a dispute or litigation over a claim for money or injury. The pictures can reflect activity twenty-four hours per day (if necessary), and the equipment can be computerized to tilt, zoom, pan, and so on in order to distinctly record crucial elements or operations. (This is also widely used for security systems.) The data are then transmitted to the company's server via an

Dates	Initials	Item	Remarks
		☐ steam	
		☐ gas	
		☐ pneumatic tube	
		☐ other	
		or obtain systems checklist from consultants. Secure Owner's approval of the list and notify consultants of approval or revisions.	
85.		Define actual occupancy density for each special area and forward to consultants. Check against program.	
86.		Have the structural engineers investigate and confirm in writing their review of applicable regulations. (HBC D-2)	
87.		Have the mechanical and electrical engineers: ☐ Contact utility companies and public authorities for all services. Request and receive written approval for all service connections. ☐ Investigate and confirm to you in writing their review of all applicable local public and utility regulations. ☐ Review architectural and structural schematic drawings to establish adequate provision for specialized systems. ☐ Prepare estimates of probable operating costs, with recommendations.	
88.		Review engineer's estimates of probable operating cost and forward to Owner. Obtain Owner's approval of the selected fuel source.	

Part C—Tasks during design development phase.

Dates	Initials	Item	Remarks
89.		Prepare site plan indicating building location(s) and site improvements.	
90.		Prepare all other necessary documents to include: plans, elevations, sections, schedules and notes as required to fix and describe the project.	
91.		Prepare area calculations (net and gross) and volume calculations. (AIA Doc. D101)	

AIA DOCUMENT D200 • PROJECT CHECKLIST • AUGUST 1982 EDITION • AIA® • ©1982 • THE AMERICAN INSTITUTE OF ARCHITECTS, 1735 NEW YORK AVE., N.W., WASHINGTON, D.C. 20006

WARNING: Unlicensed photocopying violates U.S. copyright laws and is subject to legal prosecution.

FIGURE 7.1 *(Continued).*

ISDN line and unloaded to the company's Web site for viewing by all interested parties that have compatible equipment.

These photos can also be digital. They can be drawn from the Web by satellite, wireless, ISDN, or plain old telephone connections. Professionals on the site, using PCs and PDAs, can assess the data for immediate use.

Perhaps more important to contract administrators are the newer (and still evolving) hand-held cameras that can take a picture and transmit it almost immediately to another person. This can be done to illustrate a problem without having to use

Chapter 7 Forms for the Process 211

	Dates	Initials	Item	Remarks
92.			Prepare preliminary draft of the Project Manual. Have consultants prepare their portions and coordinate. Use MASTERSPEC® Table of Contents as checklist. (HBC B-6)	
93.			Have engineers and consultants prepare layouts and drawings as required to illustrate and describe their portion of the project. (HBC D-2) ☐ Structural ☐ Mechanical ☐ Electrical ☐	
94.			Have each engineer and consultant provide a Statement of Probable Construction Cost for their respective portion of the project.	
95.			Obtain detailed cost estimate, if specifically authorized by Owner. (HBC B-5)	
96.			Prepare a further Statement of Probable Construction Cost.	
97.			Submit design development documents (drawings, calculations, preliminary draft of the Project Manual and further Statement of Probable Construction Cost) to Owner.	
98.			Confirm with Owner whether single or separate contract system will be used.	
99.			Obtain Owner's written approval of design development documents.	
100.			Obtain Owner's written authorization to proceed to the construction documents phase.	
101.			Submit end of phase statement to Owner for payment. (AIA Doc. F5002)	

AIA DOCUMENT D200 • PROJECT CHECKLIST • AUGUST 1982 EDITION • AIA® • ©1982 • THE AMERICAN INSTITUTE OF ARCHITECTS, 1735 NEW YORK AVE., N.W., WASHINGTON, D.C. 20006

WARNING: Unlicensed photocopying violates U.S. copyright laws and is subject to legal prosecution.

FIGURE 7.1 *(Continued).*

wordy and inexplicit methods of communications. In addition, if properly equipped, the receiver can transmit the solution to the problem back to the field.

There are many types of cameras from many different manufacturers. They are reasonably priced in a range from just under $1,000 to around $1,800. Picture capacity runs from 16 to 48, and many optical options can be added if desired. An audio recording feature, for example, allows verbal digital notation directly associated with the picture that is transmitted via the computer system microphone straight to a file on the disk. Additionally, when the digital picture is processed through the com-

CONSTRUCTION DOCUMENTS
Refer to AIA Handbook Chapters B-4 and B-6 and B-Series Documents

Part A—Ongoing or periodic tasks accomplished throughout this phase.

	Dates	Initials	Item	Remarks
102.			Have Owner review documents and render prompt decisions as the need arises.	
103.			Review periodically internal office budget and production schedule, and check against actual progress.	
104.			Review and update schedule of completion dates for this and all subsequent phases. Inform the project team and Owner of any revisions.	
105.			Review and update personnel time and production cost projections.	
106.			Adjust number and type of personnel as required.	
107.			Update project record book.	
108.			Update project directory. (AIA Doc. G807)	
109.			Update project data. (AIA Doc. G809)	
110.			Review and update project flow chart.	
111.			Identify all documents with project number and date.	
112.			As documents develop, confer with and obtain further review from regulatory agencies such as: ☐ Building department ☐ Fire marshal (state & local) ☐ Dept. of Health ☐ Dept. of Education ☐ Check with the applicable regulatory agencies and establish schedule for submission and/or review.	
113.			Coordinate the work of all members of the team, including consultants. Coordinate drawings with Project Manual.	

AIA DOCUMENT D200 • PROJECT CHECKLIST • AUGUST 1982 EDITION • AIA® • ©1982 • THE AMERICAN INSTITUTE OF ARCHITECTS, 1735 NEW YORK AVE., N.W., WASHINGTON, D.C. 20006

WARNING: Unlicensed photocopying violates U.S. copyright laws and is subject to legal prosecution.

FIGURE 7.1 *(Continued).*

puter software program, notes (in high contrast boxes or "bubble" areas), leader lines, dimensions, cropping, and so forth can be added directly to the picture to further explain the visual presentation. One must investigate the market to ascertain what is available to fulfill project needs, the budget, and future expectations.

Even without using computers, documentation can be enhanced by using old-fashioned photographic techniques. For example, laying a rule or tape (or other *readable* measuring device) next to the work to be pictured lets the viewer determine actual

Chapter 7   Forms for the Process

	Dates	Initials	Item	Remarks
114.			Continually update previous Statement of Probable Construction Cost and advise Owner of any changes.	
115.			Maintain expense accounting records.	
116.			Submit monthly or periodic statements to Owner for payment. (AIA Doc. F5002)	
117.			Submit monthly or periodic reimbursable expense statements to Owner for payment. (AIA Doc. F5002)	

Part B—Tasks prior to start of construction documents phase.

	Dates	Initials	Item	Remarks
118.			Review all previous phase checklists to assure yourself that they are complete and list all required data that has been obtained.	
119.			Review the program and verify compliance.	
120.			Re-check design development documents against codes and regulations.	
121.			Select additional consultants, if required, and establish contractual relationships. Obtain Owner's approval when required.	
122.			Determine the scope of the drawings, including a list of required drawings, their sequence, the information to appear on each sheet, the approximate number of sheets necessary and the sheet size. (Check requirements of the Owner and governing bodies.)	
123.			Develop title block format (check requirements of Owner, licensing laws and governing bodies).	
124.			If there are to be alternates, determine what they are.	
125.			If there are to be cash allowances, determine items for which they apply.	

AIA DOCUMENT D200 • PROJECT CHECKLIST • AUGUST 1982 EDITION • AIA® • ©1982 • THE AMERICAN INSTITUTE OF ARCHITECTS, 1735 NEW YORK AVE., N.W., WASHINGTON, D.C. 20006

WARNING: Unlicensed photocopying violates U.S. copyright laws and is subject to legal prosecution.

FIGURE 7.1   *(Continued).*

scale and size. All of the features are fully manipulable by the contract administrator and can be combined with imagination to make a distinct, specific, complete presentation or record document—and provide solutions to problems in a short amount of time, when necessary (a most valuable asset in contract administration).

The array of operations that takes place on a job site can lead to an overload of information, changes, revisions, and so on. It is far better to reduce everything possible to writing, even to written instructions of seeming consequence, and verification

	Dates	Initials	Item	Remarks
126.			Obtain Owner's instructions on insurance and bonds. (AIA Doc. G610)	
127.			Obtain Owner's instructions regarding construction agreement and bid procedures. (AIA Doc. G611)	
128.			Submit for Owner's review, copies of General Conditions, Supplementary Conditions and building criteria, or obtain Owner's specific contract requirements.	
129.			Determine what items, if any, are to be furnished by Owner, or are not to be included in the construction contract.	
130.			Obtain schedule for delivery and installation of Owner-furnished materials.	
131.			Assist Owner's legal counsel to determine the suitable type of construction contract.	

Part C — Tasks to be done during construction documents phase.

	Dates	Initials	Item	Remarks
132.			Prepare final drawings to include plans (including site and landscaping), elevations, sections, details, notes, dimensions and schedules, and have all consultants do the same. (HBC A-5)	
133.			Prepare and assemble Project Manual concurrently with preparation of the drawings. Obtain assistance from engineers and consultants where appropriate (HBC B-2, D-2 and D-5). Completed project manual will typically include the following:	
			A. If project will be bid, bidding requirements to include: ☐ Notice to bidders ☐ Advertisement or invitation to bid ☐ Instructions to Bidders (AIA Doc. A701) ☐ Sample bid form.	

AIA DOCUMENT D200 • PROJECT CHECKLIST • AUGUST 1982 EDITION • AIA® • ©1982 • THE AMERICAN INSTITUTE OF ARCHITECTS, 1735 NEW YORK AVE., N.W., WASHINGTON, D.C. 20006

WARNING: Unlicensed photocopying violates U.S. copyright laws and is subject to legal prosecution.

FIGURE 7.1 *(Continued).*

of telephone calls and face-to-face conversations (where they affect the work). Often, through the progress of a project, over months and even years of time, the information in the documents will prove to be invaluable. This does not necessarily mean only in some resulting litigation. For example, a change to the project work through an addendum or through a change order issued early on the project time line can easily be "lost" further downstream, when the work must be executed, if it is not preserved through written and/or graphic form. If the number of construction

Chapter 7   Forms for the Process

Dates	Initials	Item	Remarks
		B. Contract documents, other than drawings, to include:	
		☐ Agreement (AIA Doc. A101, A101/CM, A107, A111, A117, A171 or A177) Assist Owner's legal counsel to select the appropriate document for the project.	
		☐ General Conditions of the Contract for Construction (AIA Doc. A201, A201/CM or A271).	
		☐ Supplementary Conditions (HBC D-5, AIA Doc. A201/SC, A511, and A512).	
		☐ Specifications (MASTER-SPEC®) Have consultants prepare divisions or sections relating to their work.	
134.		Check all completed documents for coordination, compliance with program, accuracy and cross-coordination with the consultants and engineers' work, and have them make similar checks.	
135.		Revise documents as required after check and have consultants and engineers do the same.	
136.		Obtain from each consultant a further Statement of Probable Construction Cost for consultant's part of the project.	
137.		Prepare final Statement of Probable Construction Cost, or detailed cost estimate if specifically authorized by Owner. (HBC B-5)	
138.		Prepare testing and quality control program budgets, and assist Owner in selection of testing agency.	
139.		Prepare final calculations of net and gross area, and volume. (AIA Doc. D101)	
140.		Submit drawings, Project Manual, Statement of Probable Construction Cost or detailed cost estimate and calculations to Owner for review. Obtain Owner's written approval. (HBC D-1)	

FIGURE 7.1   (Continued).

materials, systems, devices, equipment, and the like can approach a half-million on even a moderately sized project, one can only imagine the number of pieces of information—valuable information—that will be produced.

The two dangerous threats toward any construction project are unnecessary claims and lengthy disputes. Both siphon time, effort, and money. To avoid these nemeses, there is a need to fully understand the pitfalls and the appropriate documentation practices involved. The key to the documentation is to ensure that (1)

Dates	Initials	Item	Remarks
141.		Review list of potential contractors with Owner.	
142.		Obtain qualification statements from interested bidders and review. If separate prime contracts are to be awarded, obtain assistance from consultants. (AIA Doc. A305 and A501)	
143.		Place Architect's and Engineer's seals on the documents and obtain any necessary signatures required by reviewing authorities.	
144.		Assist Owner in filing documents for approvals and permits.	
145.		Obtain Owner's written authorization to proceed to the bidding or negotiation phase. (HBC D-N)	
146.		Submit end of phase statement to Owner for payment. (AIA Doc. F5002)	

FIGURE 7.1 *(Continued).*

something is done and (2) whatever is done is effective. While forms and contract provisions can be relatively standard in format, their content and wording are crucial to the success of the program. Often well-intended action is rendered ineffective or aggravating merely because of faulty or inappropriate wording, the deletion of important requirements and details, or simply inadequate documentation.

The professional should ensure that all forms, notices, letters, and so on are of the highest quality as to legal parameters and technical content. Proper training of per-

Chapter 7 Forms for the Process

BIDDING OR NEGOTIATION
Refer to AIA Handbook Chapter B-4 and AIA Doc. A501.

Part A—Ongoing or periodic tasks accomplished throughout this phase.

	Dates	Initials	Item	Remarks
147.			Update project record book.	
148.			Review and update project flow chart.	
149.			Maintain expense accounting records.	
150.			Submit monthly or periodic statements to Owner for payment. (AIA Doc. F5002)	
151.			Submit monthly or periodic reimbursable expense statement to Owner for payment. (AIA Doc. F5002)	

Part B—Upon receipt of all necessary approvals from governing authorities and the Owner's written approval of the construction documents and written authorization to proceed to the bidding or negotiation phase, proceed as follows:

152.			For open bidding: Publish advertisement for bids (in some cases Owner may publish). If separate prime contracts are to be awarded, separate advertisements may be necessary.	
153.			Distribute invitation or advertisement to bid.	
154.			Obtain qualification statements from interested bidders and review. If separate prime contracts are to be awarded, obtain assistance from consultants. (AIA Doc. A305)	
155.			For competitive bidding, notify selected bidders.	
156.			For direct selection system of construction award, select a contractor for negotiation and obtain Owner's concurrence.	
157.			Prepare register of bid documents. (G804)	
158.			Distribute bidding documents to bidders and obtain deposits and bid security. (AIA Doc. A310)	
159.			Issue documents to plan rooms.	

AIA DOCUMENT D200 • PROJECT CHECKLIST • AUGUST 1982 EDITION • AIA® • ©1982 • THE AMERICAN INSTITUTE OF ARCHITECTS, 1735 NEW YORK AVE., N.W., WASHINGTON, D.C. 20006
WARNING: Unlicensed photocopying violates U.S. copyright laws and is subject to legal prosecution.

FIGURE 7.1 *(Continued).*

sonnel will ensure proper use of the various instruments at their disposal. The documentation must be practical, appropriate, well executed, properly distributed, and parallel to the course of the project.

Documentation should be made to follow the circumstances of the project. For example, it may be extremely helpful to keep a separate looseleaf binder in the job trailer that contains nothing but the drawings that result from change orders. Often, these do not filter down to the site workers. This is particularly true when the pro-

	Dates	Initials	Item	Remarks
160.			Hold pre-bid conference.	
161.			Prepare addenda record.	
162.			Prepare and issue addenda as necessary.	
163.			Return bid security to bidders who withdraw and return documents.	
164.			Return bid security to disqualified bidders.	
165.			Evaluate proposed substitutions.	
166.			Receive, tabulate and analyze bids.	
167.			Advise Owner to accept or reject bids, and obtain Owner's acceptance of one or rejection of all bids.	
168.			Notify successful bidder(s) of acceptance.	
169.			Update project directory.	
170.			Notify unsuccessful bidders and obtain return of bidding documents.	
171.			Return unsuccessful bidders' document deposits and bid security. (Hold bid security of lowest bidders until signing of contract.)	
172.			Advise Owner on selection of alternates, obtain Owner's approval and notify contractor(s).	
173.			Request and receive submission of post bid information. (AIA Doc. A701)	
174.			Assist Owner's legal counsel in preparation of construction contract(s). If separate prime contracts are to be awarded, obtain assistance of consultants.	
175.			If the construction contract cannot be executed immediately, the Owner may wish to issue letter(s) of intent. If so, provide Owner's legal counsel with technical advice. (HBC B-7 and D-5)	

AIA DOCUMENT D200 • PROJECT CHECKLIST • AUGUST 1982 EDITION • AIA® • ©1982 • THE AMERICAN INSTITUTE OF ARCHITECTS, 1735 NEW YORK AVE., N.W., WASHINGTON, D.C. 20006

WARNING: Unlicensed photocopying violates U.S. copyright laws and is subject to legal prosecution.

FIGURE 7.1 *(Continued)*.

ject is remote from the contractors' offices. For example, a major university project in Bowling Green (south-central), Kentucky, drew subcontractors from Lexington, Paducah, Henderson, Madisonville, and Louisville, Kentucky, as well as from Nashville, Tennessee; Cincinnati, Ohio; and Carmel (Indianapolis), Indiana. One major subsystem was provided by a firm from Oklahoma, which sent a traveling crew to the site for the work. It is difficult to manage and coordinate such a scenario, much less ensure that all of the pertinent information is available to the workers.

Chapter 7 Forms for the Process

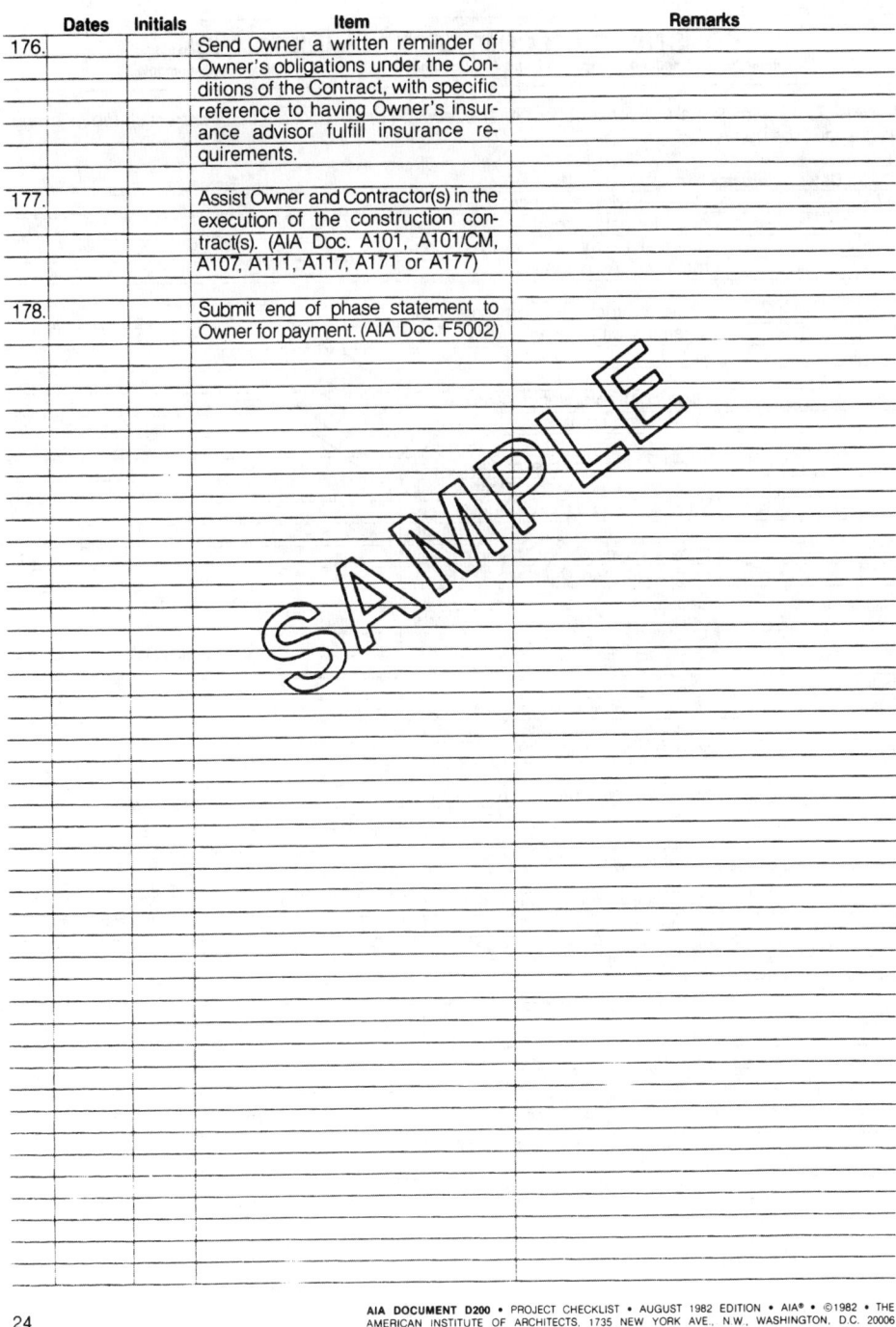

FIGURE 7.1 *(Continued).*

Often in similar situations, there is a pervasive need to upgrade even the best of communications systems and to expend extra effort to assure correct information and a smooth-running, well-timed, and coordinated project. However, this is merely an inherent part of the work of the CM *and* the contract administrator.

The basic responsibility for distributing this type of information lies with the contractors. In the general contractor configuration, this information would be more readily available on the site (in the GC's files). With a construction manager, however, the site

CONSTRUCTION CONTRACT ADMINISTRATION
Refer to AIA Handbook Chapter B-7 and AIA Documents A201, A511 and B-Series Documents

Part A — Tasks prior to start of construction. After execution of the construction contract(s), the following tasks should be completed.

	Dates	Initials	Item	Remarks
179.			Obtain, review and forward to Owner performance bond, and labor and material payment bond from Contractor. (AIA Doc. A311)	
180.			Have Contractor file certificate of insurance with owner. (AIA Doc. G705)	
181.			Have contractor secure and pay for all required permits.	
182.			Receive and approve or take other appropriate action on Contractor's list of subcontractors and notify Contractor of rejections. Obtain substitutes and approve or take other appropriate action. (AIA Doc. G805)	
183.			Obtain and review Contractor's schedule of required shop drawings and samples. (AIA Doc. G712)	
184.			Obtain, review and approve, if appropriate, Contractor's estimated progress schedule.	
185.			Furnish Contractor with required copies of contract documents.	
186.			Update project directory.	
187.			Have Owner submit applications for permanent gas, electric, water, telephone and other services, as required.	
188.			Have Owner file a copy of all property insurance policies with Contractor.	
189.			If Owner does not intend to purchase property insurance, have Owner notify contractor in writing. If Contractor elects to purchase such insurance, initiate appropriate change order. (AIA Doc. G701)	

AIA DOCUMENT D200 • PROJECT CHECKLIST • AUGUST 1982 EDITION • AIA® • ©1982 • THE AMERICAN INSTITUTE OF ARCHITECTS, 1735 NEW YORK AVE., N.W., WASHINGTON, D.C. 20006

WARNING: Unlicensed photocopying violates U.S. copyright laws and is subject to legal prosecution.

FIGURE 7.1 *(Continued).*

records will be complete, but could be much more complex and filled with material irrelevant to the site workers (cost data, for example). Retrieval could be difficult and cumbersome. Having a separate booklet for change order drawings both enhances retrieval and aids in getting the correct installation in the project. The CM is responsible for keeping all the files and this extra booklet will add little in the way of extra work.

Figure 7.2 shows specifications for the contract administrator's project site office. Note that among the other requirements, two telephone lines should be made avail-

Chapter 7 Forms for the Process

Dates	Initials	Item	Remarks
190.		If Contractor requests in writing that insurance for special hazards be included, at Contractor's expense, in the property insurance policy, have Owner purchase such insurance. Initiate change order. (AIA Doc. G701 and G610)	

Part B—Scheduled time tasks

191.		Keep owner informed of the progress of the work. Prepare field report for each visit to the site. (See checklist in HBC B-7 for scheduling visits to small projects.)	
192.		Obtain and review Contractor's updated progress schedule and advise Owner of potential revisions to anticipated occupancy date.	
193.		Prior to the first application for payment, receive, review and approve if appropriate Contractor's schedule of values. (HBC D-3)	
194.		Receive and review Contractor's applications for payment. Check against progress, retained percentage, potential claims, defective work, etc. (AIA Doc. G702, G703, G722 and G723)	
195.		At 50 percent completion, verify requirements for reduction in retainage. Where applicable have Contractor submit Consent of Surety. (AIA Doc. G707A)	
196.		If no grounds exist for withholding payment, issue certificates of payment to Owner, with copy to Contractor. (AIA Doc. G702)	
197.		Submit monthly or periodic statements of Architect to Owner for payment. (AIA Doc. F5002)	
198.		Submit monthly or periodic reimbursable expense statement of Architect to the Owner for approval. (AIA Doc. F5002)	

AIA DOCUMENT D200 • PROJECT CHECKLIST • AUGUST 1982 EDITION • AIA® • ©1982 • THE AMERICAN INSTITUTE OF ARCHITECTS, 1735 NEW YORK AVE., N.W., WASHINGTON, D.C. 20006

WARNING: Unlicensed photocopying violates U.S. copyright laws and is subject to legal prosecution.

FIGURE 7.1   (Continued).

able. One should be a regular telephone with an answering machine (if the on-site office will not be monitored continually). The other should be a fax line. The best fax service would include a function in the machine to verify a successful transmission.

Of course, a professional organization that engages in contract administration on a regular basis will have a "standardized" format for this service. Through experience, procedures, specification provisions, other contractual provisions, and the various forms can also be standardized. In a sense, the project is processed within the

Dates	Initials	Item	Remarks
		Part C—Supplemental unscheduled tasks	
199.		Obtain required test reports and review. Get assistance from consultants as appropriate.	
200.		Receive manufacturer's data, respond and return to Contractor, if submission is not satisfactory.	
201.		Receive shop drawings. Review, approve and return to Contractor without delay. Require resubmission, if necessary.	
202.		Receive samples. Review, approve and return to contractor without delay. Require resubmission, if necessary.	
203.		Maintain shop drawings and sample record. (AIA Doc. G712)	
204.		Obtain color selections and issue to Contractor.	
205.		Issue detail or supplemental drawings as required.	
206.		Visit site and have consultants visit site to observe specific events as conditions warrant. Make written field report of each visit. (AIA Doc. G711)	
207.		Review Contractor's proposals for changes and negotiate as required. Prepare change orders as necessary, obtain Owner's signature and forward to Contractor. (AIA Doc. G701)	
208.		Prepare and issue supplemental instructions as appropriate. (AIA Doc. G710, G713)	
209.		Update project record book.	
210.		Update project data. (AIA Doc. G809)	
211.		Maintain expense accounting records.	

AIA DOCUMENT D200 • PROJECT CHECKLIST • AUGUST 1982 EDITION • AIA® • ©1982 • THE AMERICAN INSTITUTE OF ARCHITECTS, 1735 NEW YORK AVE., N.W., WASHINGTON, D.C. 20006

WARNING: Unlicensed photocopying violates U.S. copyright laws and is subject to legal prosecution.

FIGURE 7.1 *(Continued).*

general context of the professional's administration program. This is not, however, rigid and dictatorial, but one that must be flexible to suit a specific project.

In creating and working within this program the contract administrator should adhere to some basic principles. These will serve the project but will not elevate the professional's liability. Some helpful hints for the contract administrator about document/record keeping on the construction job site:

Chapter 7 Forms for the Process 223

Dates	Initials	Item	Remarks

Part D—Project termination tasks—When the construction is substantially completed, the following tasks must be performed:

212.		Receive notification of substantial completion and list of items to be completed or corrected from Contractor. (AIA Doc. G704)	
213.		Inspect the project for substantial completion. Notify governmental authorities who require inspection before occupancy.	
214.		When project is substantially complete, prepare a certificate of substantial completion. (AIA Doc. G704). Obtain Owner's and Contractor's written acceptance and approval.	
215.		If certificate of occupancy or occupancy permit is required, assist the Owner in obtaining it.	
216.		Obtain from Contractor: ☐ warranties ☐ certificates of inspection ☐ schedules ☐ operating instructions ☐ keying schedule ☐ maintenance stock ☐ record drawings ☐ bonds ☐	
217.		Receive Contractor's written notice that all work has been completed.	
218.		Make final inspection of the project.	
219.		Receive from Contractor the final application for payment along with: ☐ release of liens ☐ consent of surety, if any and verify that all other conditions of the contract have been met. (AIA Doc. G702, G706, G706A and G707)	
220.		Issue final certificate for payment of Contractor. (AIA Doc. G702)	
221.		Submit final statement of Architect to Owner for payment. (AIA Doc. F5002)	

AIA DOCUMENT D200 • PROJECT CHECKLIST • AUGUST 1982 EDITION • AIA® • ©1982 • THE AMERICAN INSTITUTE OF ARCHITECTS, 1735 NEW YORK AVE., N.W., WASHINGTON, D.C. 20006

WARNING: Unlicensed photocopying violates U.S. copyright laws and is subject to legal prosecution.

FIGURE 7.1 *(Continued).*

- Even though you are not responsible for them, ensure that all agreements/contracts are in proper written form and contain all pertinent provisions, details, and approvals.
- Never write something that will impose on you any responsibility for work to be done by others.

Dates	Initials	Item	Remarks
222.		Submit final reimbursable expense statement of architect to the Owner for payment. (AIA Doc. F5002)	
223.		Assemble, analyze and file for future reference complete cost records for both construction and professional services.	
224.		If defects become evident during the one year period after completion, obtain authorization from Owner, as an Additional Service, to investigate them.	
225.		Prior to expiration of the one-year period, obtain Owner's authorization, as an Additional Service, to conduct a thorough inspection to determine if any work is required by Contractor to remedy defects.	

AIA DOCUMENT D200 • PROJECT CHECKLIST • AUGUST 1982 EDITION • AIA® • ©1982 • THE AMERICAN INSTITUTE OF ARCHITECTS, 1735 NEW YORK AVE., N.W., WASHINGTON, D.C. 20006

WARNING: Unlicensed photocopying violates U.S. copyright laws and is subject to legal prosecution.

FIGURE 7.1 *(Continued).*

- Include nothing in writing that indicates that you have any authority unless you are properly charged or under contract to exercise such authority. Be extremely careful of "authority by innuendo, implication, or assumption."
- Ensure that all specifications are job-pertinent, and no data or forms are included in the project manual that are not directly pertinent to the project at hand.
- Be sure that all required submittals, guarantees, warranties, and other representations regarding untried or substitute materials and methods are in hand, in writing.

Chapter 7 Forms for the Process 225

Dates	Initials	Item	Remarks
231.		Start-up services:	
		☐ Observe and assist in the operation of building systems during initial occupancy.	
		☐ Assist in the training of the Owner's personnel in proper operations, maintenance schedules and procedures.	
		☐ Administer and coordinate the remedial work of the Contractor(s) after final completion.	
232.		Record drawings services:	
		☐ Obtain from Contractor(s) and other parties certified information on all changes in location of concealed systems installed differently than those locations shown on the initial contract documents.	
		☐ Review for general accuracy the information submitted and certified by the Contractor(s).	
		☐ Prepare record drawings, based on information furnished by the Contractor(s), including significant changes in the work made during construction.	
		☐ Transmit record drawings and general data, appropriately identified, to the Owner and others as directed.	
233.		Warranty review services:	
		☐ Consult with Owner during the warranty periods and make recommendations concerning inadequate performance of materials, systems and equipment under warranty.	
		☐ Inspect materials, systems and equipment prior to expiration of the warranty period(s) to ascertain adequacy of performance.	
		☐ Document defects or deficiencies, and assist Owner in administering corrective action by the Contractor(s).	

AIA DOCUMENT D200 • PROJECT CHECKLIST • AUGUST 1982 EDITION • AIA® • ©1982 • THE AMERICAN INSTITUTE OF ARCHITECTS, 1735 NEW YORK AVE., N.W., WASHINGTON, D.C. 20006

WARNING: Unlicensed photocopying violates U.S. copyright laws and is subject to legal prosecution.

FIGURE 7.1 (Continued).

- Present (through proper channels) all major items that require client approval, in writing (with all ancillary data), and obtain that approval in writing, also.
- Secure, in writing, any decision by the owner regarding not requiring surety or other safeguards.
- Document all transmittal of documents, drawings, and other information to consultants and others, in writing.
- Written notices are required in instances where there may be increased cost. Include reasons for the increase and obtain the owner's acceptance in return.

Dates	Initials	Item	Remarks
234.		Evaluation services consisting of a project inspection at least one year after completion of construction; interviewing the appropriate supervisory, operating and maintenance personnel; and analysis of operating costs and related data for evaluation of:	
		☐ The project program versus actual use	
		☐ The functional effectiveness of planned spaces	
		☐ The operational effectiveness of systems and materials	
		☐ Efficiency of the design and construction delivery process.	

FIGURE 7.1 *(Continued).*
Used with permission of the American Institute of Architects.

- Keep records of all visits to the job site (or a daily report, if full-time). Include day, time, weather, and work in progress. Using a standard form is highly recommended.
- Document in writing, with appropriate detail, any rejection of work when more than minimal cost is involved.
- When previously approved work is subsequently rejected, ensure same is done in writing.

Chapter 7 Forms for the Process

```
FIELD OFFICES

Provide insulated, weathertight temporary offices of sufficient
size and number to accommodate required office personnel at the
project site. Keep offices clean and orderly on a daily basis.

    Furnish and maintain a separate field office for exclusive
    use of the architect's field (Project) representative.

    Provide, including all usage costs, adequate lighting,
    heating, air-conditioning, telephone and fax service
    (separate lines), and copy machine.

    Furnish office with a desk, 2 chairs, a 4-drawer file cabi-
    net with lock, plan table, plan rack, and 6-shelf bookcase.

    Equip with water cooler and private toilet complete with
    water closet, lavatory, and mirror-medicine cabinet.

TEMPORARY FACILITIES                                       01500-6
```

FIGURE 7.2 A specifications excerpt delineating the requirements for the facilities to be provided for the project representative. It can be modified, but the items noted are essential needs.

- All recommendations of serious consequence should be in writing and should receive a written response.
- Distribute written documentation of any fact/condition that may call for changes in drawings, specifications, and/or cost to all concerned parties.
- Always document, in written form to the contractor, your opinion regarding all work or progress that is not in compliance with the contract, and which is essential for substantial completion.
- Where interests are adversely affected by improper work or development, inform all involved parties in writing.
- Immediately after completion, keep written accounts of telephone calls, conversations, or meetings (formal or informal) in which information of more than passing interest is transmitted. Ensure that all decisions made during such activities are recorded in writing.
- In recording meeting minutes, be sure to include which party is responsible for each matter discussed. Be conclusive and press for that during the progress of the meeting. Ensure clarity on all issues and with all parties.
- Always write in a manner that is (1) direct, clear and concise—not open to misinterpretation, (2) easily understood by nonconstruction or other professional persons, and (3) able to withstand scrutiny by hostile persons, judges, juries, and/or laypersons.
- Maintain all written materials in good order, easily retrievable, no matter the volume or extent.

Think when you "put it into writing"! You may feel that all the paperwork is both a nuisance and unnecessary, but the contrary is true. It is an intrinsic part of contract administration. The keeping of appropriate records (even if they reflect a mistake on your part) will not be considered negligence. In a lawsuit, your records may be the sole recollections with any veracity (due to misunderstandings, lack of recall, "convenient" short memories, etc.), and may speak loudest and clearest on an issue, to the point of successful support of your position.

Recollections based on records contemporaneous with the event have substantial impact with judges and juries, especially when they are trying to ascertain who is being truthful. The design professional's credibility is greatly enhanced when such records are produced and shown to be part of routine practice, not hastily produced under "emergency conditions." Quite often this scenario is the solid base used in finding the professional without liability.

Timing in some cases is even more important. Recording incidents as they occur and on the dates they occur often initiates the period of the statute of limitations, and can also be the basis for successful defense (by the professional) against some claims. All project records, then, should be retained at least for the length of the statute of limitations. In the case of a problem-ridden project, one is well advised to maintain the records indefinitely.

In some instances, *not writing* is the prudent course. For example:

- Make no references to personalities. Document the facts objectively and against contract provisions. Do not defame or debase others. Record that work "is not in compliance with Specifications Section XXX," not "the contractor is doing lousy work."
- Record facts! Report; do not editorialize or judge. Do not enter conclusions or opinions about causes of incidents, accidents, etc.
- Communicate with third parties only as required by the professional services contract.

The remainder of this chapter is devoted to sample forms (Figures 7.3 through 7.30). Considerable time can be saved by utilizing such forms for recording and conveying information. First, they provide a uniform format, and second, they guide proper content, ensuring that the pertinent data are included. They are presented as suggestions and come from various sources; they are not coordinated into a system of forms that can be adopted and used directly. For the sake of continuity and uniformity, however, it is wise to establish a set of forms, similar in styling, with common information and interrelationships properly displayed. Why not use professionalism here, too, and show a businesslike image? It reveals just how the entire project is being administered!

Other forms may become necessary as the project progresses. Some forms and procedures can be found in the educational programs noted in Chapter 6. In addition to the forms included here, the Construction Specifications Institute (CSI) has a large array of contract administration forms in its *Manual of Practice* (the Construction Contract Administration Module).

EXERCISES

The following tasks are based on the concepts and information in the chapter. All of the material or information required to complete the tasks may not be contained in the chapter text. Independent research is required. Consult your instructor for guidance.

1. During a monthly meeting "walk-around," your boss verbally tells you to do several things (direct changes in the work). Some involve no added cost or time; others add cost. How do you proceed? How do you document these changes? Who receives copies of the documents, if any?
2. Discuss the value of meeting notes in the format illustrated in Figure 7–7. Why is this a preferred format?
3. A worker stops you and ask a question about the work being done. You don't know the answer. What reply should you give? When you get the answer, do you convey it directly back to the worker?
4. What is the value of writing out a form capturing the topics discussed during a phone conversation? To whom do you distribute such a completed form?

LETTER OF TRANSMITTAL
(Documents)

Easton/Reed, Associates, Inc.: Architects & Engineers

| DATE: | | REV.: |

TO: _____ DATE: _____ TRANSMITTAL NO.: _____
    _____ LOCATION: _____
    _____ COMMISSION NUMBER: _____
    _____ PREPARED BY: _____

[ ] Enclosed
[ ] Under Separate Cover
[ ] Handed

[ ] Prints
[ ] Reproducibles
[ ] Supplier Drawings
[ ] Specifications
[ ] Purchase Requisitions
[ ] _____
FOR:
[ ] Design
[ ] Construction
[ ] Your Use
[ ] Review
[ ] Approval
[ ] _____
[ ] Approved
[ ] Approved as Noted
[ ] Revise and Resubmit
[ ] Not Approved
[ ] Return to Sender
[ ] Certified
[ ] _____
BY:
[ ] Parcel Post
[ ] First Class Mail
[ ] Special Delivery
[ ] Air Mail
[ ] Express
[ ] Bus
[ ] Interoffice Mail
[ ] _____

DISTRIBUTION

WE ARE SENDING:

NO. EACH	REVISION NO.	REVISION DATE	DOCUMENT NUMBER	DESCRIPTION

COMMENTS

FIGURE 7.3 Letter of transmittal.

LETTER OF TRANSMITTAL

Emerson/Tarrs, Architects, Inc.

PROJECT:	PROJECT NO:
	DATE:

TO:

ATTN:

If checked below, please:
( ) Acknowledge receipt of enclosures.
( ) Return enclosures to us.

WE TRANSMIT
    ( ) herewith   ( ) under separate cover via _____
    ( ) in accordance with your request
FOR YOUR
    ( ) approval   ( ) distribution to parties   ( ) information
    ( ) review & comment   ( ) record
    ( ) use   ( ) _____
THE FOLLOWING
    ( ) Drawings   ( ) Shop Drawing Prints   ( ) Samples
    ( ) Specifications   ( ) Shop Drawing Reproducibles   ( ) Product Literature
    ( ) Change Order   ( ) _____

COPIES	DATE	REV. NO.	DESCRIPTION	ACTION CODE

ACTION CODE
    A. Action indicated on item transmitted
    B. No action required
    C. For signature and return to this office
    D. For signature and forwarding as noted below under REMARKS
    E. See REMARKS below

REMARKS _____

COPIES TO: enclosures

FIGURE 7.4 Letter of transmittal.

Facsimile Cover Sheet

To: _____
Company: _____
Phone: _____
Fax: _____

From: John Q. Doe
Company: XYZ architects/planners
Phone: (100) 555- 0000
Fax: (100) 555- 0001

Date: _____
Pages including this cover page: _____

Comments:

FIGURE 7.5 Fax transmittal cover sheet.

FIGURE 7.6 Meeting record form. CSI 8.0A.
Construction Specifications Institute.

FIGURE 7.7 Project memorandum form. CSI 6.0A.
Construction Specifications Institute.

File # ______

JOHNSON ROMANOWITZ ARCHITECTS
222 EAST VINE STREET LEXINGTON, KY 40507 (606) 252-6781

Name ______ Job Name ______
Company ______ Job Architect ______
Phone # ______ Job # ______ Date ______ Time ______

ACTION REQUEST

ATTENTION:

REQUEST/MESSAGE

ACTION/REPLY

Date ______

ADDITIONAL INFORMATION REQUIRED

Date ______

FIGURE 7.8 Action request form.

ARCHITECT'S FIELD REPORT

AIA DOCUMENT G711

OWNER ☐
ARCHITECT ☐
CONSULTANT ☐
FIELD ☐

PROJECT: FIELD REPORT NO:
CONTRACT: ARCHITECT'S PROJECT NO:

DATE   TIME   WEATHER   TEMP. RANGE
EST. % OF COMPLETION   CONFORMANCE WITH SCHEDULE (+, −)
WORK IN PROGRESS   PRESENT AT SITE

OBSERVATIONS

ITEMS TO VERIFY

INFORMATION OR ACTION REQUIRED

ATTACHMENTS

REPORT BY:

AIA DOCUMENT G711 • ARCHITECT'S FIELD REPORT • OCTOBER 1972 EDITION • AIA® • © 1972
THE AMERICAN INSTITUTE OF ARCHITECTS, 1735 NEW YORK AVE., NW, WASHINGTON, D.C. 20006   page   of   pages

FIGURE 7.9 Architect's field report. AIA document G711.
Used with permission of the American Institute of Architects.

FIGURE 7.10 Field inspection report. CSI 9.1A.
Construction Specifications Institute.

FIGURE 7.11 Contractor's or inspector's daily report form. CSI 9.3A. Construction Specifications Institute.

SUBSTITUTION REQUEST

Project: _____   Substitution Request Number: _____

_____   From: _____

To: _____   Date: _____

_____   A/E Project Number: _____

RE: _____   Contract For: _____

Specification Title: _____
Section: _____   Page: _____   Article/Paragraph: _____

Proposed Substitution: _____
Manufacturer: _____   Address: _____   Phone #: _____
Trade Name: _____   Model #: _____
Installer: _____   Address: _____   Phone #: _____
History: ☐ New product ☐ 2-5 years old ☐ 5-10 yrs old ☐ More than 10 years old
Differences between proposed substitution and specified product:
_____
_____

☐ Point-by-point comparative data attached - REQUIRED BY A/E

Reason for not providing specified item: _____
_____

Similar Installation:
Project: _____   Architect: _____
Address: _____   Owner: _____
_____   Date Installed: _____
Proposed substitution affects other parts of Work: ☐ No ☐ Yes; explain _____
_____

Savings to Owner for accepting substitution: _____ ($_____)
Proposed substitution changes Contract Time: ☐ No ☐ Yes; Add/Deduct _____ days.
Supporting Data Attached:
☐ Product Data ☐ Drawings ☐ Tests ☐ Reports ☐ Samples ☐ _____

SAMPLE

Page 1 of 2

July 1994
CSI Form 13.1A

FIGURE 7.12 Substitution request form. CSI 13.1A. Construction Specifications Institute.

FIGURE 7.13 Phone conversation memo. CSI 7.0A.
Construction Specifications Institute.

FIGURE 7.14   Feedback form. CSI 16.0A.
Construction Specifications Institute.

Advancement of Construction Technology

FIELD ORDER

Project: _____   Field Order Number: _____
_____   From: _____
To: _____   Date: _____
_____   A/E Project Number: _____
RE: _____   Contract For: _____

You are hereby directed to execute promptly this Field Order which interprets the Contract Documents or orders minor changes in the Work without change in Contract Sum or Contract Time.

If you consider that a change in Contract Sum or Contract Time is required, submit Change Order Request to the A/E immediately and prior to proceeding with this Work.

Description of Interpretation or Change: RE: Spec Section _____   Drawing _____

SAMPLE

☐ Attachments

Signed by: _____

Copies: ☐ Owner   ☐ Consultants   ☐ _____   ☐ _____   ☐ _____   ☐ _____   ☐ File

Page 1 of __

July 1994
CSI Form 13.4A

FIGURE 7.15   Architect's supplemental instructions form. CSI 13.4A. Construction Specifications Institute.

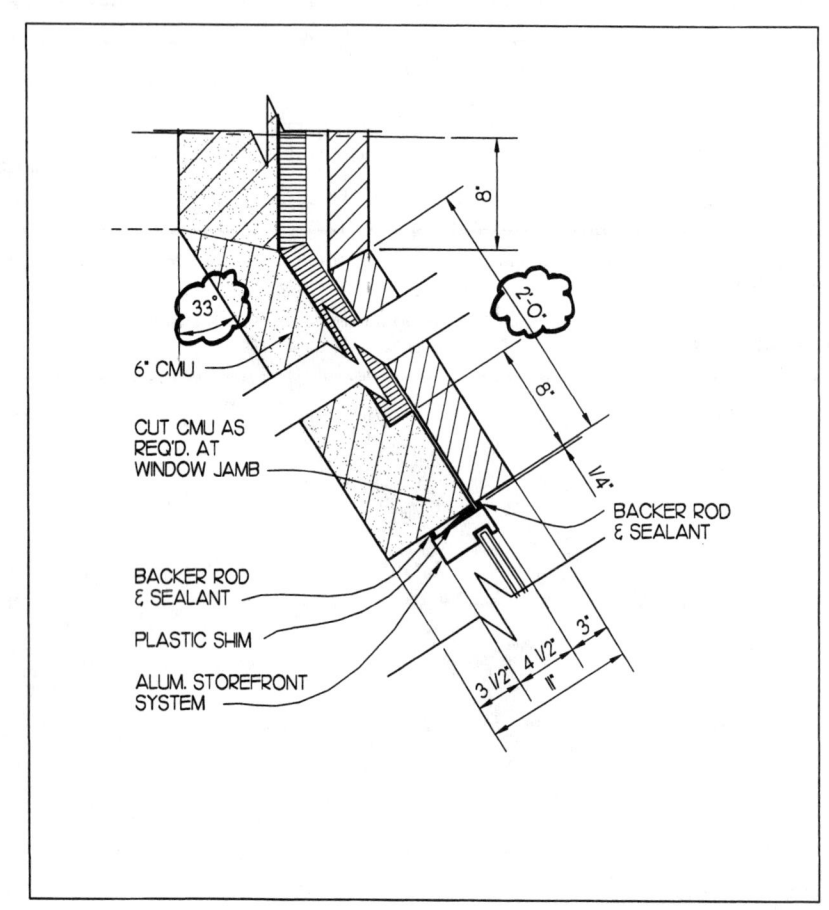

FIGURE 7.16 Addendum drawing.

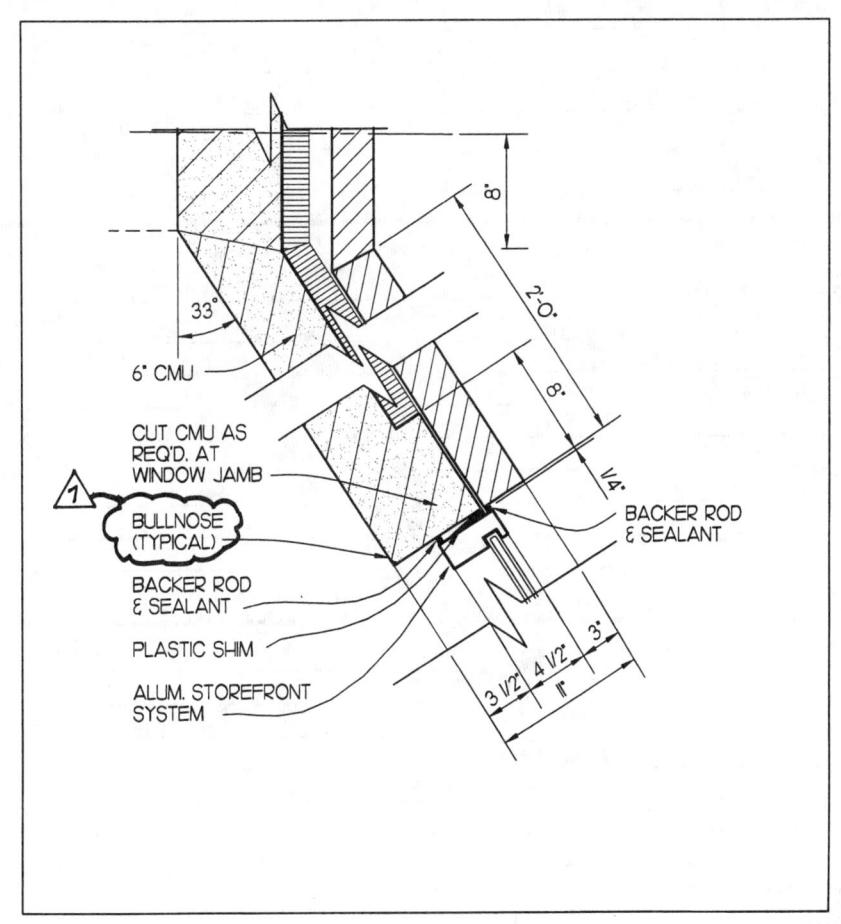

FIGURE 7.17 Bulletin drawing.

FIGURE 7.18 Notice of noncompliance. CSI 9.8A.
Construction Specifications Institute.

Advancement
of Construction
Technology

REQUEST FOR INTERPRETATION

Project: ____________________ RFI Number: ____________________

____________________ From: ____________________

To: ____________________ Date: ____________________

____________________ A/E Project Number: ____________________

RE: ____________________ Contract For: ____________________

Spec. Sec. Ref: Para: Drawing Ref: Detail:

SAMPLE

Signed: ____________________

Response:

☐ Attachments

Response From: To: Date Transmitted: __________ Date Rec'd: __________

Signed: ____________________

Copies: ☐ Owner ☐ Consultants ☐ ________ ☐ ________ ☐ ________ ☐ ________ ☐ File

Page 1 of __

July 1994
CSI Form 13.2A

FIGURE 7.19 Request for interpretation form. CSI 13.2A. Construction Specifications Institute.

Request for Information

To: _____ Date: _____

_____ Project: _____

_____

Attention: _____

Gentlemen:

We are this date requesting the following information, clarification, or direction:

_____

_____

_____

_____

The above information is needed: ⟶

☐ As soon as possible, to avoid delaying the work

☐ Immediately, to minimize delays and added costs already being incurred.

☐ Not late than _____ or the work may experience delays and added costs.

Thank you for your prompt attention to this matter:

Company: _____

By: _____
   (Signature and Title)

Response: _____

_____

_____

_____

Company: _____

By: _____
   (Signature and Title)   (Date)

FIGURE 7.20 Request for information.

REQUEST FOR INFORMATION (RFI)-SUBMITTAL/STATUS LOG

PROJECT:

RFI #	REQUESTED BY	INFORMATION REQUESTED	DATES			STATUS
			SENT	RECEIVED FIELD	RECEIVED OFF.	

FIGURE 7.21 Request for information status log.

FIGURE 7.22 Clarification form. CSI 13.3A.
Construction Specifications Institute.

5A Project Field Instruction

Field Instruction #:	C-4373	Reason:	6A Catagory:
Date Submitted:	09/26/96	● A: Design Error/Omission	○ 1 : Do Nothing
Date Approved:	10/09/96	○ B: Construction Error	○ 2 : Attach FI to EI
Work Order #:	190000	○ C: Fabrication Error	● 3 : Incorporate into 6A
Other Charge #:		○ D: Back Charge	○ 4 : Valmet
Revision #:		○ E: Information Only	○ 5 : Honeywell
System #:			

Problem:
Floor penetrations are required to be sealed as a fire precaution and as a personnel safety measure.

Solution:
Fill or cover penetrations in accordance with the attached schedule. This is per current plant standards. Materials used have plant MSDS sheet approval for use.

Remarks:

Total Estimate Amount: $16250

Attachments: (double click to Launch)

Originator:
M. Hayward
Approvals:
J. Snyder

Authorization:
J. Sathe

FIGURE 7.23 Project field instruction form.

Submittal Control Sheet

Project Title _____ Project Manager _____

Project No. _____ Contractor _____

Section No.	Article No.	Specifications Section Title (Indicate Division No. if applicable) Div.	Data Required								Date of Submittal	Date Rejected	Date Resubmitted	Date Accepted	Notes
			Samples	Shop Dwgs.	Matl. or Parts List	Descriptive Data	Mfrgs Literature	Mix Designs	Certificates	Operation Instr.	Tests					

FIGURE 7.24 Submittal control sheet.

Advancement of Construction Technology

SUBMITTAL CHECKLIST

Project: _____ Project Number: _____
Owner: _____   Contractor: _____

SPECIFICATION SECTION	PRECONSTRUCTION										POST CONSTRUCTION			
	Prod Data	Shop Dwg	Samp	Calc	Dgn Data	Test Rpt	Cert	Qualf	Mfg Inst	Conf	Mfg Fld Rpt	Rec Doc	O/M Manl	Wrty

SAMPLE

o = Required   Ø = Received   ⊗ = Disapproved/Resubmit   ● = Approved

Page 1 of ___

July 1994
CSI Form 12.1A

FIGURE 7.25 Shop drawing/sample log, CSI 12.1A. Construction Specifications Institute.

PUNCH LIST

Advancement of Construction Technology

Project: _____ From (A/E): _____

To (Contractor): _____ Site Visit Date: _____

A/E Project Number: _____

Contract For: _____

The following items require the attention of the Contractor for completion or correction. This list may not be all-inclusive, and the failure to include any items on this list does not alter the responsibility of the Contractor to complete all Work in accordance with the Contract Documents.

Item No.	Location (Area):	Description:	Correction/Completion Date:	Verification A/E Check:

SAMPLE

☐ Attachments _____

Signed by: _____ Date: _____

Copies: ☐ Owner ☐ Consultants ☐ _____ ☐ _____ ☐ _____ ☐ File

Page 1 of ___

July 1994
CSI Form 14.1A

FIGURE 7.26 Punch list form. CSI 14.1A. Construction Specifications Institute.

Color Schedule for _____   Date: _____

Color Numbers are by _____
See general notes at end of schedule.

Room Number & Name	Walls	Accent Walls	Base	Doors	Door Frames	Ceiling	Flooring		Remarks

FIGURE 7.27   Color schedule form.

CHANGE ORDER

AIA DOCUMENT G701

OWNER ☐
ARCHITECT ☐
CONTRACTOR ☐
FIELD ☐
OTHER ☐

PROJECT:
(name, address)

CHANGE ORDER NUMBER:

DATE:

TO CONTRACTOR:
(name, address)

ARCHITECT'S PROJECT NO:

CONTRACT DATE:

CONTRACT FOR:

The Contract is changed as follows:

SAMPLE

Reproduced with permission of The American Institute of Architects under license number #97016. This license expires April 30, 1998. FURTHER REPRODUCTION IS PROHIBITED. Because AIA Documents are revised from time to time, users should ascertain from the AIA the current edition of this document. Copies of the current edition of this AIA document may be purchased from The American Institute of Architects or its local distributors. The text of this document is not "model language" and is not intended for use in other documents without permission of the AIA.

Not valid until signed by the Owner, Architect and Contractor.

The original (Contract Sum) (Guaranteed Maximum Price) was . . . . $
Net change by previously authorized Change Orders . . . . $
The (Contract Sum) (Guaranteed Maximum Price) prior to this Change Order was . . . . $
The (Contract Sum) (Guaranteed Maximum Price) will be (increased) (decreased)
 (unchanged) by this Change Order in the amount of . . . . $
The new (Contract Sum) (Guaranteed Maximum Price) including this Change Order will be . . $

The Contract Time will be (increased) (decreased) (unchanged) by ( ) days.
The date of Substantial Completion as of the date of this Change Order therefore is

NOTE: This summary does not reflect changes in the Contract Sum, Contract Time or Guaranteed Maximum Price which have been authorized by Construction Change Directive.

ARCHITECT	CONTRACTOR	OWNER
Address	Address	Address
BY	BY	BY
DATE	DATE	DATE

AIA DOCUMENT G701 • CHANGE ORDER • 1987 EDITION • AIA® • ©1987 • THE AMERICAN INSTITUTE OF ARCHITECTS, 1735 NEW YORK AVE., N.W., WASHINGTON, D.C. 20006

G701—1987

FIGURE 7.28 Change order. AIA document G701.
Used with permission of the American Institute of Architects.

APPLICATION AND CERTIFICATE FOR PAYMENT AIA DOCUMENT G702 (Instructions on reverse side) PAGE ONE OF PAGES

TO OWNER:
PROJECT:
APPLICATION NO.:
PERIOD TO:
PROJECT NOS.:

Distribution to:
☐ OWNER
☐ ARCHITECT
☐ CONTRACTOR
☐
☐

FROM CONTRACTOR:
VIA ARCHITECT:
CONTRACT DATE:

CONTRACT FOR:

CONTRACTOR'S APPLICATION FOR PAYMENT

Application is made for payment, as shown below, in connection with the Contract. Continuation Sheet, AIA Document G703, is attached.

1. **ORIGINAL CONTRACT SUM** . . . . $ ______
2. **Net change by Change Orders** . . . . $ ______
3. **CONTRACT SUM TO DATE** (Line 1 ± 2) . . . . $ ______
4. **TOTAL COMPLETED & STORED TO DATE** . . . . . . $ ______
   (Column G on G703)
5. **RETAINAGE:**
   a. ______% of Completed Work $ ______
   (Columns D + E on G703)
   b. ______% of Stored Material $ ______
   (Column F on G703)
   Total Retainage (Line 5a + 5b or Total in Column I of G703) . . . . $ ______
6. **TOTAL EARNED LESS RETAINAGE** . . . . $ ______
   (Line 4 less Line 5 Total)
7. **LESS PREVIOUS CERTIFICATES FOR PAYMENT**
   (Line 6 from prior Certificate) . . . . $ ______
8. **CURRENT PAYMENT DUE** . . . . $
9. **BALANCE TO FINISH, INCLUDING RETAINAGE**
   (Line 3 less Line 6) $ ______

CHANGE ORDER SUMMARY	ADDITIONS	DEDUCTIONS
Total changes approved in previous months by Owner		
Total approved this Month		
TOTALS		
NET CHANGES by Change Order		

The undersigned Contractor certifies that to the best of the Contractor's knowledge, information and belief the Work covered by this Application for Payment has been completed in accordance with the Contract Documents, that all amounts have been paid by the Contractor for Work for which previous Certificates for Payment were issued and payments received from the Owner, and that current payment shown herein is now due.

CONTRACTOR:

By: ______ Date: ______

State of:
County of:
Subscribed and sworn to before me this day of

Notary Public:
My Commission expires:

ARCHITECT'S CERTIFICATE FOR PAYMENT

In accordance with the Contract Documents, based on on-site observations and the data comprising this application, the Architect certifies to the Owner that to the best of the Architect's knowledge, information and belief the Work has progressed as indicated, the quality of the Work is in accordance with the Contract Documents, and the Contractor is entitled to payment of the AMOUNT CERTIFIED.

AMOUNT CERTIFIED . . . . $ ______

(Attach explanation if amount certified differs from the amount applied for. Initial all figures on this Application and on the Continuation Sheet that are changed to conform to the amount certified.)

ARCHITECT:

By: ______ Date: ______

This Certificate is not negotiable. The AMOUNT CERTIFIED is payable only to the Contractor named herein. Issuance, payment and acceptance of payment are without prejudice to any rights of the Owner or Contractor under this Contract.

SAMPLE

G702-1992

CAUTION: You should use an original AIA document which has this caution printed in red. An original assures that changes will not be obscured as may occur when documents are reproduced.

FIGURE 7.29 Application and certificate for payment. AIA document G702.
Used with permission of the American Institute of Architects.

CERTIFICATE OF SUBSTANTIAL COMPLETION

AIA DOCUMENT G704

(Instructions on reverse side)

OWNER ☐
ARCHITECT ☐
CONTRACTOR ☐
FIELD ☐
OTHER ☐

PROJECT:
(Name and address)

PROJECT NO.:

CONTRACT FOR:
CONTRACT DATE:

TO OWNER:
(Name and address)

TO CONTRACTOR:
(Name and address)

Reproduced with permission of The American Institute of Architects under license number #97016. This license expires April 30, 1998. FURTHER REPRODUCTION IS PROHIBITED. Because AIA Documents are revised from time to time, users should ascertain from the AIA the current edition of this document. Copies of the current edition of this AIA document may be purchased from The American Institute of Architects or its local distributors. The text of this document is not "model language" and is not intended for use in other documents without permission of the AIA.

DATE OF ISSUANCE:
PROJECT OR DESIGNATED PORTION SHALL INCLUDE:

The Work performed under this Contract has been reviewed and found, to the Architect's best knowledge, information and belief, to be substantially complete. Substantial Completion is the stage in the progress of the Work when the Work or designated portion thereof is sufficiently complete in accordance with the Contract Documents so the Owner can occupy or utilize the Work for its intended use. The date of Substantial Completion of the Project or portion thereof designated above is hereby established as

which is also the date of commencement of applicable warranties required by the Contract Documents, except as stated below:

A list of items to be completed or corrected is attached hereto. The failure to include any items on such list does not alter the responsibility of the Contractor to complete all Work in accordance with the Contract Documents.

ARCHITECT   BY   DATE

The Contractor will complete or correct the Work on the list of items attached hereto within _____ days from the above date of Substantial Completion.

CONTRACTOR   BY   DATE

The Owner accepts the Work or designated portion thereof as substantially complete and will assume full possession thereof at _____ (time) on _____ (date).

OWNER   BY   DATE

The responsibilities of the Owner and the Contractor for security, maintenance, heat, utilities, damage to the Work and insurance shall be as follows:
(Note—Owner's and Contractor's legal and insurance counsel should determine and review insurance requirements and coverage.)

AIA   **CAUTION:** You should use an original AIA document which has this caution printed in red. An original assures that changes will not be obscured as may occur when documents are reproduced.

AIA DOCUMENT G704 • CERTIFICATE OF SUBSTANTIAL COMPLETION • 1992 EDITION • AIA® • ©1992 • THE AMERICAN INSTITUTE OF ARCHITECTS, 1735 NEW YORK AVENUE, N.W., WASHINGTON, D.C. 20006-5292
WARNING: Unlicensed photocopying violates U.S. copyright laws and will subject the violator to legal prosecution.

G704-1992

FIGURE 7.30 Certificate of substantial completion. AIA document G704.
Used with permission of the American Institute of Architects.

8

The Construction Manager

Construction manager, n. Company, firm, or corporation that performs special management services during all phases of a building project under separate or special agreement with owner. May be an architect, engineer, constructor, or other. Not part of the architect's or engineer's basic professional services, but an additional service sometimes included in comprehensive services.

Even after numerous years of successful use, there is still some controversy regarding construction management. Despite many well-intentioned attempts to properly define it, there is still misunderstanding of the process. We still see varied names, directions, opinions, and information from the professions and other interested organizations. Even more do we see variations in the tasks and responsibilities assigned the construction manager by the owner.

ROLE OF THE CONSTRUCTION MANAGER

Perhaps the best place to start is with the program designator, "Professional Construction Management (PCM)," as opposed to "third-party CM," "CM agency contract," or simply the more common "construction management" (the major misconception is that the program is active only in the construction phase of a project). Second, PCM is intended to work throughout the entire project sequence, from the inception of design to the completion of construction. Third, this refers to a firm or organization and *not* to an individual. Fourth, the best concept is that the PCM firm should have no direct connection with either the design firm nor the construction firm. Fifth, PCM involves the owner in a third services contract, in addition to those for design services and for construction.

We can summarize PCM services, from the inception of the project to finale, as participation in the following activities:

aid investigation (programming) and design process
aid/advise selection of materials/systems
review constructability; advise detailing, availability
value engineering, where implemented
advise/input to technical specifications
advise/input to administrative documents (CSI Div. 1 matters)
aid/advise creation of appropriate bid packages
biddability/bidding lists
bid analysis
aid contract award

During actual construction, on- and off-site activities involve:

aid/facilitate scheduling
review/monitor cost control
aid/facilitate coordination at all levels
administrative oversight of contracts
project (final) closeout
field operations:
    coordination of separate contracts
    fast-track (phased) contracts
    monitoring of work items/phases
    change documentation/management
    compliance of work and materials with contract documents
    arrange field/lab testing
    review/recommend progress payments
    meetings/correspondence/records

This list is incomplete, but fairly comprehensive.

The overall purview of the construction manager (and the design/build firm) is a concept called "constructability." This concept provides for the involvement of construction knowledge and experience early in the project. It provides maximum benefits when there is optimal utilization of construction knowledge and experience in planning, design, procurement, and field objectives.

This concept, more widely used on larger projects, is becoming more popular and more formalized where reviews are part of the approval process. Projects built using this concept are better, more easily built, less costly, and with fewer adverse schedule intrusions. There is no intent to have construction "drive" the project, but with this expertise on board, early and throughout the process, the project tends to be less awkward, requires less reassessment, and remains true to the design concept.

It is important to note that PCM acts in an advisory and coordinator role—not in a replacement or substitute role. This expresses the true intent of the program.

Contracts often require the CM to provide some combination of the above services as well as ancillary field services usually provided by the general contractor. Examples are:

providing/operating all temporary field construction

facilities; sheds/storage areas/sanitation

providing all temporary utilities

site security; fencing/policing/access control/parking

periodic cleaning; cleanup/dumpster service/debris removal

Obviously, the construction manager continues to be a participant in the construction process. Laws and traditions do not address the wide variety of services, duties, responsibilities, and contractual configurations that construction management allows. Owners frequently "factor in" construction management as they wish, manipulating contractual obligations and services to their perceived needs; often this means to their legal advantage. (See Figure 8.1.)

Within the wide spectrum of available services, as described by the Construction Management Association of America (CMAA), the range runs from pure agency operations (very similar to those of the design professionals) to GMP (guaranteed maximum price) or independent contractor (with operations very similar to the general contractor). Within this breadth the owner is free to dictate the terms of service that will best fit the work, finances, and timing.

Where conflicts or other legal difficulties may arise, the court would have to establish the exact relationships set forth in the contracts and how the legal obligations are attributed. With both restricted and wide-ranging legal precedents (to this date), each case must be reviewed on its own merits—a demanding and costly process.

DEVELOPMENT OF THE CONSTRUCTION MANAGER SYSTEM

In the past the vast majority of construction projects have involved a general contractor, a contractor for plumbing, one for heating, ventilating, and air-conditioning, and one for electrical work. Traditionally, four separate contracts were set up between the owner and the contractors involved.

This setup has changed through the years, mainly because it has been seen that a lack of control on the construction site was detrimental to the project. Between the traditional configuration and construction management, now very popular, there were various other contractual situations. For example, the general contractor would

ARTICLE 2   ARCHITECT AND CONSTRUCTION MANAGER

2.1 **Definition**

2.1.1 The Architect is _____ and identified as such in the Agreement and is referred to throughout the Contract Documents as if singular in number and masculine in gender. The term Architect means the Architect or his/her authorized representative.

2.1.2 Nothing contained in the Contract Documents shall create any contractual relationship between the Architect and the Contractor.

2.1.3 The Construction Manager is _____ who will provide construction management services as hereinafter described or referred to in the Contract Documents. The term "Construction Manager" means the Construction Manager or his/her authorized representative.

2.1.4 The Construction Manager will be responsible for scheduling and coordinating the work of all contractors on the Project.

2.2 **Administration of the Contract**

2.2.1 The Architect and Construction Manager will provide general administration of the construction contract, including performance of the functions hereinafter described.

2.2.2 The Architect and Construction Manager will be the Owner's representative during construction and until final payment. The Architect and the Construction Manager will have authority to act on behalf of the Owner to the extent provided in the Contract Documents, unless otherwise modified by written instrument which will be shown to the Contractor. The Architect and the Construction Manager will advise and consult with the Owner and all of the Owner's instructions to the Contractor shall be issued through the Construction Manager.

2.2.3 The Architect and the Construction Manager shall at all times have access to the work whenever it is in preparation and progress.

The Contractor shall provide facilities for such access so the Architect and the Construction Manager may perform their functions under the Contract Documents.

2.2.4 The Architect and the Construction Manager will provide a full-time representative to generally observe the progress and quality of the work and to determine in general if the work is proceeding in accordance with the Contract Documents. On the basis of their on-site observations as Architect and Construction Manager, they will keep the Owner informed of the progress of the work, and will endeavor to guard the Owner against defects and deficiencies in the work of the Contractor. The Architect and the Construction Manager will not be responsible for construction means, methods, techniques, sequences, or procedures, or for safety precautions and programs in connection with the work, and they will not be responsible for the Contractor's failure to carry out the work in accordance with the Contract Documents.

2.2.5 Based on such observations and the Contractor's Application for Payment and recommendations of the Construction Manager, the Architect will determine the amounts owing to the Contractor and will approve Application for Payment in such amounts as provided in Paragraph 9.4.

2.2.6 The Architect will be, in the first instance, the interpreter of the requirements of the Contract Documents and the judge of the performance thereunder by both the Owner and Contractor. The Construction Manager will, within a reasonable time, render such interpretations as he may deem necessary for proper execution or progress of the work.

2.2.7 Claims, disputes, and other matters in question between the Contractor and the Owner, relating to the execution or progress of the work or the interpretation of the Contract Documents shall be referred initially to the Architect through the Construction Manager for decisions which the Architect will render in writing within a reasonable amount of time.

2.2.8 All interpretations and decisions of the Architect shall be consistent with the intent of the Contract Documents. In his capacity as interpreter and judge, the Architect will exercise his best efforts to ensure faithful performance by both the Owner and the Contractor and will not show partiality to either.

FIGURE 8.1   Excerpt from typical specification delineating the duties of the architect and the construction manager. It is helpful in showing all participants the areas of responsibility and the lines of communications.

2.2.9	The Architect's decisions in matters relating to artistic effect will be final if consistent with the intent of the Contract Documents.
2.2.10	Any claims, disputes or other matters that have been referred to the Architect, except those relating to artistic effect as provided in Subparagraph 2.2.9 and except any which have been waived by the making or acceptance of final payment as provided in Subparagraphs 9.7.5 and 9.7.6 shall be subject to arbitration upon the written demand of either party. However, no demand for arbitration of any such claim, dispute, or other matter may be made until the earlier of:
2.2.10.1	The date on which the Architect has rendered his written decision, or
2.2.10.2	The tenth day after the parties have presented their evidence to the Architect or have been given a reasonable opportunity to do so, if the Architect has not rendered his written decision by that date.
2.2.11	If a decision of the Architect is made and states that it is final but subject to appeal, no demand for arbitration of a claim, dispute, or other matter covered by such decision may be made later than thirty days after the date on which the party making the demand received the decision. The failure to demand arbitration within said thirty days' period will result in the Architect's decision becoming final and binding upon the Owner and the Contractor. If the Architect renders a decision after arbitration proceedings have been initiated, such decision may be entered as evidence but will not supersede any arbitration proceedings unless the decision is acceptable to the parties concerned.
2.2.12	The Architect and the Construction Manager will have authority to reject work which does not conform to the Contract Documents. Whenever, in his/her reasonable opinion, he/she considers it necessary or advisable to ensure the proper implementation of the intent of the Contract Documents, he/she will have authority to require special inspection or testing of the work in accordance with Subparagraph 7.8.2 whether or not such work be then fabricated, installed, or completed. However, neither the Architect's authority to act under this Subparagraph 2.2.12, nor any decision made by him in good faith either to exercise or not to exercise such authority, shall give rise to any duty or responsibility of the Architect to the Contractor, any Subcontractor, any of their agents or employees, or any other person performing any of the work.
2.2.13	The Architect will review Shop Drawings and Samples as provided in Subparagraphs 4.13.1 through 4.13.8 inclusive.
2.2.14	The Architect will prepare Change Orders in accordance with Article 12, and will have authority to order minor changes in the work as provided in Subparagraph 12.3.1.
2.2.15	The Architect and the Construction Manager will conduct inspections to determine the dates of Substantial Completion and final completion, will receive and review written guarantees and related documents required by the Contract and assembled by the Contractor and will issue a final Certificate for Payment.
2.2.16	The Architect will provide one or more full-time Project Representatives to assist the Architect in carrying out his responsibilities at the site.
	The duties, responsibilities and limitations of authority of such Project Representative shall be as set forth in AIA Document B352.
2.2.17	The duties, responsibilities, and limitations of authority of the Architect and the Construction Manager as the Owner's representatives during construction as set forth in Articles 1 through 14 inclusive of these General Conditions will not be modified or extended without written consent of the Owner, the Contractor, and the Architect.
2.2.18	Neither Architect nor the Construction Manager will be responsible for the acts or omissions of the Contractor, any Subcontractors, or any of their agents or employees, or any other persons performing any of the work.
2.2.19	In case of the termination of the employment of the Architect or the Construction Manager, the Owner shall appoint an Architect or a Construction Manager against whom the Contractor makes no reasonable objection, whose status under the Contract Documents shall be that of the former Architect. Any dispute in connection with such appointment shall be subject to arbitration.

FIGURE 8.1 *(Continued).*

have the mechanical contracts assigned to him/her for administration and coordination. For this extra work the owner would pay the general contractor a separate "administration" fee. Other projects had their contracts drawn so that there was only one contract that involved all the various working trades.

The problem with the various configurations, however, was that many general contractors did not have the resources, expertise, or personnel to handle a large-scale administrative job. The general contractor, involved with all the architectural trades, had enough work controlling the various subcontractors, material suppliers, and others pertaining to his/her own contract. Because of the work involved and despite being paid an extra fee, the general contractor usually was not an effective administrator of the other contracts. For this reason, the design professionals and the owners had to come to a different solution that would provide better results. At the same time, they did not want to increase the involvement of the professional's or the owner's staff in the construction project.

The system that was devised was given the name *construction management*. At its best, construction management is exactly that—a management and expediting function. Some of the first firms or individuals who moved into construction management were general contractors. Their experience in dealing with numerous subcontracts stood them in good stead as construction managers. The element that changed, however, was that general contractors gave up their on-site working function. In essence, they were not working contractors.

Taking away participation in the actual construction and functioning strictly as managers greatly increased their effectiveness in administrating and managing the project. Most firms either became involved in construction management exclusively or expanded their operations to include a separate construction management function. Management is far less risky than general contracting, since the fee is set and guaranteed and no risk of massive financial loss is involved.

At the same time, many design professionals also moved into construction management so that they could achieve more control of the project itself. On a privately funded project the design professional-turned-manager could control who bid on the project by accepting proposals only from highly qualified and extremely reliable contractors. By maintaining control over this actual selection process and an even tighter direct control over the operatives on the job, design professionals felt that they could produce better projects—better in the sense that they met the budget with high-quality workmanship and materials within the time frame demanded by the owner.

This technique, however, must be refined and disciplined, so that tight control of the work does not bog down and literally strangle a project. Also, there is a danger of conflict of interest. Impartiality is lost when the design professional is the construction manager. Generally, the management should come from a separate organization entirely.

The two management techniques are interesting in the contrast of their approaches. Although rooted in different philosophies, both the contractor-manager and the design professional–manager can make the system work effectively. It would seem that the contractor has the advantage in attracting and dealing with other contractors and subcontractors. The design professional seems to have the advantage in overall project control, including control of the owner.

But the final analysis shows that the construction management process in many instances enhances the project by reducing costs, tightly controlling project quality, and accurately meeting time schedules.

It is extremely difficult for a superintendent employed by a general contractor to be in charge of overall supervision of the project, as well as trying to accomplish his/her own fair share of the work. Traditionally, a well-trained journeyman carpenter has been able to move into the superintendency of any construction project. Many general contractors, however, expect such people to produce as much actual carpentry work, but this is an impossible task. Usually, to protect standing as a qualified carpenter, the "superintendent" would tend to do too much trades work while project supervision and administration suffered. In the long run, then, the project would suffer more conflict and disarray. As construction projects became more complex, it became more evident that there had to be a change. It is easily seen that construction management is one answer.

Many companies saw a new opportunity to increase their workload, and of course their profits, by providing a variety of services. Usually, though, there is a distinct change in each of their operations. For example, a company acts in an altogether different manner when it is serving as a construction manager than when it is the general contractor. The methods of operation, the attitude, and the ultimate goal of the operation are different. The personnel involved will adopt different attitudes and methods when operating in the different modes. The professionals also must adjust their perspective when engaged in construction management.

The genesis of construction management is also seated in another aspect of the old traditional contract system. Under that system, most architects and engineers would design projects and include only minimal cost estimating (usually on square foot or comparison basis) as required in the standard contract forms. Owners, unless they required some better and more detailed cost analysis earlier, would not really know the cost status of their project until bids were received. Of course, where bids exceeded the budget, the owner could require adequate redesign (at the professionals' cost) to bring the contract documents to the point where an acceptable project would be produced within cost parameters.

It is easy to see how the method of construction management came into this mix. Involvement of a construction professional early in the process could provide far better costing to the owner. By even earlier involvement (in the design process) this professional could bring cost and constructability to the "method" of design, i.e., to material and system selection and the way the project was "put together." Often, when pressed for more accurate cost data, design professionals would rely on construction professionals to give cost estimate data during the design process.

Further expansion led the CM to work throughout the progress of the project, providing their expertise in every phase, to the benefit of the owner. This does not ignore the major contributions of the design professional, but reflects the change in practice, contracts, and responsibilities. It shows how CM became attractive to owners and why this system is so widely embraced today.

Agency vs. GMP Construction Management

Although it has existed for several years and has been actively used, construction management is a process that is still without a single definition. The two parameters that seem to define the extremes of the process are agency and GMP construction management.

Agency CM is more classic (and perhaps the first developed and used). It is based on the premise of a separate owner-CM contract and relationship. The CM functions are specifically set forth and, more importantly, are blended together with those duties of the design professional (who is under a separate contract with the owner and has no contractual arrangement with the CM). Usually this type of program provides for CM services in the design development phase (to overview constructability, material, and systems selection and to begin cost monitoring). It follows through the other phases in guidance and monitoring roles to ensure the best solution within the budgetary confines set by the owner. In bidding, the CM may help to develop a list of bidders or may solicit bidders for the project on behalf of the owner, to further ensure best pricing through active competition.

During construction, the agency CM will be active both on-site and in the office, ensuring contractor coordination, scheduling, product submittals, and purchase orders (to monitor that proper and timely purchasing is occurring). On-site, the CM's representative or superintendent will see that work proceeds as required and will aid coordination and problem solving, working in close conjunction with the design professional's project representative and the contractors' superintendents.

The second configuration, GMP (for guaranteed maximum price), is also known as independent-contractor CM and "CM at risk." All of these names are meant to indicate that in this mode the CM acts much like a general contractor: guarantees the price and contract time and engages in subcontracting. Indeed, here the CM is at risk in exactly the same manner as the GC (profit is at risk for faulty work, time delays, cost overruns, lack of coordination, etc.).

Since this program seems to directly reflect a GC, the variation is only in how the contract language is drawn and what benefits the owner is willing to give the CM, over and above the demeanor of a GC. The primary difference between GC and at-risk CM is that the CM is involved with the project at an earlier time (before completion of design) and provides value engineering, cost analysis and estimating, and related services.

The underlying issue in construction management, at this juncture, is the desire of the owner. It is only in the contract language (owner-CM) that one can ascertain exactly what services the CM will and will not provide; variations are frequent and across the spectrum. Basically, firms offering CM services have expanded their range of service to retain or gain market share in the face of rapidly changing owner wishes. To stay competitive, these firms have to engage in programs ("at-risk," for example) that they avoided previously.

Usually the CM has a no-risk contract (agency-type construction management) wherein a fee/profit is set into the contract. There is no threat to this money since the operation is one of administrative control (for the most part) and not one where time, material, personnel, and so on are variables. Of course, the latter elements are distinct parts of general contracting and create the risk factor for the GC or a GMP construction manager. While a responsive bid for any project should contain contingency funds and a profit level with some margin (i.e., a slightly lower percentage of profit is acceptable), there still is an inherent risk that timing, weather, deliveries, and disruptions of many kinds can cause the cost of the project (to the GC) to vary (the price to the owner, of course, is fixed by contract, unless proper change orders are executed). By being at risk the GC tends to seek rock-bottom pricing, efficiency, promptness, productivity, correct construction "the first time," timeliness in all operations, effective supervision, and so on. Of course, a hidden agenda calls for doing things so well that some of the money set aside for the actual "cost" of the project work can be converted into additional profit.

CONTRACTUAL OBLIGATIONS

The terms of the General Conditions of the Contract for Construction (in AIA document A201, CM edition) are very specific when outlining the duties of each participant in the project, including the construction manager. In the wording of the General Conditions, the manager comes by other duties by virtue of what is not written, but is inferred.

A careful examination of AIA documents B141 and B141CM reveals that the same tasks are performed under both, but it is a matter of who performs which tasks in the construction management format. The CM concept is to bring a greater technical expertise to the project for specific functions. It tends to relieve the professionals (and the owner in some ways) from these tasks, to allow them to concentrate on other aspects of the project. However, in this format the professional is still part of the project and continues to fill an important role as the owner's agent in many of the operations during construction. Usually, the CM will function in the areas of cost containment and monitoring, administration of contractors, and technical information/research/procurement. Thus, the intangibles of concept, intentions, and their physical manifestation seated solely in the designer are still available to the project

Chapter 8   The Construction Manager

as contained in the contract documents. This is important to the complete resolution of the project's contracts.

The General Conditions call specifically for the construction manager to be hired to manage the construction of the project. His/her prime obligation, by virtue of these contract documents, is the actual expediting, scheduling, and coordination of all the work and contractors working on the project or under contract to it. As has been discussed, the construction manager generally replaces the traditional general contractor, usually without doing any of the actual construction work.

Although the professional differences between architects and engineers influence some owners and some operations, the requirement for a construction manager is a different situation. Basically, this emanates from the early decision (by the owner) regarding what professional will be hired first, the project-design professional or construction manager. This decision has ramifications throughout the process of the project.

Reading the standard contract form from the professionals, one can see that the first-hired person will have the status of primary agent of the owner and will, usually by contract language, act at least as adviser to the owner about who else is hired and how their duties and functions will be delineated. This simple act sets a tone for the project and gives some insight into the owner's expectations. It is obvious in the standard professional forms that there is mutual understanding that all of the necessary tasks require attention; it is just a matter of who does what, under what kind of jurisdiction—what other tasks/duties/responsibilities are appended. Usually the design professional will "design" and the construction manager will "manage."

It is clear that the owner *must* have assistance; a legal counsel is a necessity, in addition to an open, cooperative, and expert "first hire." Care must be taken so bias, animosity, and competition are totally eliminated. The creation of the design/construction team is crucial. The team must make an integral, fully involved, fully coordinated effort in the owner's best interest.

In scheduling and coordinating the manager along with the professional is ultimately responsible for reviewing and processing the certificates for payment, including the final payment. The manager actually makes recommendations to the professional, who then issues the proper certification. Here is one of the first subtleties in the General Conditions language. Certainly no construction manager is going to make any sort of recommendation with regard to payment if he/she does not have some substantiating information for that recommendation. Basically, the contract language implies that he/she must inspect the work. In the recommendation, there must be verification of any observation/inspections with those of the professional. This means that the manager must tell the professional that certain work has been accomplished and is satisfactory (contract-abiding), and that payment is due. Although there is no specific mention of the word "inspection," the responsibility to inspect is still implied.

Further, there is specific reference to the work of the professional and her/his responsibility for periodic on-site observation. The construction manager, however, is on the site daily, much as the clerk-of-works is. Through the CM recommendation for payment and obligation to reject unsatisfactory work, the CM is part of the "inspection" team. Rejection of unsatisfactory work is a specific duty of the CM. Upon finding noncompliant work, a note is issued along with various solutions and a statement regrading impact on the project as a whole. The recommendation is to the professional, who will examine the same work and issue any further orders or recommendations to the owner. The CM is, therefore, a constant on-site "inspector" although his/her authority is somewhat restricted, by the General Conditions and the contract itself. Figure 8.2 shows how the CM's role is defined in the project manual.

Authority of the Construction Manager

In construction management in particular, placing authority in the correct hands is most important. In contrast with the conventional general contractor configuration,

ARTICLE 3—THE ARCHITECT

Add Article 3a—The Construction Manager, as follows:

"The Construction Manager will be the Owner's representative during construction and until the work is complete. The Construction Manager will advise and consult with the Owner. The Owner's instructions to the contractors shall be forwarded through the Architect and Construction Manager."

"The Construction Manager is the person or organization identified as such in the Contract Documents or referred to as CM as if singular in number and masculine in gender. The term "CM" or Construction Manager refers to either the firm or its authorized representative."

"Nothing contained in the Contract Documents shall create any Contract relationship between the Construction Manager and the contractors on the project."

"The Construction Manager shall provide the liaison between Contractor, Owner and Architect/Engineer. The purpose of this liaison is to generate continuous and current project information and bring problems to the immediate attention of the proper party."

"The Construction Manager will provide on-site administration and management services required to coordinate the work of the Contract Documents, and will endeavor to guard the Owner against defects and deficiencies in the work of the Contractor."

"The Construction Manager will not be responsible for and will not have control or charge of construction means, methods, techniques, or procedures, or safety precautions and programs in connection with the work."

"The Construction Manager will have authority during the course of the work to recommend certain changes in the means, methods, techniques, and procedures being implemented by the Contractor, if, in the opinion of the Construction Manager, the work is not being performed in such a manner as to meet milestone dates and overall project schedule."

"The Construction Manager will not be responsible for or have control or charge over acts or omissions of the Contractor, Subcontractor, or any of their agents or employees, or any other person performing any of the work."

"The Construction Manager will schedule and coordinate the work of all contractors on the project including their use of the site. The Construction Manager will keep the Contractors and Material Suppliers informed of the Project Construction Schedule to enable the contractors to plan and perform the work properly."

"The Construction Manager will review all applications for payment by the Contractor, including final payment, and will assemble them with similar applications from other contractors on the project into a combined Project Application for Payment. The Construction Manager will then make recommendations to the Architect for certification for payment."

"The Construction Manager will have authority, subject to approval by the Architect, to stop work or certain phases of work and to reject work which does not conform to the Contract Documents."

"The Construction Manager will maintain at the Project site one record copy of all Drawings, Specifications, Addenda, Change Orders, and other modifications pertaining to the Project, in good order and marked currently to record all changes made during construction, and approved Shop Drawings, Product Data, and Samples. These shall be available to the Architect and the contractors."

"The Construction Manager will assist the Architect in conducting inspections to determine dates of substantial completion and final completion, and will receive and forward to the Architect for Architect's review written warranties and related documents required by the Contract and assembled by the Contractor."

FIGURE 8.2 An excerpt from a project manual that gives an itemized definition of the construction manager role.

Chapter 8 The Construction Manager

where authority follows the obvious contract lines, the simple use of a construction manager does not provide the same shift of authority. Since the CM has a contract solely with the owner, the relationship with the various contractors can be tenuous and often leaves the CM powerless to insist on action (if a contractor is cantankerous).

One solution is for the owner to "bid the construction" by receiving bids from CM firms who, in turn, contract out the various phases of the work. At the finale, the complete package is turned over to the owner, but the owner has not been party to any construction contracts. The "lowest and best" bid can be maintained (for government clients), but the CM is able to fashion a corps of subcontractors whom he/she knows and who will perform under the CM. In addition, under a contractual relationship, the CM is empowered to deal with the contractors in every aspect of their work and can make things happen as he/she sees necessary.

With the latter scenario, and because of the complexity of modern construction, the CM firm should be restricted to traditional CM activities plus cleanup, quality control, and the like, which enhance the project overall. The CM, though, should not be permitted to perform any actual construction work, since this will blur responsibilities and open the door to other combinations of duties that could harm the project. The CM should function purely as the administrator, expediter, scheduler, and coordinator for the project.

Figures 8.3 and 8.4 show two valid configurations of construction management. Figure 8.3 shows the manager bound by a separate contract with the owner and with no other direct contractual relationships. The manager is strictly an administrator of the project. With no added authority, the manager has little with which to influence,

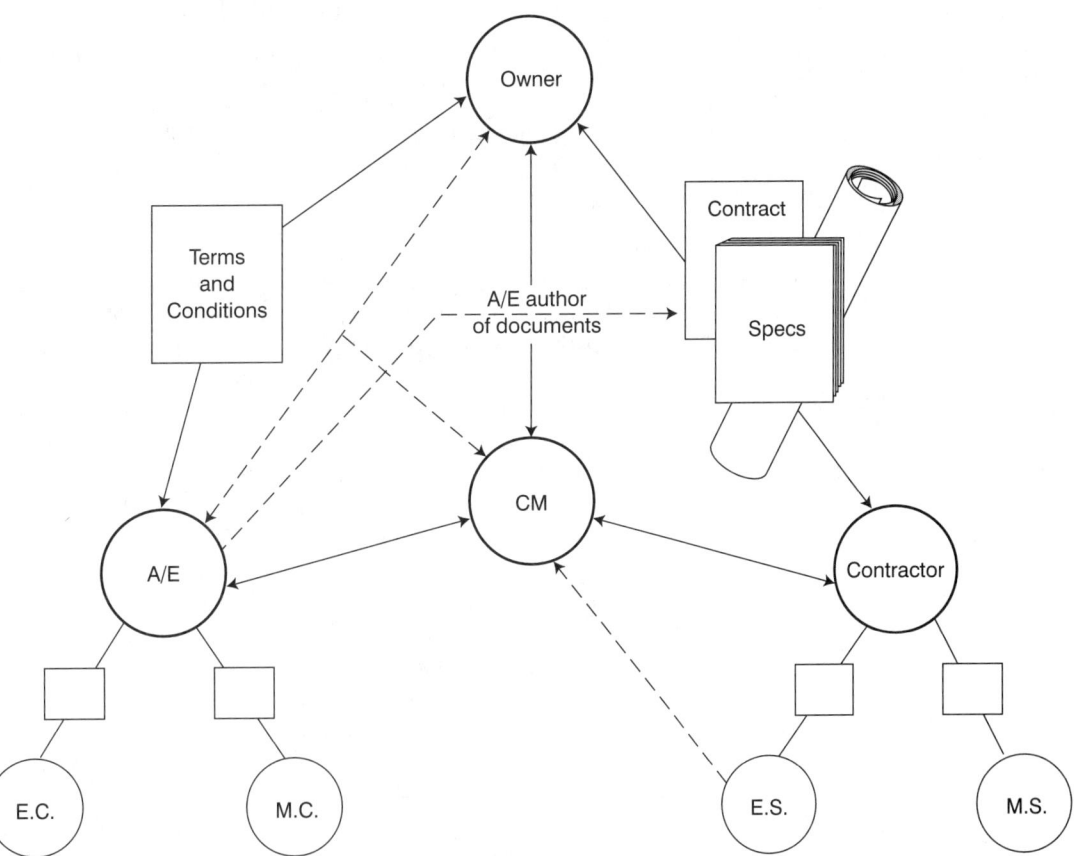

FIGURE 8.3 Construction management configuration where the CM is under a separate contract with the owner to oversee the construction process. The CM has no contractual relationship with the contractors, who contract directly with the owner.

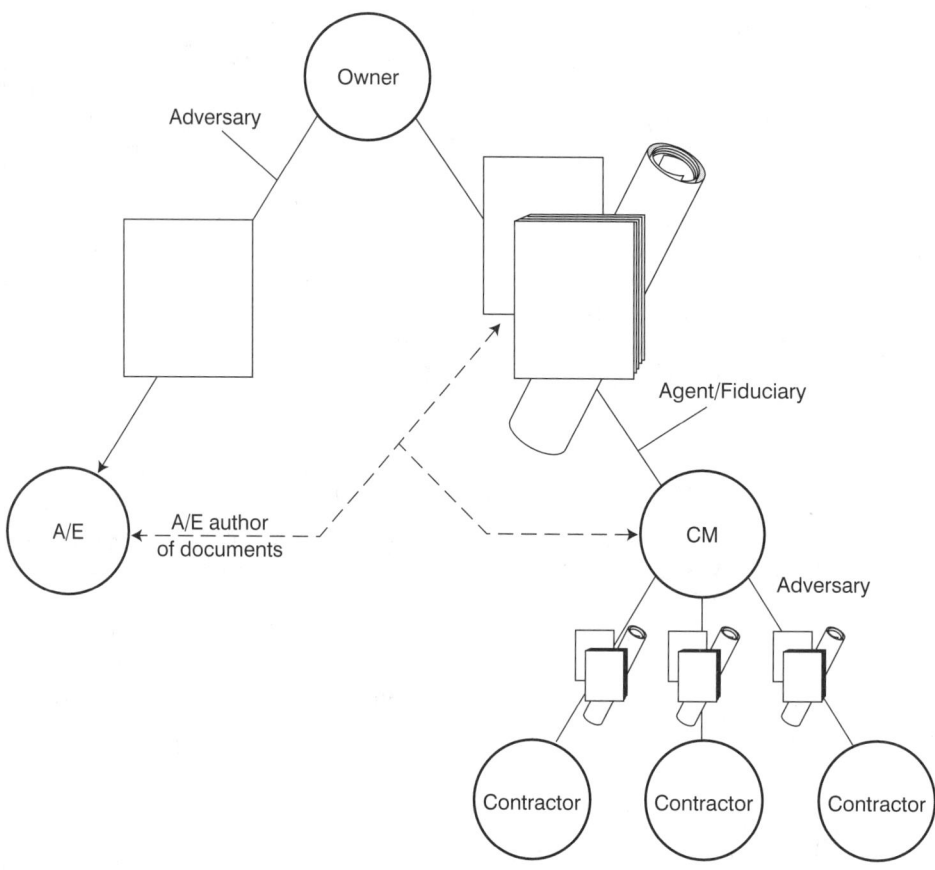

FIGURE 8.4   Construction management configuration where the CM is the agent of the owner. The contractors contract with the owner through the CM (the CM may or may not perform part of the work).

coerce, cajole, or order action by the working contractors (who themselves have separate and direct contracts with the owner). This leaves the CM in a most untenable position, i.e., any action against or toward a contractor must be conveyed first to the owner and from there to the contractor. This convolution will greatly impair a smooth-flowing project, quick approvals, and definitive action, and could lead to a reduction of the status of the CM should the owner not be fully supportive.

Figure 8.4 shows the construction manager as the operating entity of the project. All working contractors are under direct contract with the CM and report to him/her. This relates to the old general contractor configuration, with one exception. The CM here is, by contract, a nonworking contractor who is restricted to administration of the project and performs no actual construction work. Even building (or general) construction work is another and separate contract. Some general project items can be included in the CM contract where they will contribute to a better project. These could include, for example, cleanup/trash removal, site security, temporary facilities, and so on. Such assignments can lead to better coordinated efforts, elimination of duplication of efforts, and cost reduction (the same items need not be provided by several contractors).

The situation closest to the ideal would be one in which the project is under a CM and the manager is given the proper level of authority and contractual relationships with the working contractor. In this scenario, the CM is in a position to perform all of the crucial scheduling, coordinating, and administration of the project, without being burdened with the management of his/her own construction force.

Under the general contractor system a superintendent is often assigned to the project, but the duties assigned include not only administration but responsibility for some of the actual general construction work. The tendency is to do the work required (which is usually more familiar to the superintendent) at the expense of project coordination. In some situations, some contractors who bid separately are "assigned" to the GC, who is paid an added fee to administer the assigned contracts. Again, the GC will be more interested in construction work than in the "clerical" work of contract administration. Under severe circumstances, these two scenarios can become areas of controversy between owner and professionals on the one side, and the contractors on the other. Some can be left floundering, due to the lack of adequate coordination, direction, and scheduling.

Currently, unless specific new provisions are included in the contract documents, the owner is the sole entity common to all of the contracts. Where no new, specific authority is given by the owner to the CM and/or the design professional, there are no inherent enforcement "tools" in any hands but the owner's. This forces the owner into a proactive position, which may be at odds with his/her expertise or desires. It also inhibits prompt, direct action. This is not to say that the professional should seek to undercut the role or influence of the owner, but suggestions as to more even "layering" of controls and authority may serve the project and the owner well.

It must be recognized that in the prevalent construction management contract (see Figure 8.5), there is *no* contractual relationship between the CM and the various contractors. Only the owner is party to the actual contracts the contractors hold. Where no new, specific authority is given to the CM, there is no method of enforcement. The CM is restricted to a weak position, as opposed to the direct contractual obligations between the GC and the subs.

Since the design professional has no contractual relationship to the contractors either, the professional joins the CM in being at a disadvantage for taking direct action. Both must convey situations requiring action to the owner. For example, only the owner can stop work or withhold payment (with due cause), as a means of resolving problem situations. Neither the design professional nor the CM can so act, *without direct assignment of such authority to them by the owner.*

Two different examples illustrate this point further: cleanup and quality control. Each contractor may be held responsible for the trash and debris created by his/her work. But who is responsible for general sweeping and cleaning of the whole site? This job could be given to one contractor as part of a bid package, but more than likely this will be a constant source of irritation to every participant. Too much trash is generic and not attributable to one craft or another. If this was assigned to the CM, he/she could hire the necessary help and clean as required, without any "trade/craft analysis."

Much in the same light, quality control, often including minor task assignments, could be a CM responsibility. For example, drywall in a stair hall runs down past the stair stringers. Within acceptable tolerances, both the drywall and steel stringers vary, creating a slight gap between the two materials. Who fills this gap—the steel installer, the drywall contractor, the caulking contractor (if there is one), the painter? Without specific direction in the specifications, this situation can wreak havoc. Under the old general contractor configuration, it may be an item absorbed by the GC in an effort to complete the job. But under construction management the condition will languish. With some restricted quality control power, it would be proper for the CM to assign the work to the most logical contractor. This would require specific language in the contract documents. It would be proper, authorized, and out in the open for all to understand. (See Figure 8.6.)

This is the crux of the construction manager system: constantly watching each item of work as it proceeds and advising the architect as to the amount and satisfaction of the work. The CM may also require special inspection or testing of any work that does not comply with the contract documents. He/she must, of course, have reasonable cause for

---

DOCUMENT 01040
CONSTRUCTION MANAGER COORDINATION

PART I—GENERAL

RELATED DOCUMENTS

Drawings and general provisions of Contract, including General and Supplementary Conditions and other Division-1 Specification Sections, apply to this Section.

SUMMARY

This project is being constructed through the method of Construction Management. The Owner has employed the Construction Manager to provide coordination of work and on-site supervision for the duration of the project.

DUTIES OF CONSTRUCTION MANAGER

Establish and schedule work of the various contractors.

Establish on-site lines of authority and communication; schedule and conduct project meetings among:

    Owner's representative.

    Architect.

    Engineers.

    Consultants.

    Material Suppliers.

Construction Schedules:

    Coordinate schedules of the several contractors.

    Monitor schedule as work progresses.

    Observe work to monitor compliance with schedule.

Temporary Facilities:

    Allocate space for temporary structures furnished by each contractor.

    Monitor use of temporary utilities.

    Verify that adequate services are provided to comply with requirements for work and climatic conditions.

    Verify proper maintenance and operation of temporary and permanent facilities.

    Administer use of Owner's facilities.

    Administer traffic and parking controls.

    Coordinate and delegate areas for on-site storage of materials.

FIGURE 8.5 An excerpt from a project manual listing the specific duties of the construction manager. Can be used in conjunction with the listing shown in Figure 8.2.

such action, and it is subject to the interpretations and decisions of the architect. Again, this is a situation in which the architect's contract calls for a limited level of inspection. By virtue of the responsibilities given to the construction manager, he is in the position of being an inspector although not specifically called that in the documents.

    The contract between the construction manager and the owner is parallel to that between the architect and the owner. There is no contractual obligation of any sort between the architect and the construction manager. Although both are basically agents of the owner, they have specific responsibilities to the owner, and these responsibilities interlace in only a few places, primarily for overall quality control or inspection and for the expediting and proper execution of the project. This last phrase is, perhaps, the key to the entire system and the project. The contractual lines are devised so that the project is of primary importance, in full keeping with the owner's concern. The details of how the project is executed, or who executes it,

```
Changes:
    Recommend necessary or desirable changes to Architect.
    Review Contractors' request for changes and substitutions; submit recommendations to
    Architect.
    Assist Architect in negotiating change orders.
    Assist in implementation of change orders.
Interpretations of Contract Documents:
    Consult with Architect to obtain interpretations.
    Assist in resolution of questions which may arise.
    Transmit written interpretations to concerned parties.
Administer processing of:
    Payment requests.
    Shop drawings, product data, and samples.
    Field drawings.
    Coordination drawings.
    Change orders.
Maintain Contract Documents, Reports, and Records at the jobsite.
Verify that specified cleaning is done:
    During progress of the work.
    At completion of each contract.
Substantial completion:
    Upon determination of substantial completion of work or portion thereof, prepare for
    the Architect a list of incomplete or unsatisfied items.
Final completion:
    Upon determination that work is finally completed:
        Submit written notice to Architect and Owner representative that work is ready
        for final inspection.
        Secure and transmit to Architect required close-out submittals.
    Turn over to Architect spare parts and maintenance materials.
                              END OF DOCUMENT 01040
```

FIGURE 8.5   *(Continued).*

really are of little interest to the owner, who is interested in the end product, in occupying the project on time, and in not exceeding the budget.

The architect, who has many interests on a particular project, simply does not have the time to assume the role of construction manager. Although many architectural firms may have a construction manager function, this usually becomes a separate department or organization and a separate contract. Besides removing conflict of interest, this allows the proper allotment of time for each function. In the past, the architect in the classic role of master builder often tried to control a project by acting as the construction manager before this nomenclature became popular. If the architect has no other work and no other clients, this may work. Normally the architect does not have time to be the one who expedites and schedules each project. But to have someone else, with another set of eyes and another set of hands, to set all this in motion is a valuable aid for the architect.

The construction manager's on-site representative:

1. Is to be the on-site "driving force" for the project.
2. Is to act as if he/she were the superintendent of a general contractor of the project.
3. Will "call the shots" as far as schedule, coordination, readiness for the next construction activity to proceed (correctness and completeness of the work, etc.).
4. Has the authority and the responsibility to question the correctness of the work, give corrective instructions (if they are known), and contact the project manager or the design professional for the following reasons:
   a. notification that a portion of the work is not correct, or its correctness is in question
   b. request for clarification as to whether the work is correct
   c. request for instructions about corrective measures required
   d. request whether higher action (i.e., stopping of work, etc.) needs to be taken
5. Has the authority to recommend holding up payment requests of any contractor whose performance is not satisfactory.
6. Has direct line of communication with the project manager or the design professional, or, in case of emergency, the home office.
7. Has the responsibility and duty not to allow any defective or questionable work to be covered up.

FIGURE 8.6 Another view of the CM and the CM's on-site representative, which may not fit every project nor other parameters shown in other illustrations in this chapter.

The worst possible situation would occur on a project that has an architect and a construction manager who simply do not get along and cannot work together. It is the owner's responsibility to prevent this from happening. In using the architect as the basic adviser, the owner can develop a team (which includes selection of the manager) that can work together and keep the project as the prime goal. There is so much potential for controversy within any project that it is almost mandatory that the manager and the architect be the most compatible of participants on the job. This is not to say that their respective responsibilities become cloudy. They have to work together, perhaps daily, to see that the work is done properly. There is no reason the construction manager and the architect cannot talk to the owner in different ways; indeed, at times they may be in direct opposition to each other. Many different approaches can be taken to best serve the owner and his/her project, and therefore, communication with the owner, like all the communication on the job, should be open and active.

It is highly improper for the manager or the architect to go to the owner and criticize the other, in such a manner that the owner becomes suspicious of both participants. Before long, there can be pique on either part, which may become openly hostile and cause controversy on the entire project.

The owner must understand the wording and the lines of responsibility laid out in the General Conditions. It may even be proper for the architect to take an extract from the General Conditions and list the individual responsibilities of the manager and the architect for the owner. In this way, the owner knows exactly whom to approach about a particular problem. It also helps to clear up some rather vague language of the General Conditions. Although the manager has no specific inspection responsibility under the General Conditions, except in dealing with substantial completion, they imply an inspection function. The manager may not have the authority to deal directly with problems, but in producing the recommendations to the architect for payment, remedial work, or additional testing, he/she must know where the fault lies. To do this, the manager must become an inspector at least during part of the day.

Obviously, the construction manager system is not necessary for every project, especially moderate to small projects. Larger, more extensive projects benefit greatly by using it. Without the manager's constant on-site presence, certain responsibilities of the architect are enlarged. Therefore, the architect must assume some of the construction manager's subtle inspection obligations.

AVOIDING IMPEDIMENTS OF WORK PROGRESS

Because the construction manager system is more complicated, it becomes more important that all participants understand their specific roles. What are their responsibilities? What do their responsibilities exclude? The project must be kept in mind at all times. Any system that is imposed on a project should go to great lengths to avoid impeding the progress of work. This is particularly true of the construction manager system.

There is no doubt that the more participants there are in the construction and inspection system, the greater the chance for conflict, misunderstanding, and other impediments. Because of the contractual obligations of the manager and the architect and the parallelism of those contracts, it is essential that there be an extra effort to provide a smooth working relationship between the two. Communication should be comprehensive, constant, and open.

If the manager inspects some work and finds it faulty, he/she must make a recommendation to the architect, who must "pick up" the situation and resolve it. If there is to be an interpretation of the contract documents, that function belongs by contract, solely to the architect. The manager must take responsibility if there is some fault in the scheduling or the management of the work. It is not enough to understand the particular roles; one must be able to understand how the process is imposed, how the responsibilities are interwined, and where decisions must be made. There are enough problems on any construction job that the coordination between the manager and the architect cannot be allowed to deteriorate in any way. They must work together very closely in a complementary way, and they must understand and respect the process in which they are participating.

The architect's inspection facility is greatly enhanced by the construction manager system. The manager becomes an on-site inspector who can help the architect. In the end, the manager is of great help to the project itself.

Previously, the clerk of works concept was discussed, and this role, of course, can be added to the construction manager system. A very large project could easily have an architect, a clerk of works, and a construction manager. This system is more complicated and deserves much more attention by the participants. New lines of communication must be established and maintained. The owner talks through the architect to the manager, who talks to the contractors. (The owner does not talk directly to the contractors.) The architect cannot give specific instructions to the contractor through the clerk of works without informing the construction manager, otherwise the system will begin to break down.

This introduction of more personnel into the inspection system cannot be allowed to stand in the way of the construction process. A delicate balance should be achieved so the project is not administration bound. Too many people acting in similar capacities tends to cloud solutions and create communication problems, and attitudes can change because someone is bypassed or his/her functions superseded.

There must be quality control and inspections. For the most part, the contractors respect and understand this process as it involves them. Contractors become concerned only if that system impedes their work and causes time delays because decisions are not made promptly or by the proper people. In such cases, decisions must sometimes be countermanded, reissued, or delayed still further by other sorts of entanglements.

With more complicated construction projects, the construction manager system works, so long as all participants understand the manager's emerging role—primarily, his/her contractual role in the process.

EXERCISES

The following tasks are based on the concepts and information in the chapter. All of the material or information required to complete the tasks may not be contained in the chapter text. Independent research is required. Consult your instructor for guidance.

1. Discuss the changes a contracting firm must make when acting as a construction manager instead of a general contractor.
2. A medium-sized university is building a $12 million health and fitness center. The campus is located in a town of 40,000 (the school adds 16,000 when in session), about 100 miles from larger cities. Subcontractors and material suppliers from far-flung locations are needed because local labor is not sufficient for this size project. What are the advantages of using a construction manager?
3. Discuss the proposition that the contract administrator is the "prime mover" on the job site.
4. From a professional's viewpoint, a construction manager can be both an asset and a liability. Explain and give examples.

9

The Owner

Owner, n. 1. Person, firm, corporation, or agency for whom the project is being constructed. 2. Party to contract who shall make payments for work done. 3. The design professional's client.

An owner who chooses to utilize the basic service of contract administration may need to have her/his proper role defined. This definition should come from the design professional, as well as from the legal advice given the owner by an attorney.

Architectural (construction) projects start as activities between the professional and the owner (client). Contractually, the demands on both are clearly delineated. There may be some "selling" of the design concept by the professional. Disagreements may occur, but usually there are few real hassles. On rare occasions the professional is summarily dismissed by an aggravated owner, who cannot seem to get from the professional the type of project the owner has envisioned.

In the main, though, things move along quietly through programming of the owner's needs, desires, and problems, and the preliminary design. Good exchanges of information, appropriate questioning, and astute design ability are active in developing the scenario suited to the client's taste, budget, and needs. It is when the project takes a more technical turn (design development, document preparation) that it becomes quite mysterious to the owner (who fears he/she is losing control). To the professional, the conversion of the design idea to nitty-gritty details and specifications is routine.

However, it is in this conversion that the project becomes intertwined with more numerous and pervasive legal "strings" and encumbrances. What happens in this process can have a direct and lasting impact on the success of the project, the professional, and the client/owner/user. Concepts become real; lines take on new meanings as they delineate types of walls, doors, partitions, etc. It may be melodramatic but true that any one of these many, many lines (items) could be the source, at some future time, of grave liability to any of the participants. The burden of being proper and correct now becomes more than mere ego (desiring perfection) and becomes a process of ensuring compliance, safety, and stability under all conditions, no matter how variant they may be or may become.

EDUCATION OF THE OWNER

Education of the owner may sound like a presumptuous endeavor, perhaps even unnecessary. However, even if it is dismissed as a formally identified activity, *it will still occur.* The problem with it happening by happenstance is that it then becomes piecemeal, crisis-oriented, and often wrongly perceived as a cover-up on the part of the professional. This program may vary from owner to owner (and should, to meet the respective familiarity and experience of the owners in the construction process), but the effort is one that should be openly perceived. It should be planned from the very outset of the project. Of course, this is an overhead cost to the professional, but it can be woven into the overall fabric of the various phases of the project work. The more that is done and understood up front in the project time frame, the better the various relationships will be and the smoother the project will run. This concept is best portrayed in the fact that if "new" education must occur at each juncture of the job, the time required for resolution of problems will elongate and may eventually imperil the progress of the job. Here is a prime example of an "owner problem" adversely affecting not only the other participants but the project itself.

Of course, the professional must approach this effort in a tactful, diplomatic manner, but at the same time must emphasize the need, the value, and the utter importance of sitting down (perhaps several times) to discuss, explain, and understand the roles of the various parties in the different phases of the project. Most owners will be open to, and appreciative of, this effort and will cooperate. A few, though, will require extremely sensitive handling initially, and ongoing remedial education, each episode again requiring kid-glove treatment. This requires that one be aware of the attitude and personality of the owner. Usually, this will be quite evident from the outset of the contract, if not from the first contact.

Two factors are inherent in every client/professional relationship: the client's unfamiliarity with or ignorance of the construction process, and misunderstanding or false perceptions about projects and professionals. Claims against professionals are usually filed by owners/clients. Usually these claims can be traced to (1) poor communications, and (2) misunderstanding and/or ignorance on the owners' part.

The Owner as Project Leader

Unwittingly, owners are projected into a leadership role for the project. Owners, after all, have the projects built for their benefit, they pay all the bills, and the projects are their individual efforts. Generally owners are quite successful within their realm of expertise, but this usually does not extend to construction. They approach construction as the only means to provide added value, capacity, or ability for their business ventures. Owners build projects infrequently! This produces a set of lethal factors, which the owner needs to be aware of, and which must be resolved, early and fully.

It is easy to see that as the configuration of the owner changes, i.e., single proprietor, to partnership, to corporation, the ability to meet the demands of leadership is confounded. The single person will be easily overwhelmed with trying to engage the construction process and keeping the business running. As other persons become involved, decision making is complicated and slowed, and progress becomes a real problem to everyone. As decision making becomes more detached from the real "powers that be," the actual work on the project will suffer. Perhaps this is a matter of "Too many cooks spoil the broth."

It is the "mind set" of the owner that inhibits the necessary leadership required for a construction project. Most owners will have a fundamental desire to control every detail of the entire project, with tight reins on cost, scope, and quality. The perception of leadership then becomes one of demanding absolute perfection from all other participants, not one of delegating authority, instilling confidence in others (with proper expertise), and working within the system. Education of the owner, then, becomes a process of establishing the correct demeanor and position of the owner/leader, understanding of the entire design and construction process, and sharing of common effort. A leader need not be an isolated person giving orders left and right. Rather, the true leader will be part of the team, yielding to the expertise of others when appropriate and making the tough decisions when appropriate. To do this well requires an informed, willing, and understanding owner—well educated!

Development of the Education Process

At the outset the owner must come to understand that the construction process is unique and is not like other business activities. For a relatively short period of time, construction brings together dozens of parties—different in interest, expertise, perspective, intent, ability—for the design, construction, and delivery of a project; then this group disperses, perhaps never to see each other again. The construction industry has developed relationships, traditions, jurisdictions, methods, checks/balances, and even vocabulary, all unto itself. It is an intricate and complex process. Few owners understand the simple subtleties and outright mysteries of constructing a building. Owners most frequently suffer from being unfamiliar with the process and abundantly imbued with misconceptions, misunderstandings, and unrealistic expectations. They are wide open to bad advice, suspicion, shortsightedness, false economy, and making uninformed decisions.

It also may be necessary to make the owner's attorney part of the education process. Many attorneys are not familiar with the nuances of construction and construction law. Seeking the best deal possible for their client does not always produce the best deal nor the protection of the client's interests in a construction setting. Therefore, where necessary, the attorney needs to be made a party to the educational process for the construction of the project; perhaps the attorney will come to rely on the professional for assistance during the process.

No one in the construction industry likes surprises, and this certainly includes the owner. While it is not necessary to cast the owner in the part of perpetual student, during the progress of the project, pertinent education and prompt, insightful communication are a big help in having the owner be a full participant in the process. Part of the educational process is the elimination as far as possible of surprises, in both process and in the actual construction.

There is a need to address this educational process on a continuing basis, as the client's interests change at different junctures throughout the entire process of construction. Actually the process starts at selection of the professional and progresses throughout the negotiation of the professional services contract. A major test is the programming and design of the new facility and the incorporation of the owner's desires and needs. Their attention and full participation here can greatly enhance the success of the project, both in their eyes and in the way the new facility functions and serves the client and the staff.

The education process continues through the various phases of the professional's work: schematic and then preliminary design, where program elements become more defined and relationships explicit, where the owner begins to face what the facility will be and how it will function in "real" terms. Of course, the application of cost data provides yet another hallmark for the educational process. The process of producing the contract with the contractor also provides a test of the process, and invites deeper understanding of intent, ability, working relationships, nuances, details of operations, and so on.

The final stages of the process are the actual inner workings of the construction process; the time when conditions change, where adversity may occur, where glitches appear—the place where things don't necessarily go as planned. Only an understanding and fully responsive client/owner will withstand these ordeals; only good education will support and aid that demeanor.

The two primary issues of owner education still remain: communications and misconceptions. The list of troublesome issues is lengthy. In capsulized version, some examples of issues in the contract administration phase requiring resolution are:

- Tough, owner-oriented contracts, strongly enforced by the owner's attorney, will bring the owner's ultimate goal, a "perfect" project.
- Both architecture and construction are exact sciences.
- The design professional will guarantee that the project will function as intended.
- All cost estimates are accurate.
- The lowest bid, no matter how low, is still the best bargain.
- There is no need for a contingency fund, if the architect and contractors do their jobs properly.
- Items inadvertently omitted on drawings or in specifications are paid for by the design professional.
- The services of the design professional are unnecessary once the contract is signed with the contractor.
- Contractor claims are rare occurrences.
- If something in or on the new structure is incorrect or fails to work, then either the professional or the contractor is liable.
- Realistically, no design professional can act as an objective arbitrator in disputes over contract issues; the professional must always side with the owner.

Chapter 9 The Owner

This list alone is challenging. The all-inclusive list is much longer, but is well worth pursuing. It is essential that the professional, the owner, and the owner's attorney explore and resolve all such issues. The owner must be made to see that construction involves literally hundreds of thousands of decisions and details, just in the contract documents. Compound this with all of the other decisions and circumstances, and the fact that the documents in no way anticipate, much less address every problem or condition that may have to be dealt with during construction. They are not and cannot be made "perfect."

This concept, in itself, leads to other considerations, which act as further guidelines for the educational process. Owners need to understand.

- The risks involved, how they can be balanced, shared, and allocated, mainly according to the ability of the parties to control or minimize those risks
- The fact that neither construction nor architecture function in an exact manner, meeting all of the variations, the complexities, and the conflicts (in regulations, for example) of the job; also the fact that professionals are measured against standards of care and not a precise, absolute, or "perfection" standard.
- The network of safeguards that are built into the construction process to prevent and/or expose imperfections in the process, and the needless expenditure of wasteful money, time, or resources
- The multifaceted role of the architect/engineer, with all of the many nuances, subtleties, and complexities; how the role changes in various situations even to the point where it appears to be adverse to the owner's interest (but as required by contract)

A successful project (as defined in Chapter 2) is a direct result of a proper, complete and ongoing educational process, involving both design professional and owner's attorney (wherever possible). The more complete and incisive this process, the higher the degree of success, and the lower the rate of claims filed. The owner-education process is an inherent function that must pervade every aspect of the professional's work, and must be performed by every member of the professional staff, including the contract administrator, who interfaces, in any way for any period of time, the owner/client.

During the process of educating the owner, two pivotal issues should be emphasized fully understood, and consistently observed by the owner, the owner's attorney, and any staffers of the owner who become involved:

1. The owner must participate in the process of the project, in a proper, timely, complete, and correctly directed manner, utilizing the procedures established by the construction industry for the project (and over which the owner has little, if any, control).
2. While being the primary benefactor in the project, the owner cannot take the position of leader or prime mover in the construction process. This must be sought in others (from the design professional to the directly contracted contractors) and confidence placed in them. Ensure their performance by correctly enforcing the contract provisions, not by meddling in or attempting to usurp their authority within the framework of the process. They are retained for their expertise, not only in actual construction and design, but for their ability to interface correctly in the construction process.

It may be extremely difficult for an owner to accept and agree to both of these issues. Both the financial involvement and the overall managerial technique of the owner are pervasive here. However, while recognizing the astute and expert manner in which any owner runs his/her business, the construction process cannot, and will not, acquiesce to the entire array of demands the owner may make, no matter who the owner may be. Certainly, understanding, cooperation, tolerance, respect, and compromise are required. Governmental owners, followed closely by monopolistic and corporate owners, may be the most demanding, and may also be the most successful at "warping" the process to their procedures and desires.

Figures 9.1 and 9.2 illustrate the duties assigned to the on-site construction management personnel on the staff of a large corporation. It is easily seen that these persons are required to perform numerous tasks that do not directly affect the actual construction work. Again, this shows that, while owners may be interested in the process, their perspective and informational needs are quite extensive and different from those of the contractors and, most assuredly, the design professionals. Figures 9.1 and 9.2 tend to portray an extreme situation, but would be fairly typical of very large corporate efforts and complex and extensive projects, such as projects for military and other governmental agencies (at all levels). Obviously, where public funding (or stockholders) are involved, the need for meticulous documentation and accounting is required. This is usually modified as the size of the project is reduced. Home building may be the other extreme where little documentation is done (by the owner).

There is no single precedent between these extremes that works for all projects, since the vast majority of private sector owners are accountable only to a limited number of stockholders, their employees or partners, or just themselves. Also, most do not have the force of law or the leverage of a government agency. However, often some accommodation must be made in the construction process to fit the owner's specific situation; usually it is minimal, merely requiring careful detailing and explanation. More often traditional methods will suffice.

The worst scenario is when an owner, feeling that the service in place will be "see all, do all," then takes a very passive role in the project. This should not be permitted. The AIA document A201, General Conditions, describes very specifically what the owner is to do and what authority is seated with the owner; it is often vital that such authority be exercised. (AIA document B141 and others also address this issue.) Quite often, an owner will be so preoccupied with other matters that the construction is more of a necessary evil than a vibrant and interesting process. The bottom line is "get the building done, however you can, so I can get on with business." One must be extremely careful of this situation. This very businesslike person can be volatile if not pleased, yet offers no help in the process.

In other situations, the professional may find an owner's representative (a staffer) who is assigned to shepherd the project through construction (if not through the entire process). Be cautious if the same representative used during design and documentation is not used during construction. If they are not well coordinated, the situation can produce an uncomfortable position for the professional. Who has the support of the owner regarding the project?

On some projects the representative is continually undermined by the owner. This is often true in governmental agencies, particularly if elected officials are involved. Where there is no internal meeting of the minds, the professional can be caught in a real dilemma, and must take extreme care in documenting and communicating, so everyone involved is given identical information. In some cases, it may fall to the professional to draw the line and to require *all* parties to meet and reestablish authority, lines of communication, processing, and many of the fundamental ground rules regarding personnel on the project. Often, in the government context, the contracting interest will have more input to, and direct access to, the officials, more so than their own staff representative, and can manipulate the project, payments, extras, and general administration. Certainly, this is counterproductive to a good project and efficient operation, but a situation that is often repeated, unfortunately. It is, of course, necessary to try to ascertain the situation with which one will have to deal. This should be done as early in the process as possible. Along the way, be watchful for adverse changes, since they can create havoc, both in the administration of the contract and in the actual work.

Contract administration, in any event, should be an inherent part of the owner's agenda. An owner who chooses to abdicate her/his rights in the construction process

The site construction manager (CM) is the process owner for construction strategies, execution plans, and technologies for the site.

Functional Accountability

The site construction manager has functional accountability for construction execution of all projects on site, as well as general construction operations. The site CM reports through the site organization. Construction management customers include:

- Project managers
- Plant engineer or technical systems manager
- Sector construction manager
- Resident engineers (if applicable)

Responsibilities

The site construction manager has the following responsibilities.

- Manage the construction contracting strategy process for the site.
- Serve as process owner for site construction, represent needs, define accepted results, deliver the constructed product as per the drawings and specifications, interpret contract documents, and assure the construction organization has all necessary resources.
- Be a key member of the site construction team and establish an effective relationship with the contractor's project manager.
- Develop and maintain effective communications between the construction and engineering organizations, including establishing effective constructability and construction completion processes.
- Manage the interface between the construction organization and the plant; establish and maintain a Plant Construction Agreement; serve as the contractor's advocate within the plant organizations.
- Develop and maintain a good working relationship between the construction organization and the startup team.
- Serve as the interface point for work with various outside agencies, such as railroads, highway departments, utilities, building departments, and community organizations.
- Assure that the construction is being managed in the most cost-effective manner and the actual costs of the construction are being accurately reported and predicted.
- Participate in schedule planning; assure adequate coordination with engineering, procurement, startup functions, and other resources.
- Audit the contractor's use of good safety practices; provide a safe workplace for the employees at the site; protect assets.
- Assure optimum quality by working with engineering to generate the right specifications, planning for quality in contractor selection, developing a quality control program with engineering and manufacturing concurrence, having contractor perform appropriate amount of quality measuring to ensure good results, and contributing to problem solving in case of deviations.
- Assure that purchasing is appropriately involved in contractor's buying procedures; assure the contractor is using the agreed-upon buying procedures; monitor purchases to assure that they are required and receiving the best value.
- Enforce policies on scheduled overtime and support plant bargaining agreements; assure that the contractor is taking full advantage of its management rights; plan for potential disruptions because of contract expirations.
- Serve as principal leader of improvement program for the site.
- Participate in the sector network construction activities.

FIGURE 9.1 Duties of site construction manager.

The resident engineer (RE) is the process owner for construction strategies, execution plans, and technologies for *a specific project.*

Functional Accountability

The resident engineer has functional accountability for construction execution of the project and reports through the sector central organization. The resident engineer's customers include:

- Project manager
- Plant engineer or technical systems manager
- Sector construction manager
- Site construction manager

Responsibilities

The resident engineer has the following responsibilities:

- Manage the construction contracting strategy process for the project.
- Serve as process owner for project construction; represent needs; define accepted results, deliver the constructed product as per the drawings and specifications; interpret contract documents; and assure the construction organization has all necessary resources.
- Be a key member of the site construction team and establish an effective relationship with the contractor's project manager.
- Develop and maintain effective communications between the construction and engineering organizations, including establishing an effective constructability and construction completion processes.
- Serve as the interface point for work with various outside agencies, such as railroads, highway departments, utilities, building departments, and community organizations.
- Develop and maintain a good working relationship between the construction organization and the startup team.
- Assure that the construction is being managed in the most cost-effective manner and the actual costs of the construction are being accurately reported and predicted.
- Participate in schedule planning; assure adequate coordination with engineering, procurement, functions, and resources.
- Audit the contractor's use of good safety practices; provide a safe workplace for the employees at the site; protect assets.
- Assure optimum quality by working with engineering to generate the right specifications, planning for quality in contractor selection, developing a quality control program with engineering and manufacturing concurrence, having contractor perform appropriate amount of quality measuring to ensure good results, and contributing to problem solving in case of deviations.
- Enforce the policies on scheduled overtime and support plant bargaining agreements; assure that the contractor is taking full advantage of its management rights; plan for potential disruptions because of contract expirations; set up separate gates for construction.
- Assure that purchasing is appropriately involved in contractor's buying procedures; assure that the contractor is using the agreed-upon buying procedures; monitor purchases to assure that they are required and provide the best value.
- Work with the contractor and the site construction manager to mobilize and demobilize the construction organization and facilities.
- Serve as principal leader of improvement program for the project.

FIGURE 9.2 Duties of resident engineer.

will be buffeted by circumstances, change orders, disputes, and excessive costs. Plus, the owner will have no real means to control any of these aspects, or recourse to better solutions, short of litigation.

This lack of control will, in effect, relegate the owner to the role of bystander; unfortunately, a bystander who is footing the bill. This will create an adverse situation throughout the project and could lead to dismay, confrontation, conflicting orders,

Chapter 9 The Owner

refusals to pay for some legitimate items, and, in essence, no good for the project overall. The owner should be brought along, in the project sequencing, from the initial contract and programming. The actual construction should not be the point at which the owner is left out of the process. Some owners may desire to be excluded, but this is rare. However, they must be incorporated and used properly; the team concept must be stressed. The owner is an integral member of the team. Needless to say, substantial responsibilities come to the owner during this process. In fact, the owner is the party basically responsible for the project overall; a fact often forgotten or overlooked.

This may seem like a peculiar arrangement, but since the project is to the benefit of the owner, the legal responsibility lies with that person. It is the owner, in a sense, who is to see that the project is brought on-line, in full compliance with all regulations and all other parameters, including those not easily discernible, apparent, or known.

In large measure, the owner retains the professional to act as his/her direct representative on the project (the status of "agent" must be separately established). This, however, should not be perceived to be a complete release of control, whereby the owner makes no more contribution to the project's sequences. Usually, the more money involved, the more control the owner will want to retain. It is important that this retained control be clearly defined and continually exercised. The best scenario is for the owner or a designated representative (owner's staffer) to participate in all job meetings, to be part of the distribution system for correspondence, to be an integral part of any decision-making forum, and to ensure that all other participants understand that the owner desires to be (and will be) a part of the project. Of course, it is necessary for the wise owner to delineate how that participation will manifest itself.

It should never be forgotten that the project is "the owner's game." The ultimate cost and the use of the project are the top priorities of the owner, and every other participant must understand how that affects their particular area of work. This concept pervades the entire project. Not that the owner must be asked every question, must be part of every technical decision, or that suggestions are forestalled. However, the owner has the right to (and should) require that the professionals and the contractors "bend" to his/her agenda and procedures, so long as those items are reasonable and appropriate to the project.

This principle begins with the first professional contract and goes through programming, design, design development, and into the actual project documents. In the main, project procedures should be the owner's and not something superficially implanted by a standardized document or the thoughts and procedures of someone else; unless this fact is made known to the owner and permission is received to proceed in another manner that may be suggested. Here, the professional must be ready to guide the owner, where no input is forthcoming, or where the owner's thoughts may be inappropriate. Since the expertise of owners varies a great deal, we submit that there is a distinct obligation on the professional, and the other professionals and contractors, to include, accommodate, educate, and openly offer participation to the owner. The more sophisticated the owner (corporation, large developer, governmental agency, etc.), the more expertise will be brought to the project. At times, this expertise may be fragmented and even hesitant, but many times it will be demanding and quite astute. Obviously, the other participants will (must, in fact) bend to those demands and work in the environment set forth by the owner.

Where the owner is less sophisticated, or where the expertise is wanting, it is unprofessional, unethical, and a grave disservice to the client for the professionals to twist the project strictly to their advantage and self-interest. Eventually, in this scenario, the owner is going to be abused by the process, in one way or another. The relationships with the owner will then deteriorate quickly. Better to include the owner in the process, unless he/she specifically asks to be excepted. The project does not belong to the design professionals and the contractors, and they should assiduously avoid any illusions, actions, or appearance of same. They should not assume that they can act for the owner, speak for the owner, "assume" for the owner, or give others the impression that their word is the owner's and hence "final."

Early in the project, both the owner and the professional, independent of each other, must decide how the owner's participation will be effected. The wise owner will designate a representative and delegate proper and commensurate authority to that person to meet the project's demands. This is most important, since anything short of this will result in the representative being little more than a note taker. Granted, a properly authorized rep must be trained, knowledgeable, experienced, and proactive in the process. Again, large corporate or governmental owners usually have such persons, simply because this is the easiest and best way to deal with the construction process. Of course, the organization will retain the right to make any big decisions that will have a drastic effect on the project overall.

A representative with little authority and little or no support from the organization is rather worthless to the project process. There is an excellent chance that progress will be impeded, often due to the hesitant nature of the rep. Often, agencies permit the rep to make routine, technical decisions, but will either acquiesce or "stonewall" financial or self-interest decisions, in spite of the advice or input from the rep. This directly undercuts the rep and the project, and gives the other participants the uneasy feeling that several masters must be served; an almost impossible task. Also, it leads to circumvention of the rep, which only further decimates the project procedures. In some cases, corporate, institutional, or governmental situations, "owner" usually becomes a cast of persons, who collectively have a deep interest in the project, but, with diverse (and not always coordinated) perspectives. The resulting battle (anything from minor irritations, to conflicting information, to open and hostile differences) must be resolved within the owner's organization. The professional should not be drawn into any of this (or, voluntarily inserted). Obviously, taking sides is the last thing in which the professional should engage. The professional is well advised, though, to point out the shortcomings in the process, where project procedures or progress are impaired, or indications are that they will be impaired. This strikes directly at providing the owner with the project, as desired, when desired. Where the owner imperils that service track, the professional must speak out, in a factual manner, without addressing authority, personalities, or the owner's bias. In a sense, the professional is contractually bound to protect the project from the owner, as much as from other outside influences (particularly where they are adverse).

For example, a university president is very interested in the prestige of a new building. It adds to the inventory of the university and to the overall impact, service, and standing of the facility. He is not interested in details and will be involved only in the major factors of complete funding and selling the project to the Board of Regents, He will, of course, be very astute and charismatic at the ribbon cutting.

The vice president for development, too, will be involved at a very high level, but, will find some battles to be fought over funding, timing, and incorporation of the facility into the whole scheme of the university plant. Here, too, no intricate detail is necessary.

Obviously, the dean of the school to be housed in the new building will have a deep interest. How will the school's image improve? Will certification be enhanced by the new facility? What problems can be solved? What new programs can be accommodated? Again, no real details of construction have surfaced, although paint and finish colors could eventually be part of the dean's purview to some degree.

The department heads in the school will confine their perspectives mainly to that area assigned to their activities. They and their faculties will assess layouts, interrelationships, future expansion, solving current problems with new space and equipment, and so forth. This will begin to involve detail, but not so much of the construction devices and materials as the "moveables," finishes, and space relationships.

Each of the above persons may appear on the site on occasion, but they will not attend job progress meetings. Even their walk-throughs will be sporadic, increasing as the project approaches completion.

More than likely there will be a university representative who will attend the meetings and shepherd the project. The perspective of that person will be twofold:

Chapter 9 The Owner

addressing the concerns of the university overall, and addressing the needs of the project and project personnel. Usually, this person will be part of the facilities management unit and is assigned continual service in construction projects, large and small. While perhaps not part of the project planning, this person will factor the needs of the client into the operations of the project. This may begin with standardized contract documents, from distinct clauses, to required materials or systems. These will constitute the "owner's ground rules" and all participants on the project should understand that they must observe and play "the owner's games." It is far easier for owners to maintain an even and unchanging format than it is for them to move to different modes of operations on each of their projects and with varying professional involvement. Most of the other project participants, in moving continually from job to job and owner to owner, have a greater flexibility and can adapt more easily to any prevailing demands. In fact, it is incumbent on the large owner to have a standard set of document forms, procedures, formats, and perhaps even material lists, which are reused on each project. This does not inhibit creative design, but it does narrow the need for assessing new materials and being forced to accept unknowns as part of the project. In addition, it can greatly reduce the need to warehouse an overabundance of varying spare parts. All of this tends to reduce the risk and function, in the future, on the part of the owner.

It must be remembered, though, that the "facilities person" is *not* the construction manager, but certainly is the active eyes of the owner. This person may be charged with overall project control, cost assessment and recommendations, cost adjustments, assessment of design, as well as discovery of excesses, assessment of progress and staffing to establish timing, occupancy (partial), and so forth. Actually, this person often sets the tone of the entire project by sharing the owner's concerns and goals, in addition to offering unique insight and expertise.

An owner is also well advised to include some maintenance personnel in the design and construction process. This is an especially good idea for the programming and schematic design work. These persons bring well-founded "working" observations, and usually can reinforce what the facilities person will require. Since these people must service the project after the construction is completed, their input is invaluable, both to the professionals and the owner. From spare parts, to operating manuals, to the right paper towel holders, this person or persons will have distinct ideas, and must have direct input to the design documents, as well as the form in which the project is brought on-line; this dual role is essential. In this, the new project will easily be assimilated into the overall existing management scheme, and will not become an "oddball" that has unique aspects that confound purchasing, servicing, and normalized operation. Also, this is another set of eyes who can assess excesses in design, needless or improper virtues or construction, and situations or materials that will cause problems in the future. These persons know what works and what fits the management of the owner's entire complex; this is a set of eyes solely aimed at reality. Their input should be properly routed through the owner's primary representative, and not taken as a "second track" that may, for various reasons, be played off against others.

It must be recognized that where the owner is less organized and sophisticated, most, if not all, of the above exercises must still be performed. It could well be that only one person, the owner, will be involved. Obviously, the professional must be extremely careful here. While these issues must be addressed and resolved, there is less time available; the owner must continue his/her normal business operations. Meetings must be as short and as pertinent as possible; information sharing must be maximized. Issues may be more simply resolved, but the prime thought is that they be addressed, as the project requires. To do otherwise is to shortchange the project and the owner.

Unlike this single proprietor situation, and in far too many corporate situations, there is a lack of proper authority from the owner; there is a trend to retain control at the highest level. Too often, others will confuse the process, and often can cause overbudget and off-schedule projects. Hopefully, this will occur early in the project, so it can be resolved. Input is necessary, and single authorities often overlook, ignore, or forget the input of others, even their subordinates, and thereby cause problems in a smooth-running project. While the design professional can usually rectify these matters, he/she must be aware of their existence and must be ready to deal with them. The obvious thing here is to carefully and completely document everything done on the project. The professional must be ready to defend his/her position, the need for added funding and/or time, and additional fees for extra work required. The professional must be as objective as possible in assessing the malaise, if and when it occurs, and in carefully avoiding being swallowed up into it.

PROTECTION OF THE OWNER'S INTERESTS

The owner involved with any particular construction project is vitally interested in the building process but, at the same time, may be ignorant of what is happening. The owner's interest, of course, follows on the expenditure of a tremendous amount of money. Whether the owner is an individual, a corporation, or a governmental agency, none wants to pay for a project not worth the money expended.

The owner for the most part wants dearly to be involved in the construction process, but the wise owner sees how inadequate he is to the task. Except in rare instances, he would do well to involve a design professional whom he can work with, talk to, and confide in. The design professional, in serving him properly, becomes his agent, in most instances acting for him on the project and, because of contractual obligations, trying to protect his interests.

The owner, however, may very well see the design professional, even if he has full confidence in him, as just another person in the process who may try to invade the owner's interest. An error of any sort by a design professional may drastically change the project for the owner. A decision that the design professional makes may not be 100 percent in keeping with what the owner intended or anticipated. So, above all, the design professional has the responsibility to gain as much confidence from the owner as possible and to carry out the entire contractual agreement, to the letter, in a very open and complete manner.

The owner, however, still may want to be intimately involved in the project. If he is a homeowner, for his own satisfaction he often may visit the site nightly after the workers have gone, checking on progress and quality. He is really, in a sense, finding out what he is paying for. An owner who walks into an existing building and buys it as it stands does not have the problem of seeing the project in the "rough" and following the actual finishing process. He buys it as a package, much as he would buy a suit of clothes off the rack. Flaws may exist, but they are not nearly as apparent as when they are exposed one at a time during the fabrication. So it is with construction projects.

Almost everyone has some knowledge of construction and how things are put together. The owner may have helped build his childhood treehouse, but probably has not seen other buildings being built. In his adult working years, the owner has certainly seen things in buildings, some of which work for his business and some of which do not. It is to the architects' benefit if the owner has experienced a wide range of solutions or at least has some flexibility. The owner, at any rate, forms certain ideas about the best way to accomplish his project. It makes no difference whether this is a fairly modest home, a corporate headquarters, or a large governmental complex; owners have a tendency to draw certain conclusions.

The agents of the owner (staff members, not the contractual agents such as the architect) may also be on the job. They too may have preconceived ideas of what the project should be. In many instances, the owner's little knowledge can, according to the traditional axiom, be dangerous. The design professional must be able to take

Chapter 9 The Owner

the knowledge of the owner, control it, and channel it properly. This is done by preventing confrontations between the owner and the contractor, by processing complaints, questions, and inquiries, or by educating the owner—the architect literally taking the owner by the hand and explaining the process.

In many instances, the owner can become frustrated to the point that he feels he is strictly the money man. He is giving up a tremendous sum for a commodity that he doesn't know much about, and he really isn't sure what he is getting. The homeowner, for the most part, unless he wants a custom-built home with a design professional involved, has very little in the way of help to call upon. He must rely on himself and the builder or perhaps on a friend or relative who knows a little something about building. The reliable builder will take on some of the responsibility of the design professional and will helpfully educate the owner whenever an inquiry is placed. The unscrupulous homebuilder will, of course, try to gloss over, ignore, or cover up work that is shoddy or substandard. Even if the owner has his "inspector's hat" on, he may be talked into accepting some work that is not in his best interest.

On large projects, the corporate and governmental owner may assign some of their staff members to oversee the construction. This, of course, is direct control by the owner. The representatives can attend job meetings, be consulted on any major decisions that are made, and be a member of the actual control team of the project. There is really no inherent problem with this type of control. The owner, however, and his representatives must understand, as would any other participant, the system in which they are participating. They must also be willing to give up any direct communication with the contractor, communicating only through the proper channels of the control team.

It is true that the owner has a direct contractual agreement with the contractor, and it would seem that this direct line of contract obligation would allow the owner to contact the contractor directly. However, this could lead to confusion on the project and could greatly impede the progress of the work. Direct contact would completely disrupt the communication system and would leave the design professional or the construction manager in very awkward positions; they would not know what the owner told the contractor, and it is very difficult to evaluate or control something you know nothing about. An aggressive owner must be controlled. He must be told emphatically in no uncertain terms that he must work through his design professional and the established lines of communication.

A project team that has lost the owner's confidence (or never gained it in the first place) is in deep trouble. If the owner attempts to circumvent it, the entire team will be rendered ineffective, resulting in a very expensive nonentity (and a solid basis for litigation). An owner who chooses to deal directly with the contractor works to his own detriment in the long run. He should be made to see the tremendous advantage of coordinating his efforts and communications with and through the project team, starting with the design professional. It cannot be emphasized enough how important it is for the control team to have the owner's confidence and to have the owner understand that he must work through that team and not go to the contractor directly.

Policing the Contract

The owner may be party to a number of contracts on any given project. There may be one with the design professional, one with the construction manager, and one, of course, with the contractor or on some projects, contractors. His primary effort on the project is one of paying off these contracts as the terms are met. The owner is the one common denominator of all these contracts, being party to all of them. There are no contracts between the architect and the manager, the manager and the contractors, the contractors and the architect, and so forth.

The owner can be likened to a driver of a large team of horses, holding many reins in his hand, trying to control each horse so that they pull together doing the work required by the driver. This analogy means the owner must work with the architect, the manager, the contractors, and the other contractees individually, while at the same time pulling them together into a working team to complete the project. (See Figure 9.3.)

Any misunderstanding, harassment, or adversary action between any participant and the owner can only be an irritant to everyone involved and can greatly impede the work. For example, an owner who becomes disenchanted with the consulting engineer will have an extremely hard time working with this individual. However, he would also put the other design professionals in a uncomfortable position. Although the owner does not have a direct contract with the consulting engineer, the architect does, and the architect is then in the peculiar position of trying to maintain his credibility with the owner, keeping the engineer under control, and at the same time maintaining the progress of the project.

This is not to say that the participation of the owner is unwanted or unnecessary. But when the owner chooses to participate, he is well advised to understand the project as a whole and to understand what his comments and his desires might produce if they are misdirected. (See Figure 9.4.)

There is no way that an owner can be kept away from the job or out of the process. If the participation of the owner becomes a cumbersome, time-consuming impediment to the work, then he is working against himself and he has no one to blame for budget overruns and the missing of time deadlines. Of course, the owner will not always see this and will try to blame other participants. We can only advocate that the owner think carefully about just how he wants to participate. That participation should be based on his confidence in the other contract members.

Corporate and governmental owners may have a policy or statutory requirement to participate in and oversee the project. But again, there is always a supervisor over an owner's representatives. The supervisor and the representatives should all understand exactly what they can do, what they can gain and lose, and how their participation can directly affect the work.

Decision Making

In instances where the owner is represented on the project, the major benefactor of that presence is the decision-making process. With constant representation on the project and with authority to act for the owner directly, the representative can greatly expedite and simplify decision making.

The representative of the owner must, of course, be thoroughly familiar with the inspection process and must actively participate in it. The representative simply must keep abreast of the progress of the job. And it is this representative who has the luxury to reinspect the work at his leisure to be sure that there has been no deterioration and that work previously installed is, indeed, still installed properly.

Construction sites are often vulnerable to a great deal of vandalism and damage. It is good that the owner's representative is able to inspect the job over and over again, including work that has been previously completed. This is not to say that this work is totally ignored by others on the inspection team. But basically inspection is ongoing, and the other participants do not have the time for reinspections. With his proper inspection process, the owner's representative can keep abreast of the job and may be ahead of the job if he can foresee what is to come and what problems are at hand.

FIGURE 9.3 A good description of the owner's role.

> As an owner, you must establish an atmosphere where mutual support of a common goal is the only acceptable behavior, and where competition, at the expense of others, is not tolerated.

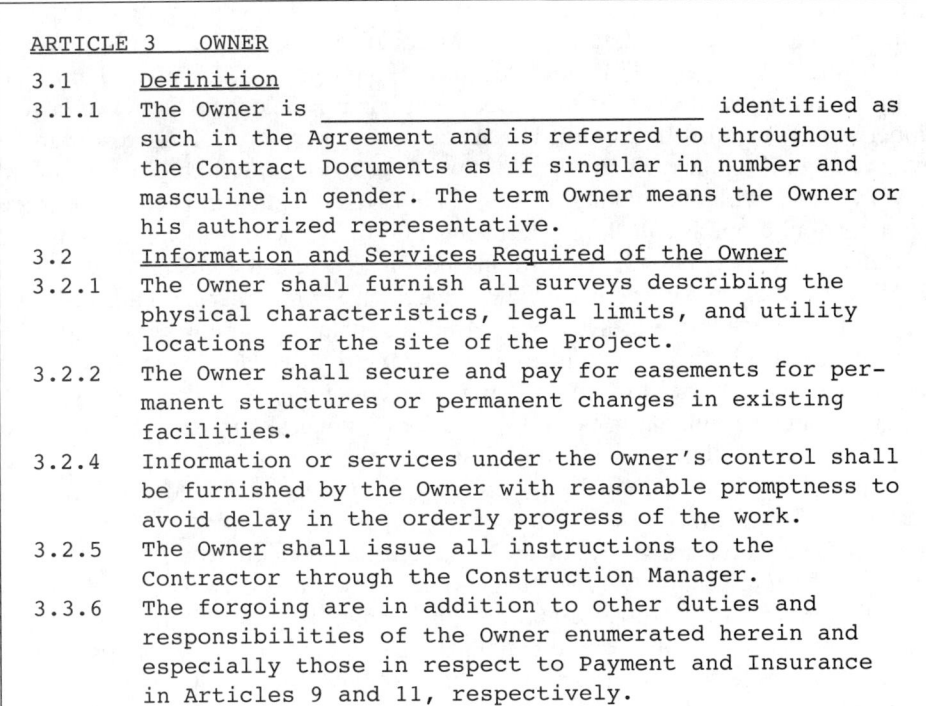

FIGURE 9.4 Excerpt from typical specification showing the areas where the owner will be active and responsible.

The owner's representative should not replace any member of the inspection team, but he is an additional member with another point of view and a different interest, serving the decision-making process as well. The delays of communicating with the owner are minimized, and a representative, if he is properly authorized, can make ongoing, immediate decisions as the job progresses, saving time, money, irritation, and involvement to all on the project. An owner who may be far away from the project can present a problem to the progress of the job. If decisions are not forthcoming immediately to meet the conditions of the job, contractors will often put in claims for extra time based on the interval needed to receive the decision.

This area can be a source of irritation, aggravation, and even hard feelings, because it is never the owner's intention to purposely delay the job. Even simple communication processes, such as telephone, telegraph, or mail, can break down, thus impeding the job. Sometimes, too, an owner remote from the project will not be thoroughly abreast of the job progress and will need time to become informed. He may have to study the drawings, previous reports, and so forth, before being able to make a good and proper decision.

Delays can occur again and again, leading to an added cost of time, personnel involvement, rentals, interest on loaned money, and so forth, to complete the job. Today, when money is extremely tight and profit margins are very small, the decision-making process must be expedited where at all possible. Perhaps one of the earliest ways is with the on-site representation of the owner.

Cost Control

The importance of cost control in the modern construction process cannot be emphasized enough. Numerous projects have run into a good deal of trouble because of the cash flow problems of the contractor, delays in decision making, a general slowdown of work, and the inspection process. The owner's representative can, of course, be one of the major forces behind an aggressive approach to the proper progress of a job. His major interests are to prevent extra charges of any kind to the owner and to deliver the job properly and on time. Keeping these two factors uppermost, coupled with the decision-making process, control of the costs can be achieved, but it requires all members of the inspection team to fully participate and to work closely with one another. It is essential that mandatory inspections be made promptly and properly, which in turn means that the work to be inspected must be executed, finished, and made ready for inspection according to schedule.

All the participants in the inspection process have some financial interest in the job itself, either for themselves or for their agencies. A building inspector, for instance, coming onto the job only to find that no progress has been made since his last visit is wasting the taxpayers' money. Job conditions will dictate changes in progress, but the building inspector should have enough knowledge of the job so that he schedules timely inspection trips. The owner is interested in seeing that things are inspected so that the progress of the job is not impeded or stopped because of faulty work, and so on for each participant.

Everyone endorses progress, but inspection must accommodate it. It is to no one's benefit to have inspection lag. Eventually some faulty condition will be found, which may then require work stoppage and remedial action. If inspection lags and allows work to move substantially beyond the faulty condition, repair work and cost factors can be greatly effected. In such cases the inspection system has totally broken down in relation to the total project.

Cost control in modern construction can be one of the most confounding factors. Many are quick to allow the design concept to dissipate for the sake of the job progress and saving money as the project proceeds. Although this practice should be minimized or not permitted at all, it is constantly at work on the construction site. Whether the project is being delivered as originally conceived depends on the dedication and commitment of the owner and his representative. Many times the representative will allow or actually force work to be changed for cost control purposes. These changes, if made without full coordination with all members of the construction and inspection team, are a tremendous source of confusion and irritation. This is particularly true of the design professional and the contractors involved. There is no indication that the project is short-changed at this juncture. It must be pointed out, however, that for cost control to be truly effective it must be an intimate part of the overall project control and decision-making processes. That, in turn, is an intimate part of the inspection system.

PROVISIONS OF AIA DOCUMENT A-201

When architects, contractors, and attorneys gather to discuss legal concerns, the content of AIA document A-201 is almost certainly a topic. As the accompanying paper indicates, not everyone is in agreement with the standard provisions in this document. The current edition of A-201, still in use, is the 1987 edition, which is discussed in the accompanying article by attorney Mary McElroy.

The AIA cautions against making changes to the printed provisions of its standard documents, including A-201. However, changes to the documents resulting from negotiations, unique circumstances or requirements, or other factors are not uncommon. However, it is highly advisable that when changes are proposed, an attorney familiar with the AIA documents and their close interrelationships be consulted. If changes are contemplated (and where no changes are proposed but a better understanding of A-201 and its effect is sought) a review and examination of the substitute provisions proposed by McElroy will be most helpful. More than a few experienced clients (and their attorneys) have developed alternative language for A-201 to fit their specific needs. But even seasoned owners, and most certainly those who are relatively new to the construction process, can benefit from McElroy's "second opinion" as to the interests that are to be served by the language found in the standard A-201 document text and in the recommended alternatives.

As with any "form contract," changes to meet the needs and desires of the contracting parties should be expected. However, as McElroy advises, the interrelationship of the language in the standard form AIA documents normally used in conjunction with A-201 require a careful assessment of the impact that any change to the standard form language will have. One common result of less than thorough modification of the standard form documents is that conflict can be created between the provisions within the same set of contract documents. This, of course, is a form of legally troublesome ambiguity.

Most would agree that knowledge is power, especially in construction contract negotiations. For that reason "An Owner's Response" is presented to balance the viewpoints that are normally present when design and construction contracts are written. Additional help and insight may be found in the following:

1. Dundin, Joseph, "A Hundred Years of AIA Standard Documents," *Architecture Magazine,* October 1988, pp. 119–121.
2. Dundin, Joseph and Ellickson, Dale, "Changes in a Dozen Basic Documents," *Architects Magazine,* April 1987, pp. 95–97.
3. Ellickson, Dale, *An Analysis of A201 Revision: Its Impact to Date,* a seminar presentation by D. Ellickson sponsored by Loyola University (New Orleans), Spring 1989 (revised July 5, 1989). Available through the American Institute of Architects Documents Program (AIA, Washington, D.C.).

An Owner's Response to the Use of the A-201 (1987) General Conditions: Major Problem Areas—Owner Recommendations[1]

Mary J. McElroy
Attorney-At-Law
Templeton, California[2]

Introduction

Few owners will take the time or will want to incur the expense inherent in the review and negotiation of the General Conditions provisions of their agreement with the contractor. Fewer still will develop their own construction contract for-

---

© 1988 Mary J. McElroy, used by permission.

[1]The author of this textbook and the author of this reprinted article caution readers that A-201 and other documents published by the American Institute of Architects are subject to revision. Therefore, the authors and the publisher refer the readers to the "Notice to the Reader" that appears on p. xiii of this textbook. Readers who have questions about publications from the American Institute of Architects or their use should contact the American Institute of Architects or legal counsel.

[2]Mary J. McElroy, former chair of the American Bar Association's Forum on the Construction Industry, is an attorney in private practice specializing in construction law. At the time this article was originally published, she was senior attorney with Homart Development Co., Chicago, Illinois.

mats. What potential problems await the owner who uses the 1987 A-201? How does an owner know if the provisions of this document are appropriate for his use?

There are two main reasons to modify the A-201 General Conditions:

I. *The accuracy with which the A-201 describes the intended roles of the parties during the Construction Phase.*

The unaltered or unmodified use by any owner of the A-201 (1987) General Conditions should vary depending on:

1. *The owner's level of construction expertise:* The knowledge and understanding which an owner has of the construction process and the relationships and responsibilities of the three primary parties: the owner, the contractor, the architect/engineer; and,

2. *The degree to which the owner intends to actively participate in the construction process.* The A-201 anticipates a relatively passive owner who relies for his contract administration services on the architect/engineer. A school board of laymen, for example, or a church group, might place more reliance on the standard provisions than a professional developer, with an in-house contract administration capability or an on-site (field) representative.

Note: When the owner intends to alter the role of the architect during the construction phase, attention should be given not only to the A-201 General Conditions to the Agreement Between the Owner and Contractor, which describes the architect's role in great detail, but also to the Agreement Between the Owner and the Architect. If you are following the AIA format this typically would be the B-141 agreement.

Where references are made to the architect's role as described in the A-201, a comparable reference will be included for the B-141. Some of the suggested language changes relating to the architect's role belong in the B-141, as well as the A-201.

II. *Legal reasons for modification of the A-201.*

The extent to which the standard provisions provide suitable protection to the owner. In this regard, I would recommend that all owners review and at least consider possible modification of at least the following major topics:

1. Definition of the Contract Documents.
2. Detailed Payment Provisions: required backup, lien responsibility and removal.
3. Insurance.
4. Indemnification Provisions.
5. Termination.
6. Warranties.
7. Assignment.
8. Arbitration.
9. Ownership of documents.
10. Changes in the commencement of the statute of limitations.

Note: No attempt has been made to provide an analysis and modification on a paragraph by paragraph basis of provisions which would be favorable to owner. Such an exercise would result in an entirely redrafted document and one which, in all probability, would require substantial negotiation with the owner's contractor. Rather, an attempt has been made to highlight major areas of owner concern.

Outline for Owner Review of A-201: General Conditions (1987)

I. *Accurate Description of the Role of the Parties.*

Not only is it important to make certain that the A-201 accurately describes the roles of the owner, architect and contractor as those parties intend to perform them, it is important to have a clear understanding of the services or work which architect and contractor traditionally perform. The more controversial aspects of the agreements are outlined below:

1. *Architect's Responsibility for Design.*
   a. Within the owner's budget. Article 5, Section 5.2, page 6, B-141.
   b. Building Code Compliance
   c. Express and implied warranties.
2. *Architect's Role During the Construction Phase.*
   a. Limited review of shop drawings, product data and samples. Article 2, Section 2.6.12., page 3, B-141. Article 4, Section 4.2.7, page 11, A-201.

   Clarifies that the architect has a sufficient period of time for adequate review.

   Clarifies that this review is "for the limited purpose of checking for conformance with information given and the design concept expressed in the Contract documents" and not "for the purpose of determining accuracy and completeness of other details such as dimensions and quantities or for substantiating instructions for installation or performance of equipment or systems designed by the Contractor . . . "

   Most owners do not comprehend the significance of this limiting language nor do they understand how detail design actually occurs in the shop drawing phase. The new 1987 language creates some practical problems for the owner, particularly when both the architect and contractor decline responsibility for shop drawing review. A prudent owner will need to consider the following protections:
   1. expand testing and inspection,
   2. institute independent peer review of key design aspects such as structural design,
   3. hire special consultants,
   4. require that all subcontractors who provide shop drawings have those drawings stamped by a licensed professional engineer who carries Errors and Omissions insurance of a stated amount.

   b. "Observation" not inspection or supervision. Limited responsibility for contractor's construction of the project in accordance with the Contract Documents. (Drawings and specifications.)

   Article 2, Section 2.6.5, 2.6.6, page 3, B-141. Article 4, Section 4.2, pages 10–11, A-201.

   c. Qualified Certifications for Payment. Article 2, Section 2.6.10, page 3, B-141. Article 9, Sections 9.4.1 and 9.4.2, pages 16–17, A-201.
   d. Architect's role as first tier arbitrator of disputes between owner and contractor. Article 2, Sections 2.6.15–2.6.18, page 4, B-141. Article 4, Sections 4.3 and 4.4, pages 11–12, A-201.

   If you accept the premise that the architect should be involved in resolution of claims between the owner and contractor, then paragraph 4.4 is a thoughtful, detailed approach to the problem. There is, however, a basic problem with the concept from an owner's perspective. Most owners prefer to see architect render decisions on technical/aesthetic matters and to provide advice to the owner when a dispute arises between the owner and contractor.

Certainly, there is a rationale for involving the architect—because he/she has first-hand knowledge of the problems involved and because the process of dispute resolution increases in cost and time when you have to turn to parties outside the triangular construction group (owner—contractor—architect) for assistance. However, it is awkward and confusing for the architect to be owners' representative *and* arbiter of disputes in which capacity he may be required to decide against the owner.

There are also other problems which arise: The provision can be viewed as one which fosters the making of claims which might otherwise be dropped, and bogs down the construction process in claims preparation and presentation.

Paragraph 4.3.2 requires claims relating to Errors and Omissions to be included in list of things on which architect is to make decisions. It is unrealistic to expect an architect to render a decision against his own interest. Paragraph 4.3.3. requires claims to be made by written notice by either owner or contractor within 21 days of the event giving rise to the claim, or within 21 days after the claimant first recognized the claim. There is also 45-day "cooling off period" during which there can be no arbitration or litigation. This appears to have the effect of avoiding an accumulation of claims at the end of the job. However, at the end of the job, parties are willing to overlook many things which might otherwise be asserted as claims, if the relationship ends up to be a good one (if parties' expectations about the contract are fulfilled). When expectations are not met, claims are resurrected and accumulated.

It remains to be seen whether this provision will encourage claims—particularly on small items and whether claims review and processing can occur constantly throughout the job and still enable timely construction to take place. Paragraph 4.5.4 provides that if the architect renders what he identifies as a final decision, a demand for arbitration must be made within 30 days or the architect's decision becomes final and binding on the owner and contractor. If a party fails to assert a claim, presumably it is barred until the 21-day rule is applied. If party fails to demand arbitration, the decision of the architect is final. (See discussion on arbitration.)

The owner should evaluate this claims process to determine whether he wishes to be bound in this fashion.

    e. Additional services of the architect. Article 3, Sections 3.3 and 3.4, pages 4–5, B-141.

3. *Contractor's Responsibility for Design Issues.*
   a. Shop Drawings. Article 3, Section 3.12, page 9, A-201. Article 4, Section 4.2.7, page 11, A-201.
   b. Drawings and Specifications—Building Code Compliance. Article 3, Section 3.2.1, page 8, A-201 and Section 3.7.3, page 8, A-201. The contractor is liable unless he recognized the Errors and Omissions and knowingly failed to report it to the architect. If the contractor performs work knowing it involves a recognized Error and Omission without notice to the architect:

   > "the contractor shall assume responsibility for such performance and shall bear an *appropriate* amount of the attributable costs for correction."

Problems with the current language: How will you ever show that the contractor knowingly failed to report an Errors and Omissions? A better

Chapter 9   The Owner

standard is that the contractor has responsibility if he knew or should have known in the exercise of reasonable care of the error or omission.

The concept reappears in Paragraph 3.7.3, page 8, A-201. The contractor is not liable for failure of Contract Documents to comply with building codes unless he knows of noncompliance and proceeds. Then the contractor assumes full responsibility for work and costs.

Owner-Recommended Provisions to Replace A-201 Provisions

Revised Paragraph 3.2.1

The Contractor shall carefully study and compare the Contract Documents and shall at once report to the Architect and the Owner any error, inconsistency or omission which he discovers or should have discovered in the exercise of reasonable care.

The Contractor shall not be liable to the Owner or the Architect for any damage resulting from any such reported, errors, inconsistencies or omissions in the Contract Documents.

The Contractor shall be responsible for any costs incurred by the Contractor resulting from errors, inconsistencies or omissions in the Contract Documents which were not reported to the Architect and Owner and which, by reasonable study of the Contract Documents, should have been discovered by the Contractor.

The Contractor shall not perform any portion of the Work at any time without Contract Documents, including, where required, reviewed and approved shop drawings, product data or samples.

Revised 3.7.3 and 3.7.4

The Contractor shall carefully study and compare the Contract Documents and shall at once notify the Architect and the Owner if he observes or should have observed in the exercise of reasonable care that the Contract Documents are not in accordance with applicable laws, statutes, ordinances, building and fire codes, or rules and regulations of governmental entities having jurisdiction over the Project.

The Contractor shall not be liable to the Owner or Architect for any damage resulting from any such reported failures of the Contract Documents to comply with applicable laws, statutes, ordinances, building and fire codes, or rules and regulations of governmental entities having jurisdiction over the Project.

The Contractor shall be responsible for any costs incurred by the Contractor resulting from failures of the Contract Documents to comply with applicable laws, statutes, ordinances, building and fire codes, or rules and regulations of governmental entities having jurisdiction over the Project, which were not reported to the Architect and Owner and which, by reasonable study of the Contract Documents, should have been discovered.

4. *Contractor Warranties.*

Article 3, Section 3.5.1, page 8, A-201.

5. *Changes in the Scope of the Work.*

(In the Contract Sum or the Contract Time.) Article 7, pages 15–16, A-201.

7.1 now requires a contractor to sign a change order and develops the concept of a construction change directive (where only owner and architect sign while the contractor is obligated to perform the work). This was considered necessary to deal with problem of performing changed work promptly—in order to maintain the schedule before the cost of the change can be determined. From the owner's standpoint, once changed work is authorized, the owner is liable to the contractor for the reasonable value of goods and services provided.

II. *Legal Considerations.*
  1. *Definition of Contract Documents.*

Paragraph 3.12.4 (A-201) specifically excludes shop drawings as part of the Contract Documents for the reason that they are prepared by contractor (or his subcontractors) to show how he intends to implement the architect's design. To the extent that they are required by the Contract Documents and to the extent that they are *reviewed* by the architect—even if not required, the prudent owner would want shop drawings either to be implemented or revised and resubmitted. An owner wants the contractor to be contractually obligated to build in accordance with those shop drawings.

Recommendation: Modify paragraph 3.12.4 to state that required, reviewed shop drawings are Contract Documents.

  2. *More Detailed Payment Provisions.*

The owner should consult his construction lender (if any) regarding the lender's requirements for review and processing of partial and final payments to the contractor. The lender's requirements should be incorporated into the agreement with the contractor. The following items are owner-oriented provisions to be added to Article 9:

  a. Coordination of application for payment with owner's construction loan draw.
  b. Obligation of contractor and subcontractors to provide waivers of lien (through proceeding payment—would apply at minimum to all applications for payment after the first).
  c. Waivers/releases of filed lien in form required by owner, any lender of owner or owner's title insurer.
  d. Greater clarification regarding retainage (amounts, times for disbursement, exclusion from Application for Payment until a separate request for release is approved.)
  e. Title to material passes from contractor to owner—upon payment *or* incorporation in work, whichever occurs first. Paragraph 9.3.3 now says that: "title to all work covered by the Application for Payment passes to owner *no later* than the time of payment."
  f. From an owner's standpoint it would also be appropriate that a contractor supply a sworn statement listing all subcontractors/material suppliers, the amount of their respective contracts, including adjustments due to the new concept of "construction change directives" as well as change orders, and the following representations:
    1. No known mechanic's/materialman's liens.
    2. All bills due/payable have been paid or will be paid from proceeds of the Application.
    3. There is no known basis for filing mechanic's lien on project.
    4. Waivers of lien have been or will be obtained from all subcontractors, material suppliers—in form proper to effect the waiver.
  g. Clarify that it is the contractor's responsibility to remove filed liens promptly (from county records).
    1. By recording release of filed lien (compromise/settlement).

2. By providing bond pursuant to state statute—the proceeds of which would substitute for and remove lien of record (if possible in that jurisdiction).

3. Also appropriate to specifically require the contractor to indemnify and defend the owner against subcontractor nonpayment claims—as, in some jurisdictions it is questionable whether the basic AA-201 indemnification provision will cover that item. It is typical now that a subcontractor with a payment dispute with his contractor will not only foreclose on his filed lien, but will sue the owner directly on one of the following theories:

   a. for failure to pay value of goods and services rendered (*quantum meruit*)

   b. breach of some statutory duty, such as a duty to hold a certain percentage of funds until the end of the project.

   In the event of such a lawsuit, the fact that a contractor was obligated to obtain a release of the lien would not be sufficient protection for the owner.

h. It is important to clarify that an owner can withhold amounts necessary to cover subcontractor's claim (Paragraph 9.5).

i. Consider possible joint check language to allow the owner to pay a Subcontractor directly without breaching its duty to pay the contractor.

Such a provision is set forth below:

"In the event bond claims and/or mechanics lien claims are filed by any subcontractors, materialmen, suppliers or mechanics of Contractor, Owner may, in its sole discretion and without prejudice to any other rights or remedies which it may have against the Contractor and without prejudice to any payment bond (a) hold money due Contractor sufficient to protect Owner against such claims or (b) make payments by joint check payable to the claimant(s) and Contractor jointly, and charge the payments against the Contract Sum due the Contractor. In the event that the Owner elects to make payment by joint check, as set forth above, the Contractor hereby consents and agrees that such payment shall satisfy the Owner's obligations to make payment to the Contractor for such amounts."

j. Paragraph 13.6.1 provides that the owner will pay interest for late payments. Owners may prefer to strike this provision from the document to avoid further grounds for disputes over amounts owed.

3. *Insurance.*

At the very least, an owner should provide the A-201 basic insurance provisions to his risk manager or insurance broker to verify that he has or can obtain the types of insurance required of him, to avoid double coverage issues, and to establish appropriate insurance limits for the contractor, which limits are established elsewhere in the document.

4. *Indemnification.*

   a. Paragraph 3.18.1 of the A-201 (page 10).

      1. note that it is extended to the architect's consultants but not to the owner's other consultants.

      2. It appears to be drafted to a comparative negligence standard.

3. It may be difficult for the owner to get the contractor to agree to indemnify the architect. Owner should seek a best efforts provision.
4. Includes attorneys' fees as an item of indemnification, but does not provide for an obligation to defend. The owner may prefer a tender of defense provision, particularly where claim is for worker injury or for a subcontractor nonpayment claim (mechanic's lien, breach of contract, *quantum meruit*).
5. Note: There is a question as to whether this indemnification is specific enough to cover mechanic's lien claims in all jurisdictions.

b. There is another very interesting indemnification provision in this revised 1987 document. Article 10, "Protection of Persons and Property," Paragraph 10.1.2, page 19 deals with the presence of asbestos or PCBs on the site. It is interesting to note that this obligation to indemnify is limited to the owner's negligence. However, it puts great burden on owner to:
1. Perform careful due diligence prior to purchase of the property, or to construction.
2. Hire competent/qualified consultants to test, monitor and treat (remove) the material.

It should be considered a fair provision for architects and contractors who are not hired to deal with hazardous materials testing, inspection or the removal process. However, it should not apply to architects or contractors hired specifically for their expertise in dealing with hazardous materials.

Paragraph 10.1.2 is a discretionary provision. It belongs in some contracts but not all. It outlines procedures if the contractor encounters PCBs or asbestos.

5. *Termination.* Article 14, pages 23–24.

a. The issue of termination for the owner's convenience is not included although there is a new paragraph addressing "*suspension* of work by owner for convenience," paragraph 14.7. This, in itself, is interesting because the Agreement between the Owner and Architect, B-141, paragraph 8.3, page 7, allows for termination by the owner if the project is abandoned. The A-201 does not contain a similar provision. From the owner's standpoint, termination for convenience is a tool of flexibility. If you consider such a tool, consider clarifying that there is no payment for lost profit or unabsorbed overhead in the event of such a termination. In order to negotiate such a provision, if may be necessary to consider payment of a termination fee.

b. Consider adding the following provisions and delete existing paragraph 14.2.4:

14.2.4.1

"In no event, whether the Owner terminates for its convenience or for cause, shall the Owner be responsible for overhead costs associated with Work not performed by the Contractor or for any profits which the Contractor might have earned if he had completed the Work."

14.2.4.2

"If the unpaid balance of the Contract Sum is exceeded by the costs of finishing the Work, the Contractor shall promptly pay the difference to the Owner upon request. The amount to

Chapter 9　The Owner

be paid to the Owner by the Contractor shall be certified by the Architect. This obligation for payment shall survive termination of this Agreement."

6. *Warranties.*

Paragraph 3.5.1, page 8, A-201 excludes from the contractor's warranty to owner a remedy where damage is due to improper maintenance or operation. A question arises about a partial recovery.

7. *Assignment.*

Article 13, Paragraph 13.2.1, page 22, A-201, restricts the right of either party to assign without the consent of the other. An owner should pay particular attention to this provision in light of his financing requirements and in light of his needs for business flexibility.

Consider adding the following provision:

> The Owner may assign its rights, duties, liabilities, and obligations under this Contract without the consent of the Contractor. From and after the effective date of the assignment, Owner shall have no further obligation to the Contractor and Contractor agrees to assert all claims and demands against the Owner's assignee. Additionally, the Owner's right, title and interest in and under the Contract Documents may be assigned, mortgaged, and/or pledged without the consent of the Contractor to any lender; and upon demand by any lender, and subject to the performance of the Owner's obligations to the Contractor under the Contract Documents, the Contractor will perform the Contract on behalf of and at the direction of any such lender.

8. *Arbitration.*

Paragraph 4.5 (A-201), page 12. Arbitration is a key aspect of the claims resolution process in the A-201. Most troublesome from the owner's perspective is the fact that the AIA documents do not require the architect to participate in the arbitration process (no compulsory joinder of parties), thereby leading to an imbalanced resolution of claims.

For the owner to view arbitration as a useful tool:

a. limit compulsory/binding arbitration to claims under a specified dollar amount.
b. require joinder—including a provision in all subcontracts and in the agreement with the architect.
c. set more detailed procedural guidelines to govern the arbitration process.
d. provide for a written opinion to be issued by the arbitrator.

9. *Ownership of Documents.*

Article 6, page 6, B-141. Paragraph 1.3.1, pages 6–7, A-201. This is expanded language which continues to recognize drawings and specifications as instruments of service and the architect as author of those documents. However, it states architect retains *all* common law and statutory rights including copyright and grants a limited license to contractor and subcontractors for purposes of constructing the project. Most owners will find the last restriction on use to be unpalatable:

> "The drawings and specifications shall not be used by the Owner for additions to the project or for completion of this project by others ... Unless architect is *adjudged* to be in default under this Agreement except by agreement in writing and with appropriate compensation to the Architect."

These revisions seek to clarify that it is a *copyright interest* that is being created, and is in need of protection. In some supporting materials, it is recommended that the architect contact the copyright office to protect his copyright interests in these documents by the formal copyright registration procedures.

In reality the owner needs to use these documents for maintenance, repair and expansion. Therefore, the following suggestions are offered:

    a. Joint title;

    b. Perpetual license granted by the architect to the owner;

    c. If the copyright remains with the architect, the owner is justified in restricting the architect's re-use of unique features or aesthetics. However, this causes problems to the extent that a feature or aesthetic becomes a motif, conscious or otherwise, in the work of that architect.

    d. Quite frankly, the biggest problem from the owner's standpoint is the question of the owner's re-use of schematics or concepts. This is tied in with owner's position that there should be termination for convenience.

more typical owner's solution would encompass the following:

    a. Ownership—copyright interest passes to owner, upon payment/completion of project or early termination with or without cause. There would be no restrictions on re-use by owner;

    b. No prior consent of architect required;

    c. Indemnification by owner to architect for damages architect actually sustains because of re-use excluding damages caused by the architect's negligence in the original design.

The following provision encompasses these points:

> Drawings and specifications are and shall remain the property of the Owner whether the Project for which they are made is executed or not. The Architect specifically agrees that Owner may use all design and construction documents and concepts developed in connection with the delivery of professional services for this Project without any further obligation to contract with the Architect for professional services in relation to said use, and without the payment of any additional compensation to the Architect provided, however, that the Architect shall not be responsible for changes made in the documents by anyone other than the Architect or for the Owner's or other's use of such documents. The Owner shall indemnify, defend and hold the Architect harmless from any and all liability, damages, causes of action, loss, cost or expense (including reasonable attorneys' fees) arising out of or in connection with or caused on account of such changes or such use of any such documents including, without limitation, such use by any third party to whom Owner may deliver such documents.

    10. *Changes in the Commencement of the Statute of Limitations.*

Paragraph 13.7, page 23, A-201. This provision attempts to trigger the statutory limitations period from substantial completion or final payment. From the owner's standpoint this creates problems with latest defects.

    III. *Controversial Provisions which are not included in the A-201 which should be considered by owners.*

      1. A no damage for delay provision restricting the contractor to an extension of time but no increased costs.

      2. A per diem liquidated damage provision in the event that the contractor is delayed in achieving substantial completion:

a. Example of a liquidated damage provision:

"Accordingly, the Owner and Contractor acknowledge and agree that, because such damages may be difficult to ascertain in precise amount, Owner shall be entitled to liquidated damages of $_ if Contractor shall be delayed more than _ (days) (months) in completion of the work (or) $ _ per (day) (month) for each (day) (month) for which Contractor has not completed the work after the scheduled date of completion.

b. Example of bonus and liquidated damages provision:

"If Contractor substantially completes the work at least 30 days prior to the substantial completion date (as adjusted), then the Contractor shall be entitled to a bonus of $_. If Contractor fails to substantially complete the work within 30 days after the substantial completion date (as adjusted), then the Contractor shall be obligated to pay the Owner, as liquidated damages for such 30-day delay, and in lieu of any actual damages for such 30-day delay, the sum of $_. Owner may seek damages for any nonexcusable delay beyond such 30-day period following the substantial completion date (as adjusted)."

3. A provision shifting the responsibility for delays and increased costs resulting from labor disputes to contractor from owner.

4. Owner's right to accelerate the work for excused and nonexcused delays.

Sample acceleration provisions are set forth below:

a. "In the event of any inexcusable delay, Owner may direct that the Work be accelerated by means of overtime, additional crews or additional shifts or resequencing of the Work. All such accelerations shall be at no cost to Owner. In the event of an excusable delay, Owner may similarly direct acceleration and the Contractor agrees to perform the same on the basis of reimbursement of direct cost (i.e., premium portion of overtime pay, additional crew, shift or equipment cost and such other items of cost requested in advance by Contractor and approved in writing by Owner) plus a percentage of the cost for overhead, general conditions and profit as set forth in the provisions governing additive change orders in the Owner-Contractor Agreement. All amounts to be paid to the Contractor due to the acceleration shall be adjustments to the Contract Sum and shall be made only by appropriate change order issued in accordance with the provisions of this Agreement. The Contractor expressly waives claims of any other compensation or damages arising out of the acceleration request of the Owner, except those set forth in the change order or otherwise agreed to by Owner in writing prior to the performance of the accelerated Work."

b. "Owner shall have the right to direct that the Work be accelerated by means of overtime, additional crews or additional shifts or resequencing of the Work, notwithstanding that the Work is progressing without delay in accordance with the established progress schedule. The Contractor agrees to perform the accelerated Work on the same terms and conditions as if the acceleration were caused by an excusable delay, pursuant to the provisions of paragraph (a) above."

EXERCISES

The following tasks are based on the concepts and information in the chapter. All of the material or information required to complete the tasks may not be contained in the chapter text. Independent research is required. Consult your instructor for guidance.

1. The owner is not obligated to use the AIA Standard A201 form of General Conditions. These can be written specifically for the owner. Is this wise? If it is done, what safeguards should be used?
2. Why do owners usually make progress payments to contractors? Discuss alternatives.
3. The owner refuses to pay the amount due on a progress payment request. The contractor complains to you, the contract administrator. What actions do you take?
4. Explain what advice every owner needs at the start of a project. Who gives this advice?
5. Discuss the possibility of an owner suing a public entity that is responsible for inspecting the work if the work does not meet the code.

10

The Contractor

Contractor, n. 1. The individual, firm, or corporation undertaking the execution of the work under the terms of the contract and acting directly or through its agents or employees. 2. A person or company who agrees to furnish materials and labor to do work for a certain price.

CONTRACTUAL OBLIGATIONS—
THE MANY THINGS INCLUDED

The contractor, by virtue of the contract, has a tremendous amount of responsibility for the execution of the project. The architect, while designing the project, has gone through a myriad of processes, selecting various materials, components, and systems for incorporation into the project. He/she has made an overwhelming number of decisions, large and small, about how the project should be constructed and what should be included. It is said that perhaps a half million different pieces of construction apparatus are involved in even a moderate-sized project.

Although there is a tremendous amount of responsibility during the design phase, the responsibility carried by the contractor begins with the architect's decisions. The drawings and specifications, which depict all the items, devices, apparatus, components, equipment, and materials, are basically a shopping list of what the contractor must purchase, install, or at the very least, coordinate. (See Figure 10.1.) An astute contractor becomes thoroughly familiar with the contract documents. He/she may be more familiar with what is not included in the contract documents. These loopholes can provide a source of additional income because of change orders, addenda, and other documents that must be issued to either clarify or make additions to the project.

The unscrupulous contractor may try to take advantage of this situation by charging inflated prices for the additional work. For the most part, though, the contractors will cooperate and provide the additional work in the normal sequence without extensive negotiations and exaggerated prices. They know it is for the benefit of the project, and in most instances they will call errors to the attention of the project team.

The contractor cannot in any way neglect what is included because those things become his responsibility. If something is required of him and does not appear in the work, it is the contractor's responsibility to correct that situation. On a large project it is an awesome task for the contractor's staff to keep track of all the material and personnel and the proper scheduling of the job. (See Figure 10.2.)

The contractor has a great many involvements in any project. He must organize his staff to follow the project from its beginning to its final completion and occupancy, in such a way that some people will be working on up-to-the-minute projects, some will be working months ahead, and some phasing out projects months after completion. In this way the contractor can anticipate situations that are coming to the fore, as well as properly executing the work as it progresses. At the same time he is following up on details and finalizing those items that have been executed and passed on. (See Figure 10.3.) Without this overall organization of the project, the contractor can become inundated and lost, and the project will suffer.

It can easily be seen that one major function of the contractor's staff is inspection. It is important that they be in a position to schedule, expedite, critique, order, pay for, store, move, and cajole as necessary. All these functions come down to the bottom line—inspection. If the contractor has not inspected the project continuously, he will not know how the project has progressed, where it is going, and where it has been, and he will lose control and be unable to execute the project to meet the terms of the contract documents.

The contractor may have several teams of people who are executing various forms of inspection. He may be involved with various subcontractors or even with the design professionals (see Figure 10.4) in a joint effort of inspection to ensure that the project is able to progress and will not become stalled because of insufficient work space or material or because previously executed work has proved faulty.

Contractors, for the most part, decry inspection. Many say it is unnecessary; many say it is an impediment; many say it is second guessing. Ironically, contractors are important participants in the inspection process. If their other cries of woe are indeed well founded, then the inspection process has broken down.

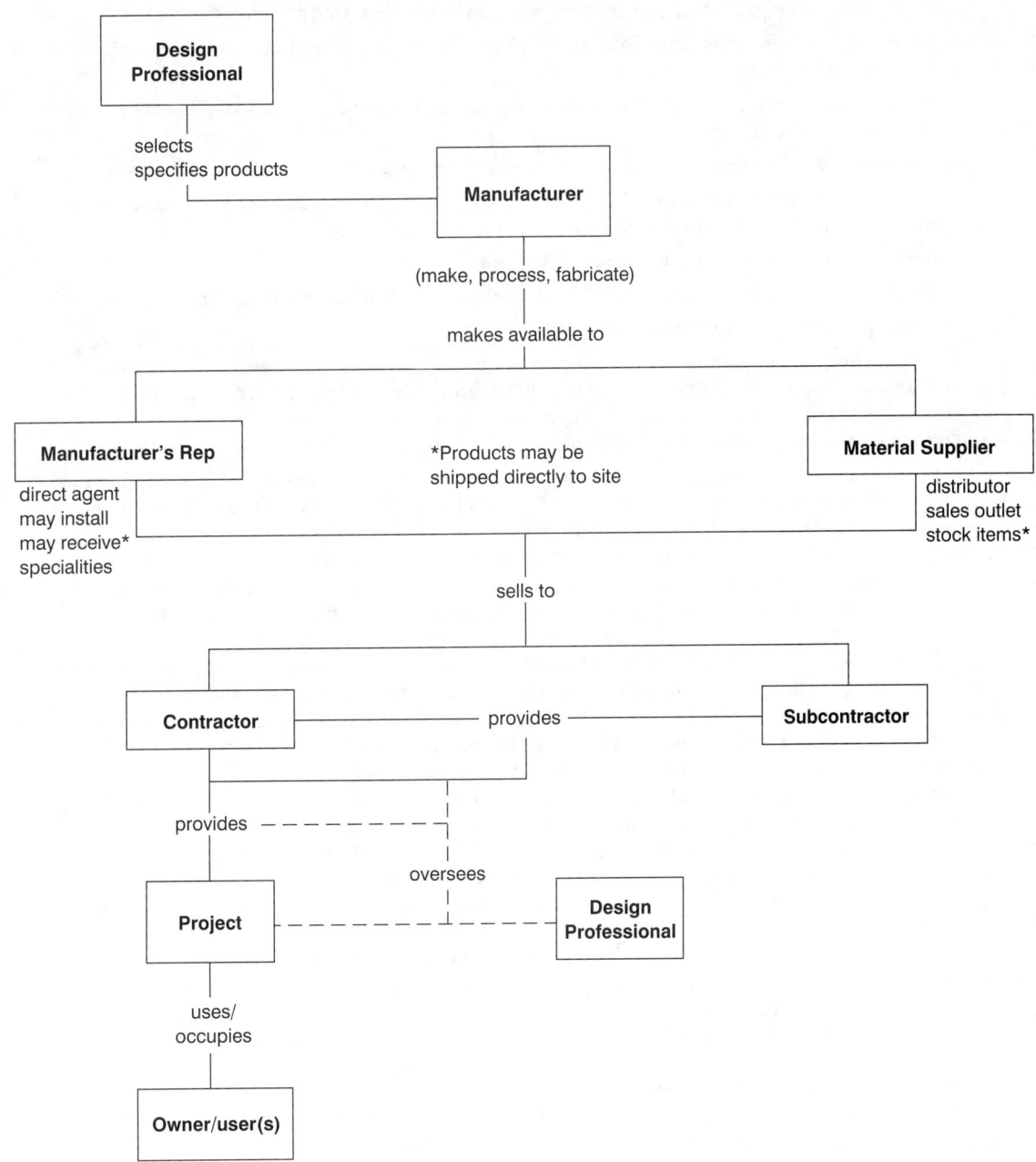

FIGURE 10.1 Chart depicting the flow of materials from/through various sources to the construction project.

DUTIES OF EACH CONTRACTOR

Each Contractor shall perform the following duties in execution of his Contract:

Coordinate work of employees and subcontractors under the Contract.

Provide adequate information and delivery dates for the development of a Project schedule by the CM.

Conduct work and expedite materials and equipment to assure compliance with the schedule.

Transmit written instructions to concerned suppliers and subcontractors.

Under the administration of Construction Manager, coordinate work with that of other trade contractors.

Cooperate with CM and Architect/Engineer.

Forward all communication to the Owner and Architect through the CM.

Attend project meetings as required.

Provide adequate manpower, material and equipment to keep the jobsite and work areas free from debris, dust and mud. Each Contractor and Subcontractor shall do daily clean-up as described in Section 01500—Temporary Facilities and Controls.

Timely completion of this project is dependent largely upon the close and active cooperation and coordination of all those involved; it is therefore expressly understood and agreed that each Contractor shall lay out and install his work at such time or times, and in such a manner, as not to delay or interfere with the carrying forward of the work of other Contractors.

In the event a dispute arises with respect to possible or alleged interference between Contractors which may retard the process of the work, the dispute shall be heard and settled by the Construction Manager, whose decision as to the party or parties at fault, and as to the manner in which the matter may be settled shall be binding and conclusive on all parties.

Any cost resulting from improperly scheduled work or caused by a Contractor failing to communicate and coordinate work activities with the Construction Manager, which affects or could possibly have effect on other trades or Contractors, shall be borne by that Contractor.

Provide construction aids and equipment required by personnel and to facilitate the execution of the work, provide scaffolds, staging, ladders, stairs, ramps, runways, platforms, railings, hoists, cranes, chutes, and other such facilities and equipment.

Contractor shall at all times enforce strict discipline and good order among the Contractor's employees. Any person conducting themselves in an inappropriate manner shall be removed from the jobsite at the request of the Owner, Architect, Construction Manager, or their representatives. Loitering on Owner's property by Contractor's employees before or after normal working hours for the project is prohibited.

Contractor shall prepare and submit safety plan to Construction Manager for review, as required by General and Supplementary Conditions.

Normal working hours for the project will be 8 hours per day, Monday through Friday each week. Any deviation from this schedule will be subject to approval of the Construction Manager and Architect.

Each Contractor shall have a qualified superintendent present at the project site whenever work of his contract is in progress. See General Conditions.

Layout of work within respective contract.

END OF SECTION 01041

FIGURE 10.2 A specification insert noting the specific duties of each contractor (and subcontractor) working on the project.

Chapter 10 The Contractor

CODES AND STANDARDS

Where codes and standards are referred to, they shall be current approved copies. It shall be the duty of the supplier of any material on this work to submit evidence, if required, that such material is in compliance with the applicable local and state codes and standards, in the method in which the material is used in this project shall be current approved copies.

Contractor shall submit his base bid in accordance with plans and specifications. If plans and specifications do not comply with any codes or utility company requirements having jurisdiction, then Contractor shall submit an alternate price on any changes necessary to comply with such codes. If such alternates are not stated in bid, it shall be assumed that Contractor's base bid includes all work necessary to comply with such codes or utility company regulations, and no extra shall be paid for any work or materials in order to meet requirements of codes or utility company regulations having jurisdiction.

PERMITS AND FEES

Before any work is started or materials purchased or purchase commitments made, the Contractor shall first take out and pay for all permits, licenses and fees as specifically detailed and required under the Building or Zoning Ordinances of the municipality or the legal authority having jurisdiction. Copies of all such permits, etc., shall be submitted to Owner with first Application for Payment. Exception: State approval will be obtained by the Architect, with Contractor reimbursing the Architect for the State permit fee.

FIGURE 10.3 Excerpt from typical specification. This insert firmly sets a basis for the work and the responsibility of the contractor, showing pointed and detailed requirements.

There is a tremendous need for contractors to meet with building officials on a continuing basis. Both have problems that involve the other, and discussion will solve many if not all of them. Semiannual seminars would be an excellent vehicle for discussion, as well as contractor membership and participation in building officials' groups. The trades should also realize that the building officials, in many instances, do not write the codes. Model codes, used widely in this country, are open to any challenge and the code writing groups are receptive to code change proposals from any person or group. We must get away from flexing "ego muscles" on both sides. We should be developing a deep concern and a dedicated effort in the design/build teams. Quibbling, hassling, and budgetary cheating over code provisions is a macabre exercise; it's gambling lives against money. Overall regulation of the construction industry is prohibitively expensive, but for the most part building code provisions are not the factors that add substantial cost to the building. In fact, the added cost is small when compared to the liability involved. Code items are not added work, they are basic work.

The contract makes the contractor walk a very narrow line. He is receiving a tremendous sum of money for the project, but little of it can be turned into profit. He and his staff are to order the material and to have it on the site in good condition, ready to be installed. He must have enough workers and subcontractors to execute the work properly. Failure in any of these areas can reduce the project to chaos.

```
1.05    COORDINATION OF THE WORK:
  A.    The contractor for general construction shall coordi-
        nate all work in progress on the site. He shall direct
        arrangements for the storage of materials; shall keep
        himself informed of the progress and detailed work of
        all contractors and shall work with the Architect in
        the coordination and expediting of all work so that
        the progress of the work shall be kept on schedule.
  B.    Contractors for all other items of the work shall
        cooperate with the contractor for general construction
        and with each other and shall keep informed of the
        progress of *all* the work and shall expedite and coor-
        dinate their work so that the progress of work *as a
        whole* is kept on schedule.
  C.    Each contractor shall confer and cooperate with all
        other contractors whose work occurs in the same area.
        If this contractor installs any of this work without
        such cooperation, and in so doing interferes with or
        prevents the installation of other work in the area,
        he shall bear all costs involved in achieving required
        compliance and coordination of *all* associated work.
```

FIGURE 10.4 Excerpt from typical specification. This provision delineates how the work is to be coordinated and who is to act as coordinator, but the design professional is still the total project coordinator.

In addition, the astute contractor will quickly realize that he has some element of the inspection system looking over his shoulder almost constantly. Each of the functions described rests with the contractor, who is the end of the line. Simply put, it is the contractor who is responsible for the execution of the work. (See Figures 10.5, 10.6, and 10.7.) Therefore, the inspection system will be looking to him to execute the work properly.

The inspection system is not intended or designed to "come down on" the contractor. It imposes on him a system of checks and balances in that he has to meet the criteria of various regulations and agencies. (See Figure 10.8.) A contractor can spend a great deal of his time involved with the inspection process, and much of his involvement can be traced directly to his philosophy. If he chooses to be a straightforward, cooperative contractor, demanding the best work of himself, his staff, and his subcontractors, he will usually have very few problems with the inspection system. If, however, he chooses to take shortcuts with an overaggressive approach that results in doing the wrong thing at the wrong time, if he is constantly in an adversary position, if he carries a chip on his shoulder from other jobs (with some of the same participants), he can very well be in trouble with the inspection system.

The system is not designed to be punitive; it is not "after" any particular person. It is there to ensure that the work is executed in a proper manner. The contractor, by bidding on the project and signing the contract, has voluntarily put himself in the position of responsibility. He has not been forced or cajoled into the position; he has accepted the contract openly and knowingly. He knows what procedures are necessary, and his philosophy will guide him in dealing with each element of the project.

It would be helpful for the contractor, just as it is for the design professional, to seek out all of the inspection agencies on the job and list them, thus having available quick reference to the proper people, telephone numbers, addresses, and so on. It would be well to have some notes about the requirements as well, such as whether notice prior to inspection is necessary, when that notice must be given, what the par-

Chapter 10 The Contractor

> COORDINATION OF TRADES
> a. Sections of these specifications set down guidelines as to the extent of work by subcontractors. These guidelines are set forth to aid in the bidding process and help with estimating trade payment breakdowns and monthly payment requests only.
> b. The General Contractor is responsible for the administration of all subcontractors and therefore is not obligated to the work descriptions listed herein. As previously mentioned the General Contractor must provide a total and complete project and thus may divide the work between subtrades as he feels is most efficient and economical to provide a complete project.
> c. The Architect is not responsible for providing complete coverage of work between subtrades in the work descriptions of the sections of these specifications. Therefore, the Architect will not arbitrate or even discuss overlapping or voids that occur in scope of work between subtrades. The General Contractor shall solely handle all problems of this type that deal with which subtrade will take care of which items of work.
> d. The General Contractor shall coordinate all subtrades regardless of the specification headings and shall make all necessary provisions for accommodation of all equipment and fittings into the building and patching after installation as necessary. The General Contractor shall coordinate all material delivery, unloading, and storage. These provisions do not, however, relieve the subcontractors from the responsibility of coordinating all work with the other subtrades.

FIGURE 10.5 Excerpt from typical specification. Another approach to the specifics of coordination, these pointed references leave little doubt as to what is required.

ticular inspector is looking for, or at what time inspection is required. It is far easier for the contractor if he can predict what the inspection system will require and when the requirements will engage the project. Planning for the inspections is wise, in that the work can be finished and examined beforehand, job sequence need not be disrupted, and proper scheduling can be maintained. Also, any faulty work can be found early and promptly remedied.

> SGC-23   EXAMINATION
> a. Before starting the work, and from time to time as the work progresses, the Contractor and each subcontractor shall examine the work installed by others insofar as it applies to his work and the Contractor shall promptly notify the Architect if any conditions exist that will prevent him or his subcontractor from giving satisfactory results in his work. Should the work be started without such notification, it shall place upon him the responsibility for replacing any of his work that it may be necessary to remove in order to correct such faults.

FIGURE 10.6 Excerpt from typical specification. Called by another name, this is a requirement for inspection by all contractors involved.

```
PART 1—STANDARD AIA GENERAL CONDITIONS (CONT'D):
  ARTICLE 4:
  4.2.1:  Delete entire paragraph.
  4.5.1:  Where the word "Architect" occurs change to read, "Developer"
          in Line 7. Delete the word "Architect" in Line 2.
  4.7.1:  Delete paragraph as written and add in its place the following:
          "The General Contractor shall secure and pay for all permits,
          unless modified by the Developer-General Contractor agreement,
          including general work building permit, and occupancy permit.
          Governmental fees and licenses, inspections, and all other
          legal fees pertaining to his trade both permanent and temporary
          which are necessary for the proper execution and completion of
          the work, which are applicable to the construction process."
  4.7.2:  Delete paragraph as written and add in its place the follow-
          ing: "The General Contractor shall give all notices and comply
          with all laws, ordinances, rules, regulations, and orders of
          any public authority bearing on the performance of the work.
          If the General Contractor observes that any of the contract
          documents are at variance therewith in any respect, he shall
          promptly notify the Developer, Architect, in writing, and any
          necessary changes shall be adjusted by appropriate modifica-
          tion. If the General Contractor performs any work knowing it
          to be contrary to such laws, ordinances, rules, and regula-
          tions, and without such notice to the Developer and Architect,
          he shall assume full responsibility therefore, and shall bear
          all costs attributable thereto."
```

FIGURE 10.7 Excerpt from typical specification. Supplements to the General Conditions modify the basic requirements for the contractor and others. These are usually imposed for special requirements of the individual project.

RESOLVING INFORMATION FROM OTHERS

The contractor, while participating in the inspection process, is placed in the position of gathering a great deal of information from other sources. This discussion deals only with the other inspection sources that are imposed on the project. Various people will be inspecting the many aspects of the job at different times. (See Figure 10.9.)

```
PERMITS AND FEES

Contractor shall secure and pay for all permits and any fees required for
carrying out the work, as may be required by governing authorities. Each
contractor shall obtain and hold in force any item required to complete
that portion of the work. The owner shall obtain and pay for general
building permit.

CODE COMPLIANCE

The Contractor shall give all requisite notices to the proper authorities,
obtain all official inspections, permits and licenses made necessary by
the work and shall comply with all laws, ordinances, rules and regulations
pertaining thereto.
```

FIGURE 10.8 Excerpt from typical specification. The beginning of the inspection process. A requirement to meet the code with payment by owner for complying work.

> 1.08 REQUIREMENTS OF REGULATORY AGENCIES:
>   A. It shall be the responsibility of each contractor or subcontractor to apply for and obtain any and all permits and inspections which may be required for this work by local laws, ordinances, rules and regulations.
>   B. Copies of all such permits and inspection certificates shall be filed with the Architect.
>   C. The fees for such permits and inspections will be paid for by the contractor securing same.
>
> 1.09 INSPECTIONS BY GOVERNING AGENCIES:
>   A. Contractors for plumbing, heating, ventilation and air-conditioning, electric, and for any other work requiring special inspection, shall arrange for the inspection and test of the installation as required by the governing authority and/or by the specifications and shall provide all necessary tools, equipment, and personnel to conduct the required tests and shall notify the Architect at least three (3) days in advance of such scheduled inspections and tests and shall submit approved certificates of inspection or copy thereof, from the governing agency prior to request for final payment.

FIGURE 10.9 Excerpt from typical specification. These provisions incorporate governmental inspection into the total project.

It would be helpful if the inspection team and all its participants could move about the job periodically, side by side in a joint effort, but this is not always possible.

Although large groups may prove cumbersome and ineffective, several inspectors can review the job at the same time, allowing immediate decisions to be made and adjustments to be documented by all present. Preferably, such groups should include the design professional, the construction manager, and the contractor. Any such group should proceed in a controlled fashion and should not be allowed to become a "grand tour" of the project with little if any inspection other than some cursory viewing.

The contractor, then, with a myriad of information from various sources must begin to sort out all this information. It is possible that this information could be contradictory, out of sequence, possibly out of date, or completely unfounded. A participant in the inspection system may try to force the contractor to execute some work not in keeping with the contract documents, but in the way that the participant wants it done, whether it meets the documents or the minimum standards or not.

Here the contractor must be not only astute, but also firm and flexible. He must be able to stand his ground and say "This is not in keeping with my contract; therefore, there must be some adjustment." If he is played the fool too often, the contractor can become very upset and require that certain conditions be met before he continues work. These conditions could be that all information must be in writing and verified by others or otherwise properly documented. It is far easier, however, if all participants recognize where their responsibilities stop and stay within them. This is particularly true with the requirements of the contract documents, which really are the parameters of the entire inspection system.

The contractor may have to do some fancy footwork from time to time in an effort to resolve or placate the various inspection sources. Again, he must be able to come forth, show the contradictory elements that are involved, and ask for guidance on how properly to proceed. In this event, the contractor should see this informational

---

    1.02   WORK COORDINATION & JOB MEETINGS

           A. Coordination: The Contractor(s) and subcontractor(s)
               shall, when so required to expedite the general
               progress of job, temporarily omit or leave unfinished
               any part of their work in order to facilitate work
               of others and Owner shall not be charged for such
               accommodations. Job meetings will be every ten work-
               ing days, or when otherwise scheduled by Architect.

FIGURE 10.10 Excerpt from typical specification. This is a good requirement for all contractors to communicate and participate in discussions about the job.

boondoggle as symptom of problems on the project. He may have to keep a constant watch for false, irrelevant and contradictory information, or else the problem may recur. If he has to deal with it on several occasions, he should take a very hard line with the inspection system and call it to task so that it performs properly. Proper communication should be reestablished. (See Figure 10.10.) The contractor and the project should not be placed in an adverse position.

ENSURING GOOD WORK

One of the major features of the contractor's inspection is to ensure that all work is properly executed; that is, to ensure quality control. The specifications require a certain level of workmanship on almost every aspect of the work. The specifications should not be ambiguous; they should be precise and enumerate exactly what is expected: the quality of the work, the finished appearance of the work, and the workmanlike approach for installing the work.

The contractor should be aware of these job requirements. He must, therefore, be actively engaged in an inspection system that views all of them objectively. If something is not done in compliance with the contract documents, the contractor should correct it on his own initiative. Faulty work should not be ignored or allowed to remain uncorrected until another part of the inspection system catches it and takes action.

Many contractors, though, do not want to inspect the work before the more authoritative inspections. It may be a matter of time or priority, but more likely than not the contractors simply do not want to criticize the work of others. They feel the subcontractors know their work and should perform it properly; if they don't, then any remedial work or slowdown in job progress rests with the subcontractors. In the end, the project suffers from this lack of initiative, no matter who is responsible. All contractors should be made fully aware of their contractual requirements and should be made to perform accordingly.

A good contractor on a medium-to-large job can best serve himself with a good superintendent, preferably a nonworking one whose job is only administration and coordination (see Figure 10.11). While this is another overhead cost that the contractor must absorb, it can enhance the profit. A carpenter or carpenter-foreman who is required to produce a full share of the work himself cannot really be a proper overall superintendent of a construction job. Hence, he serves neither his employer nor the project well. A proper superintendent must be able to move around the job, checking on all phases of the work as it progresses. He should be intimately aware, of course, of all the contract document requirements, and he should be an active inspector. He should have direct communication not only with the contractor's staff, but with all subcontractors.

Chapter 10 The Contractor

> CONTRACTOR'S SUPERINTENDENT
>
> The contractor shall keep the same superintendent on the job during the duration. Superintendent shall not be required to work with tools, or perform any other work not related to administering, expediting, or coordinating the work under this contract. He shall have previous experience in this type of work and shall maintain progress schedule and be authorized to make field decisions in the absence of the contractor.
>
> SGC-33   REQUIREMENTS FOR ALL CONTRACTORS:
>
> a. The Contractor shall keep on his work, during its progress, a competent Superintendent and any necessary assistants, all satisfactory to the Owner. The Superintendent shall not be changed except with the consent of the Owner, unless the Superintendent proves to be unsatisfactory to the Contractor and ceases to be in his employ. The Superintendent shall represent the Contractor in his absence, and all directions given to him shall be as binding as if given to the Contractor. Other directions shall be confirmed on written request in each case.
> b. The Contractor shall give efficient supervision to the work using his best skill and attention. He shall carefully study and compare all drawings, specifications and other instructions and shall at once report to the Architect any error, inconsistency or omission which may be discovered.

FIGURE 10.11 Excerpt from typical specification. This is an excellent statement about job superintendents and the work to be performed, requiring proper and constant inspection.

This type of inspection should be ongoing from the first day of the project until the very last item on the punch list is completed. The contractor should actively and aggressively participate in this quality control inspection. He can save himself a lot of headaches by looking at the work and having it corrected if improperly done. Of course, if there is any question in his mind as to the propriety of the work, he can call in any of the other participants to help him make a decision. Too often with the tight scheduling of construction projects, work is done and forgotten. It is never really inspected for quality control, and unfortunately too often the attitude is "Oh, it will be okay if nobody else catches it."

The inspection system is by no means perfect, and not all faulty work can be caught. It certainly is refreshing to have the contractor on the job constantly looking at the work to ensure that it is done properly. Such a procedure can ease the finalization of the project, and callbacks to the project will be minimized because the work was done properly in the first place.

COORDINATION AND RESPONSIBILITY FOR SUBCONTRACTORS

On a traditionally organized project, the general contractor will have several subcontractors directly under contract to him. These subcontractors have no direct dealings with the owner, all of their business affairs being handled through the general contractor. They submit their bids to the general contractor, who reviews and analyzes them. Usually the lowest subcontractor's bid will be included as the price for that particular work in the overall bid submitted to the design professional and the owner. The contractual lines of the subcontractor system make it mandatory that the general contractor be responsible for and coordinate all the "subs." No one else has any direct leverage.

Although general contractors look for low bids, they also try to work with certain subcontractors. From experience, they know which subs are reliable, businesslike, and good mechanics. However, a formal contract should be drawn each time a subcontractor is given work. The benefits of using the better subcontractor should not be allowed to disappear because of informalities or uncertainties in contractual relations. Even good friends (contractors) should have businesslike contracts.

The general contractor is often confounded by the subcontractor, and will throw up his arms in frustration and lament that there is nothing he can do, that the subcontractor is a stubborn so-and-so and will not do his work as specified. The shrug and the lament do not, however, relieve the general contractor of his responsibility. Controlling the subcontractors is part of the general contractor's contractual obligations, and he cannot ignore it. He has contracted to provide that particular work, and he alone has chosen the subcontractor. According to Gerald G. Weisbach, FAIA, architect/attorney in San Francisco, CA,

> Changes in the construction industry have transformed contractors into brokers for the trades, rather than being builders as they formerly were.

This responsibility is not simply to have the telephone number and a contract in the subcontractor's office; rather, a good working relationship is the intention of the contract. The two parties must work together and must schedule their work properly. The subcontractor should understand that his work, first of all, will be scrutinized by the contractor. The contractor should never feel that he's being intimidated by the subcontractors. He should retain control and ensure that all work is proper before the sub is in any way released from the project or to another work area.

The general contractor needs a good working knowledge of all phases of the construction process. He knows what is required from the drawings and specifications, and he has the first-line quality control over the subcontractor. This calls for inspection, plain and simple. The contractor should be willing to take on this inspection, and the subcontractor should understand that this is the first of several inspections to which he will be subjected.

Inspection should never be punitive. It should be a process of ensuring that the work is done correctly and completely and left in good order, meeting the requirements of the contract documents. The general contractor should be an active and willing participant, because if the work is not correct, it will be the general contractor who will be called to account. He, in turn, will have to try to deal with the subcontractor. A proper relationship and inspection at the moment the work is completed can lead to good feelings, respect, and an enjoyable project atmosphere. The alternative can drastically affect the project and the future relationship (and reputations) of both the general contractor and the subcontractor.

In a system where the general contractor no longer is the main functionary, the construction manager is responsible for the subcontractors or, more properly stated, for the various contractors. The manager must participate in the inspection sequence to ensure that the work is ready for inspection by others within the system. If this is carried out, it greatly expedites the work and can substantially reduce the amount of inspection time needed on the job. One of the greatest by-products of the system is, again, the elimination of hard feelings and of extended periods of time to have corrections made. It is far easier for the contractor or the subcontractor to replace or repair work while he is still on the job site than to have to come back to the job for a few hours' work to make an adjustment. Not only does this expedite the work, but it also reduces the cost to almost everyone involved.

Chapter 10 The Contractor

TEMPO OF THE JOB

The general contractor, or the construction manager, has an unwritten responsibility that involves the general handling or the tempo of the project. Nothing is written into the contract, however, that says the job must be run in a certain way. The contractor, through years of experience, will know that a well-run, well-scheduled, well-inspected project will progress very well. For the most part, it will seem to run itself, and, in the final analysis, it will be a well-executed project.

Many people can work very hard on the tempo of the job, and yet some can eminently succeed and others can fail miserably. There are no real guidelines on how to run a job successfully (however, see Figure 10.12). Good management practices, of course, are necessary as well as a good knowledge of the construction industry, of current labor and economic conditions in the community, and of the system under which the project is being worked.

The inspection system plays a part in the running of the job. If the system is allowed to deteriorate and becomes a matter of "I'm going to do the work, catch me if you can," or if it becomes full of conflict, the tempo of the job is going to be slowed. It's going to be erratic, jerky, and generally unproductive. With this in mind, the contractor is best advised to take the inspection system into account and anticipate it at every turn. In this way, the inspection system becomes a known quantity.

The inspection system is not intended to place project members constantly on the defensive. It is a simple matter of maintaining the checks and balances. If the job becomes notorious because of its failures, the inspection system is going to have a more direct and frequent influence on the project. If the job is well executed, well run, and coordinated with the requirements of the contract, the inspection system will become a fringe system doing its job, but in no way impinging on the jobs progress.

The tempo of the job is set by the contractor or the construction manager in the initial stages. A cooperative spirit, an all-encompassing communication system, an open-

---

1. Our objectives are safety, quality, efficiency, scheduling, cooperation, and honesty.
2. Treat all of those associated with the project on an equal basis and the way you like to be treated.
3. Always give the owners 100 percent of what they have paid for.
4. Maintain quality levels above the norm. Do not be satisfied with the norm.
5. Lead by leading, not always pushing.
6. Put aside preconceived ideas in regards to a contractor.
7. Put aside personality conflicts in order to better our objective.
8. Use the project schedule as a tool.
9. Bring potential problems to the table as soon as they arise.
10. Keep your project manager fully informed of all important matters.
11. Report any accident, no matter how small, to your project manager immediately after emergency action has been taken.
12. Be thoroughly familiar with all aspects of the project. If you don't know or understand, ask someone who does.
13. Check, check, and recheck. Don't take anything for granted. Mistakes caught early are easily corrected. After mistakes are compounded, corrections are harder and more costly.
14. Maintain a clean job site at all times.

FIGURE 10.12 Project superintendent's objectives.

ness, and a willingness to listen all contribute to the tempo of the job and the successful incorporation of the inspection system. To be devious or to try to ignore the inspection system will lead only to the downfall of the general progress of the job. The system exists to satisfy certain requirements, but it should never be allowed to become so important that the job suffers. However, if the job is poorly executed from the beginning, the inspection system will have a marked impact on the job. There is no implied threat in this. It is a matter of evaluating the work against job and code requirements.

The great majority of construction projects are well run and well executed, and all participants walk away friends, proud of their joint accomplishment. There is no reason this can't happen on every job. No inspector, whether a design professional, an owner, or from a governmental agency, is on site to harass the contractor, impede the job, inflate his own ego, or seek his own self-interest. He is there for the good of the project. The inspection system itself can, for the most part, be self-enforcing. One element of that system will not allow another element to become too pushy or to impinge on progress.

Someone, though, must start the ball rolling, and this responsibility falls on the contractor or the manager. The job ideally begins with proper scheduling, proper communications, and an allowance for the inspection system. This allowance does not mean that on certain days the entire job will just shut down so everything can be inspected. This means not only that the inspection system is acknowledged, but also that it will be called in when required, that work will be ready for inspection when required, and that the work will have had some preliminary inspection prior to the major inspection.

With this sequence, the inspection system will add to the tempo and help the job. It will stay out of the way, allowing progress and a well-executed job, meeting the budget and schedule without any problems.

EXERCISES

The following tasks are based on the concepts and information in the chapter. All of the material or information required to complete the tasks may not be contained in the chapter text. Independent research is required. Consult your instructor for guidance.

1. Explain why cash flow during a project often is more critical to a contractor than the profit at the end.
2. Explain why a contractor usually needs credit to perform a construction contract and how this affects the cost and performance of the work.

11

The Subcontractor

Subcontractor, n. A secondary contractor who performs some part of the prime contractor's obligation under the contract.

CONTRACTUAL OBLIGATIONS

The subcontractor on a construction project is contractually the "end of the line." He/she is basically responsible for a specific, limited amount of the work to be done on the project. Although the scope of this work can vary greatly, the subcontractor's interest in many ways is extremely narrow.

In the current system, the subcontractor may indeed subdivide his work and have a series of sub-subcontractors below him; even further subdivision is possible. In addition, there are contractual agreements with material suppliers, manufacturers, distributors, and manufacturer's representatives. This entire group of participants in the construction process, however, can be categorized under the general heading of subcontractor.

All these participants have the same basic contractual obligations and responsibilities. Each has a limited, narrow responsibility to the overall project. His responsibility lies in having a smaller organization than that required to produce the entire project, and his work is highly specialized. He could be in charge of an aspect as large as the entire masonry or bricklaying contract, or he might provide only a portion of the carpeting for the project and nothing more.

Basically, the subcontractor provides the actual hands-on work force for the project. The administration of the subcontractor is fairly small and is confined to the overall administration of the business and work under contract. The subcontractor has only a small portion of the administrative responsibilities on a large project. He must keep in touch with each project to know when his work is required and what time schedules he must meet. Although he may be only a small cog in the machinery of a large project, he is nonetheless very important to the overall quality, progress, and completion of the project.

All of the requirements he must meet should be enumerated in the contract the subcontractor has with the general contractor. Of course, the general contractor, having overall administration, inspection facilities, and functions on the project, will be looking very carefully and objectively at the subcontractor, wanting to ensure that the subcontractor performs well and that the work will not be a source of either irritation or callbacks on the part of the contractor. Also, he will want to see that the subcontractor performs in a timely manner—getting onto the job, doing the work, and then leaving in an expeditious manner when the work is finished. (See Figure 11.1.)

RESPONSIBILITY TO EVERY MEMBER OF THE DESIGN/BUILD TEAM

It is important that each subcontractor feels a part of the project. He should be included in the job meetings when he is on the job and performing his work. If he is ignored, he may feel that he is merely putting in his time and that he owes no responsibility to the overall project. In fact, he should never feel that he is being ignored, bypassed, or taken for granted; rather he should be respected for the knowledge and ability that he brings to the project. His responsibility to each member of the design/build team, of course, is to produce the best work possible, utilizing the best mechanics and workers and always using the material and methods listed in the specifications for the project.

The subcontractor is also a full participant in the inspection system. (See Figures 11.2 and 11.3, which refer to inspection and compliance.) He should determine that the standards applicable to his work are fully met and that all work is installed in keeping with the manufacturer's printed instructions. His work force should be trained in the minute detail of all the products they provide and install, but inspection of the work is still essential. (See Figures 11.4 and 11.5.)

By performing in this manner, the subcontractor will have a long business life, in that he will be treated with respect and will gain in stature as he takes on more

Chapter 11 The Subcontractor

The prime contractor is responsible for:
- all subcontracts, including sole source
- management of all subs as part of project management
- coordination of work
- financing work, including that of subs

The owner must:
- not interfere
- use prime as main contact (do not bypass prime)
- stay out of prime/sub disputes
- inspect work of prime, not subs, even when they are the same
- be protected against sub-liens, claims, and bonds
- be aware of contract provisions that call for notice from subs
- be cautious about subs pulling out of the project

While no rights are given by the owner and sub to each other, there still must be an enforceable contract for the sub. The owner has a measure of control, primarily in the pre-award phase. This is retained in (1) listing of the subcontracts, (2) determination of responsibilities, clearly and completely, and (3) the possible use of bonding. In all of this the owner is wise to address the following issues:

1. Is the scope of the subcontract well defined?
2. When and how is the sub to be paid?
3. What ensures the sub's timely performance of the work?
4. How and when will the prime contractor default a sub?
5. How will the disputes between prime and sub be resolved?
6. How can the sub's financial instability be limited?

FIGURE 11.1   Parameters, sometimes unstated, that delineate the relationship between the owner and the various contractors. For proper control and administration, these guidelines should be followed.

responsibility. He will distinguish himself as he performs in a good and workmanlike manner, extending himself in the best way possible and keeping in mind that the project, of course, should be placed above all else.

PROTECTION OF SUBCONTRACTOR'S RIGHTS, INTERESTS, AND BUSINESS

The inspection aspect of the subcontractor's work involves more self-interest than that of any other inspection participant. His inspection should be aimed directly at his own interests to ensure that the quality of work he is producing meets the specifications and the requirements of the project. At the same time, it should not become burdensome to his own work.

Being the "bottom-line, do-it" contractor, the burden to perform weighs heavily on the sub. However, the smart subcontractor performs only the work for which he is engaged, and should not allow any outside influence to sway him from his work or to convince him to do work for which he is not trained or qualified.

This is not to say that the subcontractor should be constantly looking for loopholes or shortcuts. He has some basic business sense. He will know that his contract price is valid and that his materials are as specified. He should then in a businesslike way be ready to perform to the best of his ability. His inspection should involve his self-interest in that he should look at the work of others in relationship to his own. (See Figure 11.6.) For example, a contractor under contract to build the structural

> 6. **COMPLIANCE WITH LAWS:**
>   a. Each subcontractor shall assume all responsibility and hold the Architect and the Developer harmless for compliance with all applicable laws, ordinances, rules, regulations, and orders of any public authority bearing on the performance of the work, including furnishing and installing of all labor and material required to obtain the occupancy permit.
>   b. Each Subcontractor, including all employees, shall comply with OSHA to the extent required by the Department of Labor, Occupational Safety and Health Administration. The General Contractor shall notify the Developer and the Architect of any changes required by OSHA.
>   c. Each subcontractor warrants that all construction and installation work shall be performed in accordance with the requirements, orders, and limitations of all local, state, or federal departments and bureaus having jurisdiction. Upon completion, the premises shall be in full compliance with all governing requirements for the use which the Developer may make of them, and that all necessary certificates of inspection shall be obtained by the General Contractor and submitted to the Developer.
>   d. Each subcontractor shall carefully inspect all divisions of these specifications and the construction drawings, prior to the submission of his bid, and notify the General Contractor of any noncomplying materials or erection methods. If the subcontractor furnishes any work which is not in conformance with such laws, ordinances, rules, and regulations, and without notice to the General Contractor, he shall bear all costs arising from the correction thereof.
>   e. All permits, licenses, certificates, and necessary insurance required for the work shall be obtained and paid for by the subcontractor, as hereinafter specified and in other contract documents.
>   f. All electrical equipment and installations shall be approved by the Underwriter's Board and local electric company; all gas equipment and installations shall be approved by the local gas company.
>   g. Compliance with laws, etc., shall not be used as a means of justifying the installation or application of parts, assemblies, or methods inferior to those specified.

FIGURE 11.2 Excerpt from typical specification. A listing of all requirements imposed on the subcontractor implies inspection in several instances.

frame of a building should investigate and inspect the foundation work that has already been done by another contractor. It would be foolhardy to begin work without doing so and find that problems with his work are created by faulty foundations.

Similarly, a flooring contractor should carefully inspect the concrete floor slabs before installing his material. Rough surfacing, cracks in the slab, or other problems with the concrete could cause problems for the subcontractor, costing additional money to remedy and to ensure that the work is finished in a complete and proper manner. In most specifications the writer will specifically call for these inspections and place the responsibility for them on the subcontractor for the new work. In other words, the contractor for the structural frame is given the responsibility to find any faults that may exist in the foundation system. So, too, the flooring contractor is responsible for any faults in the concrete floor slabs.

If a subcontractor finds anything wrong with the work that preceded him, he should report it in writing, through the proper channels so that it can be remedied prior to the installation of any new work. This, of course, is the reasonable approach and this procedure should be documented in the subcontractor's contract, as well as in the construction documents.

Although this may seem like self-preservation on the part of the subcontractor, it is only right and proper that the responsibility for any faulty work should not be

Chapter 11   The Subcontractor

1.08   ADHERENCE TO CODES AND REGULATIONS:
a. Before proceeding with the work, each contractor shall thoroughly review the drawings and specifications to assure the design is in accordance with all laws, ordinances, regulations, and building code regulations applicable to the specific work hereunder.
b. Should a discrepancy to such applicable laws, ordinances, codes or regulations be observed or determined, the discrepancy shall be brought to the attention of the Architect who will provide corrective instructions.
c. If the contractor performs any work knowing it to be contrary to applicable laws, ordinances, rules and regulations, and without notice to the Architect, he shall assume the full responsibility therefore and shall bear all costs attributable thereto.
d. The "Ohio Basic Building Code" as administered and modified by the Division of Public Works, City of Cincinnati, Ohio shall govern all the work in addition to any other code authority indicated in the specifications.

1.09   APPLICABLE STANDARDS:
A. Where reference is made in the specifications to published standards, the issue of such standards current to the date of the invitation to bid shall govern the applicable work.
B. In case of conflict between the published standard and project specifications, the latter shall govern; standards required by codes shall supersede all other standards.
C. References to known standard specifications shall mean the latest edition of such specifications adopted and published at date of invitation to bid. Reference to technical societies, organizations, or bodies is made in specifications in accordance with [accepted] abbreviations.

FIGURE 11.3   Excerpt from typical specification. Contractor requirements ensure that the project meets all applicable laws and codes, ensures understanding of the project, and allows input from the contractor, who may have information different from that of the design professional.

shifted from one contractor to another. Future litigation or problems that develop later with the user of the project can be extremely hard to resolve if it is difficult to trace the lines of responsibility. Hence a proper contract, proper documentation, and proper inspections are essential.

The duration of legal liability of subcontractors, and indeed of all contractors, varies from state to state. In about a dozen states, the courts have struck down the limit of liability for defects in a project, so that in some states the limit that had been set at ten years now lasts as good as forever. About two-thirds of the states still

---

GENERAL GYPSUM BOARD INSTALLATION REQUIREMENTS:
A. Pre-installation Conference: Meet at the project site with the installers of related work and review the coordination and sequencing of work to ensure that everything to be concealed by gypsum drywall has been accomplished and that chases, access panels, openings, supplementary framing, blocking, and similar provisions have been completed.

FIGURE 11.4   Excerpt from typical specification. This excellent requirement brings the subcontractor into the project. It coordinates the subcontractor's work and helps him/her understand the project as a whole.

```
1.03  QUALITY ASSURANCE:
   A.  Job Mock-Up:
       1.  Prior to installation of masonry work, erect sam-
           ple wall panel mock-up using materials, bond, and
           joint tooling required for final work. Build mock-
           up at the site, where directed, of full thickness
           and approximately 4' X 4', indicating the proposed
           range of color, texture, and workmanship to be
           expected in the completed work. Obtain architect's
           acceptance of visual qualities of the mock-up
           before start of masonry work. Retain mock-up dur-
           ing construction as a standard for judging com-
           pleted masonry work. Do not alter, move, or
           destroy mock-up until work is completed. Use sam-
           ple panels to test proposed cleaning procedures.
           Provide separate mock-up for the following:
           a.  Typical face brick wall.
           b.  Typical interior exposed CMU wall.
```

FIGURE 11.5 Excerpt from typical specification. This requirement establishes a model against which other work can be evaluated. The model should be executed in keeping with all standards, for it is easier to change a faulty model than a large portion of the work in place.

retain laws that set definite limits, but a trend may be developing that would demand even closer inspection by every contractor, because of extended liability.

The subcontractor, having a smaller organization, is not in a position to financially assume extra work on the job. He is not in a position to have his own work force cover a large amount of faulty work. Therefore, he must be very aware of all aspects of the work he is concerned with.

The subcontractor must be sensitive to the job requirements and to his own position. He should resist pressures that try to involve him in work unfamiliar to him or that is beyond his expertise and thus extends his liability. He must be aggressive but reasonable in his approach, doing his work properly and watching every condition around him.

His inspection carries over into his own work. He should be concerned that his forces are, first of all, meeting the requirements of the specifications. (See Figures 11.7 and 11.8.) Of course, he will be inspecting his work to see that his workers are performing according to his time schedule and are not extending his contract exposure time on the project to the detriment of both his finances and his reputation.

Ideally, every subcontractor would like to get the job, perform his specialty without interruption, leave the site, and get paid. Unfortunately, not all jobs proceed in this smooth sequence. Unforeseen time delays, return trips, long waiting periods, and the like, can lead the subcontractor to ruin.

A good contractor can get in, do the work, and get out without being detrimental to the project. If he has good workers, has his material on hand, and has a well-coordinated and well-administered organization, he should be able to perform in this manner and provide the best work possible.

His inspection, of course, is specialized to whatever materials or systems he is working on, and he should be well trained in the inspection of those materials and systems. He should know his work literally inside out and should prepare it to meet the requirements of the job. A subcontractor working on the job in a relatively late stage can be asked to do additional work to cover the mistakes of others. This, of course, should be added to his contract. He should not be forced to do this work without extra remuneration.

Chapter 11 The Subcontractor

PART 3.00—EXECUTION
3.01 INSPECTION:
    A. Applicator must examine the areas and conditions under which painting work is to be applied and notify the Contractor in writing of conditions detrimental to the proper and timely completion of this work. Do not proceed with the work until unsatisfactory conditions have been corrected in a manner acceptable to the Applicator.
    B. Starting of painting work will be construed as the Applicator's acceptance of the surfaces and conditions within any particular area.

3.01 INSPECTION AND PREPARATION:

Inspect all surfaces to be waterproofed and remove all material that may be detrimental to bonding of waterproofing. Report to Architects any conditions which are not satisfactory for waterproofing application. Application of material constitutes acceptance of surface conditions by waterproofing contractor.

PART 3—EXECUTION
3.01 INSPECTION:
    A. Installer must examine the areas and conditions under which resilient flooring accessories are to be installed, and notify the Contractor in writing of conditions detrimental to the proper and timely completion of this work. Do not proceed with the work until satisfactory conditions have been corrected in a manner acceptable to the Installer.

FIGURE 11.6 Excerpts from typical specification. Three examples of inspection by the subcontractor for his own interest properly applies where some work is contingent on other work.

The subcontractor's inspection of work done previously should always be done in a spirit of cooperation and in the best interest of the project. The subcontractor should never be made to feel that he is betraying his fellow contractors by reporting work that was not performed properly. Unless the subcontractor's work is extremely limited in scope, it is hard for every minute detail of every piece of work to be properly examined and inspected. He should inspect his work as others inspect theirs. Overall, though, he will allot a rather limited time for inspection. His people working in a limited area should be intimately aware of the details of this work. Some follow-up inspection is essential to see that the work fits properly into the project as a whole. Many times things will slip through and affect other contractors.

    C. Fire-Rated Assemblies:
      1. Wherever a fire-resistant classification is shown or scheduled for hollow metal work, provide fire-rated hollow metal doors and frames investigated and tested as a fire door assembly, complete with type of fire door hardware to be used. Identify each fire door and frame with recognized testing laboratory labels, indicating applicable fire rating of both door and frame.
      2. Construct assemblies and provide for installation complying with NFiPA Standard No. 80, and as herein specified.

FIGURE 11.7 Excerpt from typical specification. To ease the burden of the inspector, code-abiding data must be displayed on the work.

```
1.02   REQUIREMENTS OF REGULATORY AGENCIES
       A.   All work included under this Project Manual shall conform to the
            National Electric Code as approved by the Inspection Bureau,
            Inc., and regulations of the Cincinnati Bell Company and the
            Cincinnati Gas and Electric Company. Include requirements of all
            codes whether shown on drawings or not. Differences between the
            Bid Documents and local codes and regulations shall be cor-
            rected, as required, without any cost to Owner.
       B.   The Contractor shall obtain all required permits, temporary
            releases, and inspections, and include the cost thereof in his
            Bid. Certificates of approval shall be submitted prior to
            request for final payment.
```

FIGURE 11.8 Excerpt from typical specification. Detailed requirements for the mechanical subcontractors state that all work on the project must comply with all applicable regulations.

The spirit of cooperation is necessary among all participants, from the prime contractor to the person who supplies the smallest item on the project. If this is not inherent in the project, there could be serious problems, and some people will try to avoid direct responsibility. This is another of the basic reasons for a coordinated inspection system.

EXERCISES

The following tasks are based on the concepts and information in the chapter. All of the material or information required to complete the tasks may not be contained in the chapter text. Independent research is required. Consult your instructor for guidance.

1. Discuss the difference between a subcontractor and a supplier. Can they be one and the same?
2. A material supplier informs a subcontractor that the specified material is no longer available. What procedure should the subcontractor follow? As the contract administrator, when informed of this (via the proper procedure), what do you do?
3. A subcontractor performs a rather small, but vital and expensive portion of the work. Two weeks after this subcontractor leaves the site, work finished, you begin to see faults developing in the work. What do you do? What needs to happen and when?
4. You hear that a sub refuses to come back on the site to complete her work, because she has not been paid properly by the general contractor. What do you do, when, and why?
5. During the development of the punch list, you find defects in a sub's work. However, although the work is complete, you know that the sub has gone out of business. How do you resolve this situation?
6. A subcontractor becomes a problem when he does not appear as necessary, comes in at odd (off-work) hours, and performs just small portions of the work required for good progress of the job. What can be done? By whom? What part does the contract administrator play?
7. The painting contractor "subs" out finishing of the doors. Your boss, in making a periodic visit, answers some questions of this subcontractor, but does not pass the final information on to you. The work proceeds but is not as specified. What action do you take?

12

The Building Inspector

Building inspector, n. A representative of a governmental authority employed to inspect construction for compliance with applicable codes, regulations, and ordinances.

BACKGROUND

Building codes were born of disaster, human tragedies, and violations of public health, safety, and welfare. For over 5,000 years, governmental officials have found it necessary to control the construction and placement of buildings within civilized communities. As soon as the human race found it necessary to construct shelter for protection from the elements, hazardous situations were encountered. The lack of knowledge about natural laws and the capabilities of various materials that could be used for building were initially a danger. Earthen walls and roofs, as well as light wood structures, could collapse because the most basic engineering requirements and principles were not known. The only "method" was that of trial and error, which could easily lead to disaster. Even when fire was controlled, providing warmth and comfort, it could also be a tremendous hazard to the user's well-being. This is the basic root of the building code and inspection system that exists today.

The system of construction now enforced is a tug-of-war between the best interests of the owner and developer and the best interests of the general public. The best interests of the owner are not always, and not necessarily, in the best interests of the public, and this is responsible for the imposition of governmental inspection on construction projects. Governmental inspection includes building inspection, inspections by fire-prevention officers, health and sanitation officials, plumbing inspectors, electrical inspectors, and inspectors of the various other engineering disciplines. This chapter discusses the building inspector primarily, although all governmental inspectors act in basically the same way. The contract administrator must understand the need for inspections, must anticipate them, and must make them part of the overall scheme of the project. They are an important factor in the success and completion of any project.

Simply put, construction drawings and specifications are a commitment to comply with the standing law of the jurisdiction, the building code. The building inspector and the inspection system are a check mechanism to verify that such compliance is achieved. It could very well be that the owner or the use of the building will, at some time, imperil the building and its occupants. It is imperative, then, that the building inspector, speaking for the governmental jurisdiction, see that the finished building turned over to the using public is constructed to abide by the code as closely as possible.

The use of the building must begin with the code in mind in order to prevent future misuse, the downgrading of protection, and the addition of fire loading to the structure. But here the interests of the owner and developer can directly conflict with those of the public. The owner, by virtue of his financial commitment, feels very deeply that he has a right to develop and use his building in his best interests and in the way that he wants to use it. But just as any right has a definite limit, so too this right of use has a limit, that limit being the imposition of the best interests of public health, safety, and welfare.

Well-intentioned, expensive construction assemblies frequently are rendered ineffective by lack of attention to significant detail in the design process, poor execution of the work, lack of proper in-progress inspection, or improper renovation after final inspection. These errors include the following:

Within Buildings

Improper or lack of fire separations (vertical and/or horizontal).

Lack of proper protection for such floor and wall openings as stairs, doors, windows, pipe penetrations, shafts, ducts, chutes, conveyors, elevators, dumbwaiters, and escalators.

Inadequate fire-stopping or divisioning in concealed spaces above ceilings or in walls.

Application or installation of combustible interior finishes, protective coatings, and insulations.

Fire walls supporting combustible framing members, such as joists, beams, and girders.

Poor anchorage of structural members into masonry walls.

Improper explosion venting.

Improper venting facilities for fire gases.

Violation or improper construction of floors and walls installed to act as fire barriers.

Between Buildings

Lack of adequate fire separation distance.

Inadequate fire resistance of exterior walls.

Improper opening protection.

Lack of adequate fire walls between buildings.

Combustible construction and materials in roofs, roof coverings, roof structures, overhanging eaves, trim, and the like.

Inadequate protection at connections between buildings: utility tunnels, passageways, ducts, conveying systems.

Collapse of protective exterior walls due to explosion or fire progress.

In very few cases is there a direct attempt to circumvent the provisions of the building code, although some, of course, have planned their buildings to achieve something outside the code. But on the whole, the design professionals and the owners want their buildings to abide by the code and the law.

The problem today is that neither the owner nor the design professional wants to take the time to explore and understand the details of the code or the rationale behind its development. They are simply more interested in developing the structure and imposing the owner's program than they are in seeing that every detail of the building code is met. Neither the owner nor the design professional is trying to circumvent minimum standards; rather, there is such a massive array of information touching on the construction of a building that it is very hard to be familiar with all the details. Figure 12.1 shows the format of a list of applicable standards in a typical specification.

The building code and inspection system set up a program of review to see that the documents show compliance with the code, and the inspections verify that the construction itself abides by the code. This is the basic concern of the existing inspection system.

Although the design professional, the owner, the construction manager, the contractor, and the subcontractor all have an inspection function, they also all have a varying degree of self-interest in the project. Each has an axe to grind, which has a tendency to color inspection capabilities and intentions. By contrast, the governmental inspector comes to the job with no self-interest. His/her basic charge is to check for compliance with the applicable codes. He is acting for the general public, and its interest is his prime mandate.

In years past the governmental inspector was suspect, mainly because of his credentials (or lack thereof), his tremendous power, his ability to influence a project by approval or nonapproval, and the manner by which he got his job. With more complicated projects and an enlightened electorate, the basic protection of the code officials system has caused the system to move toward a much more professional level. Still, the governmental inspector is the only truly impartial party on the project, his basic interest being the protection of the using public and not of any individual participant on the job. It is amazing that, with all the research done before a project is started, the code compliance inspection is needed more today than ever before. Designs are being produced in a very short time, and most shortcomings can be traced to hasty and inadequate document preparation. They are not being produced to purposely circumvent the code.

> 1.06 SCHEDULE OF REFERENCES:
>
> A. Abbreviations and Names: Where acronyms or abbreviations are used in specifications or other contract documents they are defined to mean the industry-recognized name of the trade association, standards-generating organization, governing authority, or other entity applicable to context of text provision.
>
> B. The following listed acronyms or abbreviations as referenced in contract documents denote the [are defined to mean] associate names. Both names and addresses are subject to change, and are believed to be, but are not assured to be, accurate and up-to-date as of date of contract documents. Refer to "Encyclopedia of Associations," published by Gale Research Co., available in large libraries, for verification or update of information:
>
> 1. **AA** Aluminum Association, 818 Connecticut Ave., NW, Washington, DC 20006; (202)862-5100
> 2. **AAMA** Architectural Aluminum Manufacturers Association, 2700 River Road, Suite 118, Des Plaines, IL 60018; (312)699-7310

FIGURE 12.1 Excerpt from typical specification. An elaborate listing of the many standards that apply to construction work provides basic, standardized, well-researched material. Most of these standards appear in the building and other codes; others are imposed by the design professional. The list in this figure should be completed as appropriate for the project requirements. See Appendix E for a list of possible insertions.

The design professional is not taught how to deal with codes, learning of such regulations only through experience with them on the jobs, and has a tendency to ignore the more refined and involved details. He may have an intuitive feeling that he will be caught and that at that time he can remedy any violations of the building code. Moreover, he feels that his time is better spent attending to other parts of the project. He knows that his drawings and specifications will be reviewed by the building department prior to issuance of the building permit, and what is not caught in the way of code compliance at this juncture surely, he feels, will be caught during the inspection sequence.

Similarly, the contractor is not aware of all the details of all the basic reference standards contained within any code. The contractor has a tendency to work on a project in his own way and in the manner in which he was trained. His training may or may not have been in coordination with, or with the knowledge of, the minimum standards contained in the building code. How, then, is code compliance achieved?

Basically, information is gathered about various aspects of the project. Some are compared with code requirements. The plan review, prior to the issuance of a permit, usually reveals discrepancies with the code. The inspection system is in place to review work in progress to ensure compliance with the final approved contract documents.

There will always be a need for an inspection system, but it is becoming extremely important that code compliance become part of the basic design sequence. More and more design professionals are seeing that it is virtually impossible to separate the owner's design program criteria from the criteria of the building code. The building code contains massive, wide-ranging, and very complex regulations of a broad scope. Many consider these to be restraints, restrictions, or impositions that destroy the basic design concepts. This is not true. The design professional who is initially aware of the code provisions can, with the help of the building official, come to an understanding that will allow for accommodation of the design and still provide for basic compliance with the code. An increasing number of design professional offices are developing code search documents that are filled in during the initial design stages of each project. This code search provides the "public" requirements of the job. When coupled with the private owner's requirements for the job, the full parameters of the project are established.

It is a simple fact that the design professional is putting himself in the position of a baseball manager. No matter how many times the manager has had his team play in a certain baseball park, at home plate before every game there is a review of the ground rules for that park. Design professionals are beginning to realize that each project, no matter how similar it may be to previous projects, requires a review of the "ground rules," that is, the basic code provisions.

Unfortunately, building codes currently being enforced contain long lists of reference standards. The codes would be impossibly large documents if all of the material contained in the reference standards were enumerated directly in the code. The standards are incorporated by reference only. Very few design professional firms, or even building departments, have all the standards immediately available to them in their offices. In fact, it is not very hard to find some standards contained in the codes are now out of print. These requirements must still be met, even though they are no longer readily available.

The design professional and the owner must realize that once there is an invitation to the public to enter a structure, whether it is for a social visit or to do business, they both have a responsibility for the protection of the public. That protection begins when entrance to the structure is achieved, continues during the occupancy of the structure, and requires proper exiting. The responsibility is compounded when there is some sort of an emergency inside the building, for the public must be protected until everyone can safely exit the building.

An owner should realize, or be made to realize, that he may have to spend money that will not directly benefit his interests. If he intends to do business, he must follow a prudent, thorough, code-abiding sequence for the protection of his visitors and customers. Every year there is an alarming loss of life and injury due to fire, building collapse, people falling, and the like. Most cases are litigated, and in many instances massive payments are made to the survivors or the maimed.

An owner can be a prudent and sharp businessman, but he becomes suspect when he begins to misuse his building. If he has not permitted his structure to be designed and built properly in the first place, he becomes even more suspect when somebody begins to look at how prudently his entire building development was achieved.

That owners, developers, builders, and design professionals do not always act for the good of the public necessitates governmental jurisdiction and is the basis of the building inspection department. So many factors are at play that some must act in the public interest—not to intrude or prohibit and not to restrict or impose, but to set down a prudent and cautious course to be followed so that the business will have a long life and anyone touched by that business will have a chance for a long life.

PARAMETERS OF THE JOB

At the outset, it should be understood that the function of the building inspection department, the code administrator, and any building official or inspector is not one of design or construction. No properly concerned building official will delve into the basic design premise of the structure in any way. He/she must have a knowledge of construction and knowledge of the code requirements. The best statements that he can make to a design professional with a problem are perhaps the suggestions of two or three solutions to the problem. But the code official should not dictate the solution. He should point out the area of distress, indicate which provisions apply, and what latitude the code allows in this situation. He may then go on to suggest various general solutions that he has seen elsewhere, or which come to mind.

Private citizens building their own homes will often try to extract information from the building official regarding the design and construction. Any official trying to act in a friendly, cooperative manner can easily fall victim to such a request. The best answer, given in a friendly manner, is to have the owner seek the advice of a design professional who is familiar with the solution to the problem. Too often a solution may be offered, only to have the citizen become dissatisfied later on and distort the facts when telling others, including the news media. The owner may say that he was "made to do it this way" by the building official, or that the official threatened that approval would be withheld if it wasn't done this way and with this material.

Too often in the past, building departments have been graft-ridden for one reason or another because people wanted to circumvent the code. Sometimes code officials were not professional and chose to seek remuneration in the wrong way by suggesting certain materials, certain salesmen, and so forth. These practices are rapidly disappearing as professionalism becomes more the key word in building inspection.

It is no longer possible for a building department to be founded or allowed to continue in an uninformed, "hip-pocket" context. Construction is too complex and hazards too great to permit merely a passing stab at code enforcement. Every jurisdiction, no matter how large or small, should demand the best from its building department. This is not popular work, but history shows time and again the tragic results of inconsistent, incomplete, or lackadaisical enforcement. Good programs must be developed, proper codes adopted, well-trained and well-informed staffs hired, and proper financial and moral support given by the citizens and the elected officials.

The seat of knowledge in any building department lies within the office organization. It is here that, in many jurisdictions, the plan examiners are required to be licensed professionals, and it is here that the library of standards and information is kept and updated. There is no way that any jurisdiction can possibly require its field inspectors to carry all of the standards information in their automobiles. The basic concept of construction inspection lies in the fact that plan review done in a methodical, systematic manner in an office surrounded with all available information will lead to the discovery of the major problems with code compliance. Plan reviewers can then, in conjunction with the design professional, remedy the situation and have the documents revised and redrawn as required so that the approved drawings will properly reflect code compliance for the project involved.

The field inspector then utilizes these approved documents during the numerous inspections. The documents should be either with the inspector or on the project every time the inspector is on the project, and his inspection of the job should check the actual, in-place construction against the approved drawings. Any deviations he finds should be resolved among the contractor, the design professional, and the plan examiner.

Although the field inspector is given latitude and discretion, it is done in a limited manner. The basic approval is given by the plan examiner. Any deviations or changes should be coordinated with the examiner and should be fully understood by everyone involved, and properly documented.

It is absolutely essential that a complete and accurate job history be kept and placed in the department's files when the project is finished. This requirement does not vary one iota from project to project. Just because one project is smaller than another does not mean that fewer problems can develop and there is less need for a proper history. These files are invaluable to the department and all parties to the project in the event of problems or litigation. Laws vary, but many departments must retain such files for ten years or longer. Many now use microfilm to reduce bulk.

THE NEED TO UNDERSTAND VARIOUS CONSTRUCTION SYSTEMS

Each year thousands of new materials are introduced into the construction industry. The vast majority are well-researched, well-founded, well-engineered products, which have a valid place in the modern construction system. Some, however, are produced with little research; some may even be plagiarized copies of other material systems and be poorly manufactured. As long as there is a variance between the basic concept and the production of new materials, there will be a need for an inspection system. The prudent materials manufacturer, in an effort to increase the marketability of his product, will find the necessary funding to research the basic design of the product, to properly develop the product, and to have the product tested to prove its reliability. The testing of the basic virtues of the material is vitally important to the building official and should be just as important to each of the participants in the construction system.

New material may appear on a project at any time. It is not the function of the building inspector or the building code to restrict and prohibit the introduction of new materials. It is an acknowledged right of the owner, contractor, or design professional to introduce new alternative materials. The problem becomes one of equivalency. To introduce a new material or a new system of construction into the project while it is in progress presents a problem to the building inspector. Does this material qualify as a direct replacement for the other material? Does it have the same attributes? Are the same virtues present whether they be structural, fire resistant, or decorative? The introduction of this material without data that substantiates equivalency becomes a problem. Careful, meticulous material selection takes place early in the design process, and the introduction of an alternative material that does not have the same attributes as the original begins to cause the deterioration of the project. If the particular product is to be fire-rated to a certain level and an alternative is introduced that does not carry the same rating, the project must be downgraded. Either the alternative must be prohibited or other adjustments must be made within the fire protection system to maintain the level required.

How can all of this take place in the middle of a muddy construction site? The building inspector must know construction materials and systems at their basic, fundamental level. He cannot possibly carry with him the entire array of test results and the attributes of all materials and their alternatives.

The building inspector must be able to look at a particular system and evaluate whether it meets the minimum standards of the building code. Many times a building inspector, trained and active in a construction trade for many years, will have to swallow very hard to accept a piece of work. If he were to do the work, he would do it in an entirely different manner. But the building inspector must use the yardstick of the minimum standard in accepting or rejecting the work he sees. The inspector simply cannot impose his will or his method on the project, since neither of these may be in keeping with the minimum standard. He cannot ask for more than what the standard requires, but he certainly cannot tolerate less.

Prudence and flexibility are two tremendous attributes that the inspector must have. He must be strong enough to say, "Stop, we must investigate this situation." Often it's very hard to get a contractor to cooperate in a situation like this and the building inspector then must retire to his firm-but-fair manner of operating. He simply must take the time, even at the expense of the project, to investigate the alternative being introduced so that he does not contribute to the degradation of the project.

A building inspector cannot be strictly a tradesman or strictly a design professional. He must have wide-ranging interests, and should have experience that will have exposed him to many different situations. He must be interested in continuing his education, learning about the new materials and systems, and keeping abreast of the changes in the building codes.

Most educational programs for building inspectors are continuing education opportunities. A few schools now offer full certificate or degree programs in building inspection, but the vast majority of inspectors are hired from the trades or professions. Basic initial instruction is required, with the other training sequences acting as booster or reinforcement devices. The building inspector should be exposed to and trained in:

1. Basic code usage and interpretation
2. Periodic code updating
3. Techniques of inspection for each major material system
4. Public relations
5. Written and oral communications
6. Energy code provisions
7. Relation to other codes of the jurisdiction
8. Newly introduced material and construction systems
9. Department or regional operating procedures

Several efforts are being made for the formal certification of building officials and inspectors. The first system is in place in New Jersey, and recently some inspectors have lost their jobs because they were not able to achieve the certification required by the state. Other states are moving in this direction, while others have chosen to use a voluntary system. Both initial and recertification require education sequences. The certificate must be renewed periodically, with continuing education being a requirement. Although demanding, this will place a great deal of emphasis on professionalism in the ranks of the inspection corps.

Basic to construction inspection is the principle that such inspection should be progressive, that is, in keeping with the progress of the job. Once an item has been inspected and found to comply with all regulations, it is in effect given approval and can be so recorded on the inspection records. Subsequent inspections need not include these approved items unless there has been some damage, addition, or alteration to them. It would be virtually impossible and a useless exercise to inspect every item on every inspection trip in addition to inspecting the new items just recently installed. It may well be that an item can be installed early in the construction and receive approval, only to have a finishing procedure, material, or item added later in the construction. A final approval, or at least another approval, would be necessary. However, between the initial approval and the next "inspectable" feature, there would be no inspection except as noted.

As with the other inspection participants, the building inspector must keep pace with the project and must impartially and thoroughly inspect each project for which he is responsible. In other words, the inspector cannot allow certain projects to become his pets, receiving easy approval or extra inspection time simply because he is intrigued by the construction of the project. He must keep pace with the progress of each job. He must make sure that all the necessary requirements of each job are met, and he must see to it that all the inspections are properly carried out. This must be done day in, day out for each project. Methods vary among departments, but it is essential that each project receive proper and timely attention.

It is an imposing list of characteristics that the inspector should carry into his inspection techniques. Basically, he should be consistent, continuous, timely, well founded, well documented, firm but fair, relevant and proper, thorough, persistent, courteous, and knowledgeable. An inspector who is interested in his work and who has confidence in his superiors and in the basic system itself can easily accomplish all these aspects of inspection, but he must somehow have these characteristics as part of his nature.

An inspector who is knowledgeable but at the same time abrasive in his dealings with others can become quite a problem to the building officials and the governing body of jurisdiction. This is not to say that every inspector should be an easygoing, happy-go-lucky person who may be short on knowledge; then nothing is gained because the basic system or proper construction is not being served. The inspector's demeanor and action should be proper for each condition he faces.

The inspector must be able to change as the situation at hand warrants. For the most part, projects will run smoothly, and the contractors will prove cooperative. However, in some situations a contractor feels that his rights are being denied or his method of operation is being challenged. Here the inspector must be able to draw upon every resource at his command to resolve the problem as quickly but as satisfactorily as possible. A well-founded (properly based), knowledgeable, and soft-spoken approach will solve the vast majority of problems, but the inspector should never be afraid to use other tactics if necessary. He can involve other personnel, supervisors, formal orders, the appeals process, and, of course, formal legal action. The wise inspector keeps these options in proper perspective and in proper priority so that he does not use "an elephant gun to solve a rabbit-sized problem."

THE AUTHORITATIVE BYSTANDER

The building inspector, since he does not participate directly in the design or the construction of the project, is a bystander to a large extent. To many people, this posture is a source of basic irritation. "Who is he to tell us how to do something?" they say. The intent and basic charge of inspection is not to tell others how to build the project, but to see that the applicable minimum standards are met. The inspector must know the code. Everyone involved in the project is well trained in his discipline, but little formal instruction is available regarding the principles, intent, and operation of the various codes.

Traditionally, code experience was gained through trial and error, work stoppage, and field changes. More recently design professionals have recognized the constraints of the code and make a code search before beginning their project design. In this way, the code provisions can be more easily accommodated, and the design will not be disturbed by a late revision. Design professionals can help themselves by making early contact with the building department plan examiner to establish basic understandings and to receive the necessary interpretations of the code provisions.

This attitude may also carry over to the field inspector. Contact between the design professional and the inspector can be mutually beneficial. The field inspector can quickly carry any problems back to the office, have them discussed, and promptly present solutions to the design professional. This is certainly a much more positive and prudent approach than to try to deal with the codes as an afterthought or when a problem is discovered. No one in the inspection system, particularly the building inspector, is interested in posting a stop-work order on any project. It is as much in his interest to see that the job progresses as it is for anyone else. When he does not find the design professional or the contractor cooperative, he may have to resort to more drastic actions. The inspector must be able to analyze the situation and apply the proper personal and inspection techniques to resolve the problem. No one can predict the reaction of another; a design professional can often be as hard-nosed as the most aggressive contractor. Proper training and experience will stand the inspector in good stead under any circumstance.

The problem with some building departments is that either through the personalities and reputations of the people involved or through their efforts to reflect the attitude of their governing body, they become so restrictive, bureaucratic, and uptight that they become obstructions not only to an individual project but to the overall development of their community. The reputation of a community is on the line each time a building inspector opens his mouth. If the reputation gets too bad, larger developments, expansion, and possibilities of increased tax bases may vanish because the new developers simply do not want to deal with the people in that jurisdiction. This creates a drastic situation for everyone involved.

Although the building department may be seen as a bystander, it does bring a certain expertise to the project that would otherwise not be there. Therefore, it not only behooves the design professional to extend himself to the building official; the building official and all of his personnel should extend themselves to the design professional and the contractors. The spirit of cooperation that ensues can greatly enhance the entire development picture of a community, and, of course, it can greatly ease the construction of an individual project.

Professionalism within the ranks of the building inspection community has become a daily watchword. It is no longer considered good practice to allow an unqualified but well-liked individual, who has been an active participant in the community, to serve out his few remaining working years as a building inspector for the county. This system saw building departments manned by people who were physically incapable of making inspections, who were inflexible and unknowledgeable and who, for the most part, collected their paychecks rather than becoming viable public employees.

The complexity of construction, the development of communities, the continuing hazard of one use to another, the introduction of new materials and systems, and the entire economic and social life of the community are changing so rapidly that only a professional can handle the situation. More and more systems of certification and continuing education not only for building inspectors, but also for code administrators, are coming to the fore. Basic criteria for hiring personnel are becoming more and more selective and restrictive. Simply put, a person must bring much more to the job than was required previously.

Continuing education is an absolute must. Remedial education, simply going over the existing policies and procedures of the department, must also be an ongoing process. There is still a tremendous need for early education in the principles of the code, such as exiting, fire prevention, fire protection, and the like. Once the basic provisions of the code are understood, they can easily be incorporated into an inventive design. Not only must the design professional be educated, but the building inspector must be as well, so both can see how problems can be solved.

Today, though, many dislike some provisions of the code and will insist on trying to cheat on these items in every design. An architect may simply detest stairs, and his designs always merit a very close look because his exit stairs are always suspect in one way or another. What has he basically gained or achieved? What is he really trying to do? Certainly, there is an acceptable feature in each design that can accommodate proper exit stairs. Besides, a beautifully designed project can be as disastrous as a poorly designed one if basic safety requirements are ignored or toyed with.

The design professional, like the building inspector, is always and intimately involved in public health, safety, and welfare. In many states, registration laws place this burden on the design professional. Such laws demand that the professional design and execute the entire project within code limits. Further, he is to report to "the building official" any acts he knows of or observes that are contrary to the code. Basically, the design professional is placed in the "public health, safety, and welfare" business, and he can't avoid this responsibility without violating the very law that allows him to practice. What is gained by trying to shortchange the occupants of a project, simply for the sake of a design? Nine times out of ten, any shortchanging will be caught, and the remedy for this noncompliance could destroy the design. However, chances are excellent that with a little extra effort the design concept can be maintained while still adhering to the code.

Codes are not meant to impinge or inhibit design. Each code provision must be met, and it can be met in any one of numerous ways. Here again the flexibility of the building inspector, the code administrator, and the design professional comes into play. The more interplay, the more understanding; and the better the relationship between the design professional and the building officials, the easier it will be to find a satisfactory solution for any design problem.

However, the presence of the building inspector on the job is mandatory and is still the major function of the inspection system. Recently building officials have seen an erosion in their immunity to liability (personal tort liability). There seems to be a movement away from governmental immunity, which will make the official walk a fine line. Each jurisdiction must establish its own position in this regard, but the inspection system should be ready to be evaluated according to the level of compliance, the timeliness of inspection, the quality of inspection, and the volume of work accomplished.

The entire process of construction in the United States has some counterproductive aspects. The bidding process, for instance, which demands that the low bidder be hired for the job, can play havoc with any project. The contractor can make an error in his bidding calculations. His normal tendency is to try to recover the money reflected in the error so that he will not lose money during the construction of the project. In trying to do this, he often will, in one way or another, adjust his methods or "cheat." As long as there are people who have this attitude, there will be a need for inspection at all levels.

In some instances, complying with the code requirements adds expense to the project. In trying to adjust his cost a contractor may try to cut out some of this added cost. Again, this points up the need for the governmental inspector to see that the basic minimum standards are achieved. If there must be cutting on the job, if there must be adjustments, it will have to be done in some area other than code compliance. Of course, the architect must police the contract to see that the construction documents are abided by and that the owner is given full value for the job, including the code requirements. In Europe the prevailing bidding process whereby the high and low bidders are eliminated and the award is made to the bidder who comes closest to the average of the remaining bids produces a contractor whose contract price is comfortable. The bid is one the contractor can live with, without fear of encountering gross errors in his estimate later. The bid is the right one for the project.

Surely this system produces fewer adjustments and less need for an inspection system or a policing of this contractor. He knows that by doing the work in abidance with the code and the other documents, the cost he has calculated will allow him to make a profit, barring any unforeseen problems on the job.

In the private sector, the owner often will have a private list of bidders—people recommended to him or people with whom he has dealt in previous projects, who can be relied on to give proper pricing and at the same time will produce a code-abiding project. This system tends to eliminate the "bad" contractor and gives the owner some assurance that his project will proceed and will be completed in good fashion. Although this system is not 100 percent foolproof, the owner is more comfortable in this situation.

The building inspector, by not participating in the bidding process, many times will not know the intimate details of the cost of the project. He will, through his experience, find good contractors and bad contractors—people who are cooperative and produce good work without any coercion, as opposed to those who are always looking for an "out." He must adjust his inspection technique to the character of the contractor. If he is known to be a devious, cheating person, the inspector will have to keep a closer watch on the project. He will have to stand firm in enforcing any orders that are given to this contractor.

Faced with a very bad situation, the inspector must be fully prepared to follow the proper channels in resolving the problem. If a mild approach proves fruitless, he must contemplate issuing formal instructive orders, stop-work orders, or using appeals, injunctions, or other court actions. He must involve his supervisors in the proceedings and seek the advice of the local legal counsel as necessary. A good department will train its personnel in the proper court procedures and the proper method of testifying in court. The inspector should issue proper oral and written orders and have a good job history to introduce in court. All this is time-consuming and often frustrating, but the inspector must remember his basic charge and must be ready to do whatever is necessary to carry it out in a precise and professional manner.

All of this becomes part of the job of the inspector. The building inspector should allow nothing to separate him from his basic charge—the protection of the public interest. He must understand that he is doing something the general public cannot do for itself. That's why he is hired; that's why he is paid by tax money.

The building inspector's function is not all one-sided. There is a possibility that the building department can, in fact, also be of benefit to the contractor or the developer. If someone is injured or dies while on the premises, or if there is some sort of litigation, the records of the building department may be subpoenaed by the court. An owner, developer, or contractor who can produce documentation that shows that he followed proper procedures and that his project is code-abiding (that he did, indeed, take the public interest into account) will find that those records may well stand him in good stead with the court. A person who has a record of constant code violations, constant conflict with the inspection system, or shutdown of his job because of deviousness and poor construction is not going to have the same good reputation to present to the court. Therefore, one aspect of building inspection that cannot be denied and that can never be underestimated is recordkeeping. Every project that comes into the building department and is under its control should have a thorough history: a record of all reviews and approvals; a record of any variations or revisions; a record of all inspections; a record of any orders and any extenuating circumstances; and a complete chronological history of the project from start to finish. Statutes usually require the building department to keep the documents for a number of years. They are public documents and are open for review, and they can be used on either side in court proceedings.

To be of value, the inspection history should be free of guesswork, offhand comments, opinions, and irrelevant information. It is simply not enough to record that on a certain day the inspector was on the job and looked at a particular portion of the construction. Behind his initials and the date of his inspection record should be a checklist enumerating all the items that were checked and found to be in compliance. The inspection system cannot be arbitrary, imposing more on one job than on another. Each project must have the same items checked in the same manner as any other project. Flexibility must be built in for items that do change from project to project, but there are basic construction elements that are the same or very similar on every project. In this manner the opinion of the building inspector is taken out of the picture. He has no choice in what he is going to check; his choice comes down to how does this comply and does it comply.

Of course, it's possible for fraudulent documents to be filed. Over a period of time, the type of person who would engage in such activity is going to be found out, and he should be summarily dismissed. An inspector who has lost sight of his charge, who has simply chosen to "windshield" inspections (from his car), does not want to get his clothes dirty, does not want to "get up" into the project, and does not want to follow the basic checklist and the operating procedures of his department is of no value to the department. He simply is not doing the job.

There may be other problems with building department personnel. A plan examiner who is extremely lenient with interpretations of the code is just as bad as a field inspector who closes his eyes to substandard construction. The basic charge of the department—the protection of the public health, safety, and welfare—must be met. Leniency fosters disaster; further advantage will be taken of lenient personnel. The public cannot enforce codes for itself; the government hires the building inspectors to do that work.

It is baffling that some people involved with construction try to build projects that don't meet the code restrictions. Everyone has the attitude that "it can't happen to me." But it does! We have learned little from the tremendous building disasters in our history. We still build to meet the law (in most cases) but forego, for the sake of convenience, the protection it offers. Despite codes and inspection that are better than those of the past, lives are still being lost due to causes that would be silly if they weren't so tragic. There must be constant improvement in attitudes and an intensification of interest among the building department staff, and the staff's education along job lines must be upgraded.

No building department should take on a pious, know-it-all attitude, assuming that contractors or others are wrong until they prove themselves right. The staff is hired by the citizenry; its charge is to protect the citizenry. The building department must be acutely aware of this in its handling of all inspections and all other departmental functions. The inspection procedure can become very sensitive because it regulates homeowners in their "castles." It forces them to take action to meet the code, where they feel it is not necessary. It makes them spend money for additional or remedial work or for additional inspections. Inspection is as much a job of public relations and public education as anything else.

Perhaps one of the greatest problems to any building inspection department is the lack of an educated public. It would be very helpful if students were given some basic instruction about the laws of their community. The schools best serve students by instructing them in how to protect themselves a little more or making them more aware of the laws and the protection the laws provide.

Students are well aware of the speed limit in their area and of the law that requires schooling to a certain age; but we could build a better society if we could also make students aware of more of the laws affecting their daily lives. They would become more aware citizens and might demand more of their laws, elected officials, and enforcement agencies. Further, they might support these efforts in a larger measure.

Until the public is taught about the laws that affect them, an inspection system, primarily the governmental building inspection system, is an essential part in all construction. Only through public education and public awareness will meaningful changes be made in public support and funding for proper inspection activities, which can directly protect the citizenry from hazards in all structures. They must demand excellence from the inspection program and demand a system of severe penalties for code violations. As summarized by Edward C. Vavreck, legal counsel to the Department of Inspections in Minneapolis, MN,

> You, Building Officials, carry with you a public trust of the highest order. You share along with police and fire officials, a number of important duties. You enter the private homes and the private lives of many of our citizens. Fire and police personnel usually arrive on the scene *after* something has gone awry. You get on the scene in an attempt to head off and avert trouble and disaster. It is for this reason that you serve on a higher plane than most public servants. To most of our citizens, you are the only nonemergency personnel in public service they ever meet. You deal in more than service. You provide the understanding and the expertise which most of the citizens do not possess, but which each sorely needs. You are servants of the public of the highest calling. You are the couriers and carriers of the public trust, and being that, you are the synthesis of government in action and the epitome of our democratic processes, personified.

EXERCISES

The following tasks are based on the concepts and information in the chapter. All of the material or information required to complete the tasks may not be contained in the chapter text. Independent research is required. Consult your instructor for guidance.

1. Write a short essay on the need for government building departments.
2. Select a local political jurisdiction and make a list of all laws, ordinances, and regulations that directly affect construction in that area.

3. Explain why building codes are not an adequate substitute for project specifications.
4. What public authorities have the right of entry to inspect construction projects?
5. On what specific grounds can a public official reject work?

13

Conclusion

The secret of joy in work
  is contained in one word—
  EXCELLENCE.
To know how to do something well
  is to enjoy it.

—Pearl Buck

The basic value of the ongoing presence of the design professional on the construction site is indisputable. With the detailed, unique information that the professional has, no other person or system of control is a proper substitute for contract administration. Different words may be used to describe the function of the various parties, but *contract administration* and *design professional* are linked inseparably. Benefit, of course, goes not only to the professional, but to the project itself, to the owner, and, more importantly, to the public. The slight reduction in fee that may result from deleting this basic service is small in comparison with the potential value gained with it in place. An owner may fail to recognize this, so it falls to the professional to pursue it from an educational as well as a professional perspective.

It must be remembered that this system recognizes and respects the roles of all of the other participants on the project. It leads to a "better" project process and, in the end, a "better" project, overall. The final project will realize all of the expectations initially imbued in it. To ensure this end, there is a prevailing need for the constant presence of some party in every phase of the work. Many participants come and go throughout the process. The owner, while "in residence" throughout, more than likely is not expert enough to carry through every detail, function, and process. Obviously, the design professional can provide exactly the service required to address the needs, whims, and idiosyncrasies of the owner in construction terms—written, verbal, and pictorial. In the same fashion, only the professional can address all of the intangible, political, aesthetical, and technological aspects of the project, their interface, and their solution, with insight and information available from no other source. According to Aristotle,

> We are what we repeatedly do.
> Excellence, then, is not an act, but a habit.

This text has explained contract administration. The following discussion shows the importance of contract administration as an integral part of constructing safe structures. It underscores the point that the public is the ultimate and most important "client" for any design professional.

BUILDING FOR SAFETY

Despite sound, sophisticated construction technology, the United States continues to see an intolerable number of lives lost to fire and other building-related incidents. The technology is not failing—people are! Figures continue to indicate that almost 50 percent of all fires are directly attributable to human error, which accounts for 75 percent of the monetary loss due to fire (several *billions* of dollars annually). We are our own worst enemy. Years may pass without a major disaster, but they still occur, all too often. This is repugnant to a society that places great value on the individual life.

From 1913 to 1994, we have achieved a reduction in fire deaths in the United States from the high of 8,900 to a low of 4,275. We regret the loss of each of these lives lost each year in this needless fashion; yet we continue to see a tragic parade of fires that result in multiple deaths. We continue to use facilities that can be patently unsafe. Entering a building, like crossing a street, is an everyday gamble. Only when the general public is aware of this will we have a chance to reduce the loss. Recognition should trigger calls for help, safety, and compliance. Many would not recognize potential hazards, since the public is not well educated in safety codes. We are careless; we even expose our children to unnecessary risks, losing far too many.

The design and construction professionals need to understand that owners, occupants, and users of buildings tend to "abuse" and degrade the buildings; some by unwitting misuse, others driven by greed. Owners in particular tend to modify their buildings to suit their purposes, often obviating or disrupting valid protectives (some required by

Chapter 13   Conclusion

code). Professionals must do their best, within the limited realm of architecture and construction, to ensure proper selection and installation of protection devices and systems.

With the lunacy and blind fanaticism of terrorism (arson, bombings, etc.) being introduced into American life, mindlessly targeting "things" and principles with utter disregard for human life, and in particular innocent persons (including children), new perils appear. So long as terrorism is coupled with random, unpreventable criminal acts, our buildings and people will be at risk. Architecture and construction cannot solve this problem—they can only provide protection within current technology. By innovatively stretching their minds professionals can create situations that save lives by providing added time before tragedy occurs (barriers to illegal/"attack" entry, security devices, convoluted entries, alarm/detection/notice systems, etc.) or "escape time" for avoiding the result of the intrusion. We must come to understand that no building is bombproof, just as no building is truly fireproof. Elements of buildings may withstand the attack, but lives can still be lost.

In like manner, still other aspects of private and business life have the ability to affect us and our buildings. An increasing number of chemicals bring wondrous new products and tremendous opportunities. However, in all too many instances they have a downside: "hazardous!" They can pose explosive, fire, health, or radioactive threats. We are only vaguely aware of what is in the tank truck next to us in traffic, or on the rail line near us. And the more imposing threat from these materials is when they occupy portions of the same building that we are in. The building code will require increased protective construction, but still, periodically, there are leaks, "accidents," evacuations, fires, the need to decontaminate, and unfortunately, death.

There is only a small core of talented visionaries who understand the full potential of these materials and who, daily, think about every imaginable situation and the results (usually adverse) that could occur. This group includes the local fire service personnel, who must know how to deal with any of a myriad of chemicals. Here, as with fire, we must approach this situation with our best design and construction efforts, and with diligence to ensure that everything reasonable can be done for protection. We can try to prepare for any event, but absolute prevention eludes us. Neither can we eliminate the quirky nuances of the chemicals and their potentially tragic results. We can only try to minimize the human loss first, and property loss second.

After each major tragedy, the public is outraged and then sensitized. They become aware of codes, exits, and wall finishes; speculate about construction and warn their children. But far too soon we're back to our routines, going about our business unconcerned about the buildings we live and work in. We don't like to install, maintain, or pay for preventive measures. Deviant minds will always force the creation of a better "mousetrap"—a better, more reliable, less costly device or system, to prevent reoccurrence of the problem. When this is done, more than likely another way of producing problems will have been concocted. Hence, there is no choice but for each person to be ever vigilant. We need to demand that buildings provide what shelter and protection professionals can design into them, and then see to their proper/complete installation.

All of the social ills that plague our civilization cannot be resolved within the building and design processes. The professionals must continue to take advantage of their single opportunity to provide some measure of protection (as the clientele deems necessary). This needs to be a conscientious, informed, and proper effort, with absolute enforcement of the requirements for the construction of buildings.

FIRE SAFETY

Each year a list similar to Table 13.1 is produced. The National Fire Protection Association (NFiPA) maintains a large array of statistics about fires, and in particu-

TABLE 13.1 Partial list of major multideath fires in U.S. structures.

Date	Structure	Place	Deaths
November 28, 1942	Night club	Boston, MA	492
June 5, 1946	Hotel	Chicago, IL	61
December 7, 1948	Hotel	Atlanta, GA	119
December 1, 1958	School	Chicago, IL	93
December 8, 1961	Hospital	Hartford, CT	16
November 23, 1963	Nursing home	Fitchville, OH	63
December 23, 1963	Hotel	Jacksonville, FL	22
February 7, 1967	Restaurant	Montgomery, AL	25
April 5, 1967	Dormitory	Cayugo Heights, NY	9
February 25, 1969	Office building	New York, NY	11
January 9, 1970	Convalescent home	Marietta, OH	31
March 20, 1970	Hotel	Seattle, WA	19
September 13, 1970	Hotel	Los Angeles, CA	19
December 20, 1970	Hotel	Tucson, AZ	28
January 14, 1971	Elderly housing	Buechel, KY	9
October 19, 1971	Nursing home	Honesdale, PA	15
January 16, 1972	Hotel	Tyrone, PA	12
November 30, 1972	High-rise building	Atlanta, GA	10
June 24, 1973	Cocktail lounge	New Orleans, LA	32
September 13, 1973	Nursing home	Philadelphia, PA	11
November 15, 1973	Apartment building	Los Angeles, CA	25
June 30, 1974	Discotheque	Port Chester, NY	24
June 9, 1975	County jail	Sanford, FL	11
December 18, 1975	Night club	New York, NY	7
January 10, 1976	Hotel	Fremont, NE	20
January 30, 1976	Nursing home	Chicago, IL	24
February 4, 1976	Apartment building	New York, NY	11
August 12, 1976	Oil refinery	Chalmette, LA	12
October 24, 1976	Social club	Bronx, NY	25
December 20, 1976	Apartment building	Los Angeles, CA	10
December 22, 1976	Department store	Brooklyn, NY	12
December 23, 1976	Apartment building	Chicago, IL	12
January 28, 1977	Hotel	Breckinridge, MN	17
May 28, 1977	Supper club	Southgate, KY	165
June 26, 1977	Prison	Columbia, TN	42
October 24, 1977	Cinema	Washington, DC	9
December 10, 1977	Hotel	Bay City, MI	10
December 13, 1977	Dormitory	Providence, RI	10
December 22, 1977	Grain elevator	Westwego, LA	36
December 27, 1977	Grain elevator	Galveston, TX	18
January 28, 1978	Hotel	Kansas City, MO	20
November 5, 1978	Hotel	Honesdale, PA	12
November 5, 1978	Department store	Des Moines, IA	10
November 26, 1978	Motel	Greece, NY	10
December 7, 1978	Tenement	Newark, NJ	12
December 29, 1978	Mental institution	Ellisville, MS	15
January 20, 1979	Apartment building	Hoboken, NJ	21
April 1, 1979	Boarding house	Connellsville, PA	10
April 2, 1979	Boarding house	Farmington, MO	25
April 11, 1979	Boarding house	Washington, DC	10
July 31, 1979	Motel	Cambridge, OH	10
November 11, 1979	Boarding house	Pioneer, OH	14
December 27, 1979	Jail	Lancaster, SC	11
July 26, 1980	Motel	Bradley Beach, NJ	23

TABLE 13.1 *Continued.*

Date	Structure	Place	Deaths
November 21, 1980	Hotel	Las Vegas, NV	84
December 4, 1980	Motel	Westchester County, NY	26
January 9, 1981	Halfway house	Morganville, NJ	31
February 10, 1981	Hotel	Las Vegas, NV	9
March 14, 1981	Hotel	Chicago, IL	19
April 7, 1981	Grain elevator	Corpus Christi, TX	9
January 11, 1981	Residence	East St. Louis, IL	11
April 17, 1981	Apartment house	Kansas City, MO	8
October 15, 1981	Hotel	Paterson, NJ	8
October 24, 1981	Business/residence	Hoboken, NJ	11
March 6, 1982	Hotel	Houston, TX	12
September 4, 1982	Hotel	Los Angeles, CA	24
November 8, 1982	Prison cell block	Biloxi, MS	29
March 5, 1982	Hotel	Lowell, MA	8
April 30, 1982	Hotel	Hoboken, NJ	13
May 19, 1982	Residence	Huntsville, AL	11
April 24, 1982	Residence	Aiken, SC	8
May 14, 1982	Residence complex	Baltimore, MD	10
July 5, 1982	Hotel	Waterbury, CT	14
November 21, 1982	Residence	Quincy, FL	8
August 31, 1983	Halfway house	Lawrenceville, GA	8
December 9, 1983	Residence	Baltimore, MD	8
December 18, 1983	Hotel	Detroit, MI	8
May 11, 1984	Amusement center	Jackson, NJ	8
July 4, 1984	Hotel	Beverly, MA	15
October 18, 1984	Hotel	Paterson, NJ	15
December 25, 1984	Clothing store in hotel	Waukegan, IL	8
February 8, 1984	Residence	Philadelphia, PA	10
February 4, 1984	Residence	Tacoma, WA	9
September 20, 1984	Residence	Meadville, PA	8
November 11, 1984	Residence	Waterbury, CT	8
December 19, 1984	Apartment house	Orangeville, UT	27
July 23, 1985	Coal mine	Sweet Valley, PA	10
December 15, 1985	Elderly housing	Southfield, MI	8
December 16, 1985	Sanitarium	Glenwood Springs, CO	12
May 13, 1985	Auto repair/paint shop	Philadelphia, PA	11
May 29, 1985	Residence	Youngstown, OH	9
	Hazardous Chemical Manufacturer		
April 4, 1986	Hazardous Chemical Manufacturer	San Francisco, CA	9
May 8, 1986	Apartment complex	Philadelphia, PA	9
January 26, 1986	Residence	Washington, DC	9
February 4, 1986	Furniture store	Crystal Springs, MS	8
June 21, 1986	Residence	Tacoma, WA	8
October 25, 1986	Two-family residence	Minneapolis, MN	8
December 7, 1986	Residence	Bamberg, SC	8
April 20, 1987	Two-family residence	Camden, NJ	8
September 30, 1987	Residence	Milwaukee, WI	12
October 20, 1987	Hotel	Indianapolis, IN	9
February 2, 1988	Residence	Silverton, TX	8
August 20, 1988	Apartment	Jersey City, NJ	8
October 5, 1989	Elderly housing	Norfolk, VA	12
December 24, 1989	Hotel	Johnson City, TN	16
January 1, 1989	Residence	Remer, MN	10

TABLE 13.1 Continued.

Date	Structure	Place	Deaths
January 15, 1989	Business/residential	Yonkers, NY	8
February 11, 1989	Residence	Augusta, AR	8
April 11, 1989	Two-family residence	Peoria, IL	9
September 13, 1989	Coal mine	Wheatcroft, KY	10
November 18, 1989	Residence	Maxton, NC	9
October 23, 1989	Plastic manufacturer	Pasadena, CA	23
March 25, 1990	Nightclub	Brooklyn, NY	87
April 6, 1990	Hotel	Miami Beach, FL	9
July 5, 1990	Industrial chemicals manufacturer	Houston, TX	17
February 14, 1991	Dwelling	Pennsylvania	6
February 16, 1991	Three-family dwelling	Connecticut	10
February 18, 1991	Rooming house	Ohio	6
March 4, 1991	Board-and-care	Colorado	10
March 9, 1991	Dwelling	Tennessee	6
March 9, 1991	Refinery	Louisiana	6
March 13, 1991	Apartment house	New York	6
March 24, 1991	Group home	Texas	4
April 10, 1991	Dwelling	New York	7
May 8, 1991	Chemical plant	Louisiana	8
June 7, 1991	Chemical plant	South Carolina	9
July 12, 1991	Hotel	Minnesota	7
September 3, 1991	Food processing	North Carolina	25
September 14, 1991	Prison	Missouri	4
November 11, 1991	Repair garage	Ohio	4
December 9, 1991	Apartment house	Illinois	10
December 19, 1991	Dwelling	Washington	5
February 6, 1992	Restaurant	Indiana	11
June 2, 1992	Board-and-care	Michigan	10
December 7, 1992	Coal mine building	Virginia	8
January 29, 1993	Apartment	Mississippi	8
February 2, 1993	Apartment	Michigan	9
March 16, 1993	Hotel	Illinois	20
April 19, 1993	Dormitory	Texas	47+30 non-fire deaths
May 3, 1993	Apartment	California	12
March 29, 1994	Falling debris from air crash	North Carolina	23
April 27, 1994	Plane into building	Connecticut	8
July 3, 1994	Ground fire	Colorado	14
November 14, 1994	Debris hits auto	Wisconsin	6
April 19, 1995	Bomb government offices	Oklahoma	168
April 21, 1995	Pharmaceutical plant	New Jersey	5

lar multideath and "catastrophic" fires (where more than five deaths occur in residential structures). In the years 1990–1993, record low numbers were recorded, along with record high numbers (for both fires and related deaths):

Year	No. catastrophic fires/deaths	No. residence	No. commercial
1990	60 fires/385 deaths	25/145	35/240
1991	52 fires/342 deaths	27/160	25/182

Chapter 13   Conclusion

1992	35 fires/176 deaths[1]	17/93	18/83
1993	51 fires/316 deaths[2]	31/240[3]	20/76
1994	51 fires/307 deaths	32/193	9/42
1995	45 fires/384 deaths	21/116	11/208

The litany of each fire, whether a "catastrophic" incident or a single-death fire, contains the same sad commentary: the same violations of codes, the carelessness, the needless loss of these lives. Inoperable doors, inadequate or locked exits, lack of smoke detectors, smoking in bed, children playing with matches or lighters, children left unattended, metal security grilles over windows that inhibit exit and access, no sprinkler systems (or inoperable), overcrowding—the same year after year.

The implication, in this context, is that the design professionals, including the contract administrators, all have an obligation and responsibility to aid in the prevention of such deaths. Obviously, the professionals are not available during the operation of the building, or during periods when unwise and noncompliant acts are committed. Too often abuse, neglect, selfishness, immediacy, greed, carelessness, thoughtlessness, and forgetfulness on the part of owners, users, occupants, tenants, and guests mitigate the professionals' efforts. All the construction professionals have is their expertise, insight, and skill in incorporating all of the required systems and devices that promote fire prevention, extinguishment, and safety, and ensuring their proper installation and function, even to the extent of tamper-proof fixtures/devices, backup (redundant) systems, and as much fire-resistance that the project's budget will stand. According to J. Russell Groves, Jr., AIA, associate dean at the School of Architecture, University of Kentucky,

> Observation after occupancy can demonstrate continued commitment to the client, help to minimize present and future problems, and generate future commissions.

No one advocates that people be fearful whenever they walk into a building and that they feel safe only in their own homes. But we must seek and demand proper protection. Why don't we support fire department levies? Why don't we try to understand the inspection system and even demand more of it? Arson is a crime, but maybe we should make fire a crime; perhaps code violations should be crimes rather than minor offenses. Why don't we get strong court penalties, especially since only the worst cases are getting into the courts anyway? We don't seek these changes because not enough people are touched by the disasters and there is no organized effort by a large segment of the population to effectively demand change. Even a recent incident where a fire claimed the lives of several top executives from one corporation did not effect a massive change, although it may trigger some procedural studies by other corporations in an effort to prevent other such calamities. These small, isolated actions, however, may help in the long run.

Lives are lost every day in fires started by unknowing children playing with matches or candles. But lives are also lost, and at a more rapid rate, because of fires promulgated by code violations: holes in fire walls, flammable finishes, material producing toxic fumes, and locked exits or, worse, inadequate exits. We recycle and refurbish older buildings for new uses, and in the process old hazards may be glossed over or covered by new materials or may have additional loads imposed on them.

---

1. First year with less than 50 fires since 1971; fires down 33%, deaths down 49%.

2. Death rate lower than in 1988–91.

3. Death rate highest in last 7 years partly due to 47 deaths in Waco, TX. complex

Although the hazards are unseen, they still exist, and eventually they may claim some lives. In all these situations unthinking humans are to blame.

INSPECTION AND CODE COMPLIANCE

Although the profusion of single-death fires is still repulsive, two major American disasters within the last few years vividly point out the need for the imposition of proper construction inspection in occupancies other than residences (primarily in mass lodging and places of assembly).

The 1977 fire at the Beverly Hills Supper Club, in Southgate, Kentucky, which took 165 lives, seems to point up the problematic situation of the owners' interests apparently being allowed to take priority over the interests of the using public. This facility, open to the public, was constructed after being designed by a professional. However, the documents were incomplete (not intended for construction), and the construction was not supervised by the design professional; nor were the permit system and the governmental inspection agencies properly involved. A grand jury investigation (reviewed by a special prosecutor) produced no indictments and, indeed, vindicated the owners. There was no conspiracy to violate the codes, it was found, and apparently the owners had done all they could to comply with the permit and review sequence. The jury, however, scored the inspection system of the governmental agencies (both local and state) for lack of understanding, communication, action, and initiative in following the construction. Basically, no one could be held responsible for the fire; it was just something that "slipped through a crack."

The owners built what they needed when they needed it, with the assumption that all work was code-abiding, under permit, and inspected. However, the result was a terrible disaster. Investigation points to the possibility of improper and insufficient exits; highly flammable interior finishes; toxic smoke production; inadequate fire stopping; little if any fire suppression systems, fire fighting equipment, or alarm system; improper wiring; concealed voids between segments of construction (with combustibles present); and overcrowding of the entire facility. Any one of these discrepancies could foster trouble, but in combination they caused disaster. Elegance, convenience, and mere accommodation replaced the top-priority items of proper code-abiding construction and basic safety systems.

Not only had the inspection system obviously failed, it was never properly triggered. This was a massive facility in a very small town, which had no real handle on the hazard in its midst. The state jurisdiction was located some distance away, and most of the construction was carried on within the old walls. Given the confusion among inspection agencies, the lack of distinct jurisdiction, and the antiquated code that was the only tool available, it was amazing that the disaster was not of even greater proportion. A valiant effort by the town's fire department and those of many surrounding communities was the real "saving" grace. A far better and easier alternative to fire suppression and rescue lies with initial and continued compliance, inspection, and upgrading, along with owner and public consciousness of safety.

Inspection of existing facilities is extremely superficial at best, but the more obvious deficiencies could have been observed and orders issued. The basic construction, though, is hard to evaluate, especially when it is continual and old construction is covered over.

Overcrowding is a big problem in every jurisdiction, but to convict and penalize those responsible is difficult. The law and courts simply are not strict enough in these situations. If a liquor permit can be denied because of obscene performances, how about denial because of failure to meet and observe codes or for violating posted occupancy limits? When violations are allowed to stand, basically unchecked, lives can be lost.

Chapter 13   Conclusion **347**

The 1980 fire at the MGM Grand Hotel in Las Vegas, Nevada, which took 84 lives, was a major disaster, but with a different twist. The building, built in a high-powered, extremely active market, well protected by codes, was built to meet the law in effect in 1972 when construction began. How then could this facility be the site of a disastrous fire? Major changes in thinking have produced entirely new concepts of fire-rated construction, fire suppression systems, fire alarms, evacuation of buildings, compartmentalization, refuge areas, and the like, within the last eight years. But the MGM Grand was behind the time; it met an old code. There was no provision, however, that demanded that a code-abiding structure be constantly upgraded to meet the latest code provisions. Simply, the MGM Grand could not be built today as it was in 1972.

Oddly enough, though, the cause of the disaster was again human in nature. Investigation points to code violations in the building that allowed the fire to spread and become a killer. Apparently, holes in fire walls, flammable finishes and furnishings, limited use of sprinklers, and lack of proper venting contributed to the disaster. Although the building met the code of its day, it was beyond the technology of that day, and the violations allowed a marginal situation to become bad.

Even with a thorough inspection system, few jurisdictions have the clout to demand that all buildings keep in line with the kaleidoscopic code provisions. This process would be endless and prohibitively expensive. Who could afford to own a building that was constantly "under construction," installing the latest equipment? Yet doesn't the public health, safety, and welfare warrant this? At best, initial construction must comply, and periodic inspections (by any agency) are necessary to ensure continued compliance.

In one eight-year period little more than a handful of fires (on average, eight per year) accounted for an average of 164 lives lost each year. But these were just some of the 249 multideath fires that took some 1,100 victims annually. This carnage was but a minuscule 0.009 percent of the total fires (3,070,600 in 1978, for example) but accounted for an alarming 13 percent of the yearly 8,780 fire deaths. Since this period, the figures have varied (see Table 13.1) downward, for the most part. Although this can be taken as a hopeful sign, one can easily see that the number of fires is outrageous. One can understand (to some degree) death in ones and twos in relatively unregulated units. But there is no reason or excuse for the high death rate in highly regulated buildings. Needless to say, the compliance and inspection system continues to be inadequate to the task, since often the best of protective systems is obviated by the occupants. Certainly, no absolutely foolproof system is available to eliminate fire deaths. To the contract administrator, as an extension of the responsible design professional, this serves to implacably reinforce the need for diligence in both design *and* construction execution.

Construction and construction inspection are both imperfect sciences. The perfectly constructed building, with all elements being code-abiding, may never be built. As long as unthinking individuals feel the need to take shortcuts, there will be built-in imperfections. These imperfections sometimes allow problems to start, sometimes allow problems to continue and worsen, and perhaps result in a disaster. The "innocent" scrapping of fire-proofing from a steel column to allow a tighter fit of another material is an imperfection that led to disaster. The building caught fire and the other material burned away, exposing the column to the flames. The heat twisted the column, which was on the thirty-second floor of the building. In an instant, there is a major problem resulting from a small act.

Whatever measures are necessary must be employed by the governmental agencies and their inspectors to ensure compliance. Fees should be adjusted to provide adequate funding for a professional staff, proper departmental operations, and a diligent, complete inspection sequence. The penalties should be harsh, and the judiciary should recognize the hazard of light "wrist-slapping" as a symptom of disaster. A person in violation should expect to be caught and fined. Don't most people observe the speed limit for these reasons?

Much research has been done over the years, but still there are differences of opinion on just how the codes should be written and what techniques should be used to protect the public. There is, currently, no clear path that will result in maximum protection and still allow for modern construction techniques. This is not to say that all new construction methods and materials are faulty, but society has not yet solved the multifaceted problems of construction, energy saving, fire prevention, life saving, and economics. Our research must be broadened, preferably under the sponsorship of one agency.

Building and fire codes should be set forth in such a positive and creditable form that any person who is involved with them can easily and clearly see his responsibility and the consequence of noncompliance. Ideally, voluntary enforcement (and compliance) is the result being sought through the coding and inspection program. Great, though, are the pressures to do otherwise, unfortunately!

The inspection system is locked into the current code provisions, which, of course, tend to reflect the latest thinking and technology. Some time lag is normal between introduction of new systems and materials, and code acceptance, but everyone is party to seeking approval if it is to their benefit. Simply, the code must be enforced as it stands. In some instances owners will advance their protection on the advice of their insurance experts. This is no particular stroke of genius; it is amazing that every business venture does not seek maximum protection at least through compliance if not through more extreme measures. This would minimize liability, especially where the public is directly involved. Remember, the inspection system only sees that the current minimum provisions are not undercut.

Interesting situations sometimes occur. Within a short time after the Beverly Hills Supper Club fire disaster, two new clubs were built. They were located within miles of each other, and not far from the ill-fated Beverly Hills Club site. One was strictly a supper club, and the other was a club and discotheque. The supper club owner refurbished an old building and maintained a high level of public relations about meeting and, indeed, exceeding the required code provisions (for instance, sprinkler heads six feet apart throughout). All furniture carried a one-hour fire rating. Small placards were placed on each table, with instructions on fire reporting and evacuation, as well as the following information:

For Your Safety, This Building Is Constructed with:
1. The most modern fire and safety devices.
2. Complete sprinkler system.
3. Adequate fire exits and handicap ramps.
4. All employees trained to assist in the event of an emergency.
5. An alarm system connected to fire and police departments.
6. All doors closing when an alarm is pulled, but they will not lock, so you may exit, if necessary. This is to contain a fire at its origin.
7. All doors and walls fire rated 1-hour.
8. Furniture and furnishings fire resistant and nontoxic.

The club/discotheque was built by the owners of the Beverly Hills Club. They built within the corporate limits of the city of Cincinnati (within a few blocks of City Hall and the central fire station, as some have pointed out). They met every code provision and every requirement of a fully certified building department.

It was interesting to follow these two projects and see how they were accepted by the public. Public awareness was still high (as a result of the Beverly Hills Club fire) when the new ventures were built and first used. After changing hands several times, both businesses are now closed. The safety issues remained for a short period, but the other aspects of the businesses soon overwhelmed the public's impression. No major safety instances were reported during their relatively short tenures, but the approaches taken still merit some review. The cyclical awareness of the public and the continual demands on owners and business people to make things easier or more convenient (for both themselves and the customers) will always combine to reduce the safety systems that may be required. Those systems, by their passive nature, degrade the impression that they are, indeed, needed. Many times over, there are reports of sprinkler systems that failed (due to lack of maintenance or shut-down services), or even smoke detectors that were inoperable because someone removed the battery.

Many times a building, as constructed, will meet the code, but the use of the building (or rather, misuse) will produce massive adverse results, as happened at the MGM Grand. Narrowness of mind and trying to solve the immediate need at the earliest time, in the cheapest way, cause people to negate the best fire safety features and to add substantially to the fire loading of the building. Building codes cannot prevent this. Fire codes are just now emerging that try to handle these situations. Obviously, stiffer regulation is at hand; it is hoped that the full support of the public and the courts is forthcoming, as well.

In a 1977 instance, the stockpiling of material to extreme heights in a West German warehouse canceled out the effectiveness of the sprinkler system. The result was a massive loss ($100 million to the building and stored materials) because the water from the sprinklers could not reach the fire. Human error, again, and taking a shortcut to solve a problem, were responsible. The building, though, met the codes.

Oddly enough, building occupancy and use are the basic hazards to a building and its occupants. A structure built to be 100 percent safe would be highly undesirable (one could not live comfortably in it) and would demand constant supervision of the occupants by an unyielding, uncompromising authority. Bad habits and any tendency toward bad housekeeping, clutter, and taking shortcuts are the real hazards. These are not controlled by building codes, although some aspects are now being controlled more by fire codes.

Since it is well established that owners and occupants are going to "abuse" buildings, it is imperative that the inspection system demand absolute adherence to approved documents.

Contract administration is the professional's lasting contribution to this effort. It is true that *a building is never safer than at that moment just prior to occupancy.* In that is the foregone conclusion that the structure will be downgraded by the occupants, maneuvered or "warped" to make the owner's lot easier, or simply neglected. All of this either involves or affects the safety systems built into the building during construction. The professional really has but one opportunity to fulfill his/her obligation to protect public health, safety and welfare. That is to ensure that the proper systems are properly designed into the building, and then, through contract administration, ensuring that those systems survive the construction process and are available, if and when needed. Beyond this, the professional is without remedy or authority, and can only stand by if degradation of the building occurs. However, by shepherding the project through the entire building process, and closely following progress through adequate and proper contract administration, the professional has fully met his/her moral as well as contractual obligations. No one can ask more.

In our current economic climate, maintenance and upgrading are relegated to very low priorities, while convenience, attractiveness, and money-saving ventures are pushed at every turn. Fires continue to claim lives. Youngsters will still take batteries from smoke detectors, so they can furtively smoke a cigarette; some people will smoke in bed, and equipment will be poorly maintained. Sick minds will continue to set fires, and antiquated and deteriorated materials will erupt in flames. But we must try to achieve the best possible solutions, even though the deck may be stacked against us.

The tremendous pressure of time often forces acceptance of conditions that are changed or that are not in compliance, particularly with regard to material substitution or systems deletions: "We can't get it." "It's in transit and lost." "They're on strike."

"It's not available here." "Out of stock." "It has to be a special run." "This is just as good as . . ." "Thought we had it in our warehouse." "Can't afford to wait for the right stuff." "Who will know the difference?" "Can't stop now." "It'll cost me a bundle." These are real problems, but hardly grounds for not following code provisions.

The pressure of the economy on the contractor and developer is great indeed. No job can afford a slowdown or suspension because of the intolerably high interest rates on construction loans. Poor planning and unrealistic scheduling, which promises project delivery without consideration for lost time, tend to foster shortcuts, illegal substitutions, and shoddy, noncomplying work, but these are not viable alternatives to proper code-abiding construction.

One of the best instruments at the command of the inspection system is the certificate of occupancy (see Figure 13.1). This document can be withheld (with proper legislation) until all approval agencies have inspected the work and deem it safe for use. This program must be strongly enforced, and both occupant and contractor held liable for improper occupancy, thus ensuring that the public receives the best effort of the inspection system.

Many things must happen. There is a basic and vital need for cooperation and understanding. Sniping gets nowhere; laments about the past produce nothing. We need open discussion among the design professional, the contractor, and the code official. This should foster even better cooperation and coordination.

Although concern about building safety is reaching federal levels (reflected in congressional investigations and hearings), nationally mandated regulations are not needed. They would only confuse the code situation and provide unenforceable provisions, since there is no national building department. Local control should remain, but the local codes must be reviewed, updated, strengthened, and *fully enforced*. The same recommendations are made over and over:

1. Make thorough inspections during and after construction.
2. Enforce posted occupancy limits and impose strong penalties for violations.
3. Install direct alarm systems to fire departments.
4. Adopt revised national model codes in a timely manner.
5. Retrofit old buildings with safety systems (now or when remodeled).
6. Upgrade inspection system with more professionalism, proper budgets, and other support.
7. Make unannounced inspections during peak and off-hours of buildings used for public assembly.
8. Obtain personal liability protection for inspectors.
9. Permit only minimal and sparing grace periods for compliance.

The public must be educated about safety. Some building departments are using public service television announcements to let the public know what is required in

FIGURE 13.1 Excerpt from typical specification. A listing of the terminal documents required ensures that all work is properly touched by the inspection system, right up to initial occupancy.

```
1.04   CONTRACTOR'S CLOSEOUT SUBMITTALS TO ARCHITECT:

       a.   Submit evidence of compliance with requirements
            of governing authorities:
            1.   Certificate of Occupancy
            2.   Certificates of inspection:
                 (a)   Freight elevator
                 (b)   Electrical system inspection
                 (c)   Plumbing system inspection
                 (d)   Sprinkler system inspection
```

the way of permits and inspections, and how the services can be received. Manufacturers' associations are producing more literature explaining new research, new products, and new uses for old products. More and more building officials' groups are forming active community groups. We must tap every resource at our command.

Although budgets are being restricted and cut, the charge remains unchanged and perhaps is more urgent now than ever. We must do for the public what it can't do for itself. The inspection system must demand the most from every element of the imperfect construction process; it must be flexible and accommodating, yet ever unyielding and diligent. *There is no other way!*

TASKS AND EXERCISES

The following tasks are based on the concepts and information in the preceding chapters. Much of the material or information required to complete the tasks is found throughout the texts of several chapters. Some material is not in the text. Independent resources material and research is required. Consult your instructor for guidance.

1. Safety is a major concern on any project. What are the responsibilities of the contract administrator regarding safety?
2. A contract administrator has the authority to make any necessary changes in the project so long as they are documented eventually. True or false? Defend your position.
3. The contract administrator is ready to leave the site for a holiday weekend; the job construction manager has already left the site. The CA receives a report from the contractor that the swimming pool is leaking down into the pool equipment room. What action does the CA take?
4. What are the facilities the CA requires on site for use during visits? Why are they required?
5. Work has progressed, and is being pursued in a manner that requires evaluation and perhaps revision. There is a need to stop work for a period to make necessary adjustments. Who can stop the work? Who recommends such stoppage?
6. When is the best time to bring the construction manager onto a project team?
7. It is two days before Christmas. Upon entering the car, the CA sees an envelope on the seat. Inside the envelope is a $100 bill; there is no note or identification on the envelope. What action should the CA take?
8. List and describe 10 major, specific reasons why construction is particularly hazardous.
9. Discuss the impact of refusing or slow processing a payment request.
10. Explain how a cash allowance works, from the perspective of the contract administrator.
11. A contractor comes to you and explains that a series of required tests are running much more than the cost allowed in the estimate. What advice do you give?
12. What specific part does the design professional play in making changes in the work, necessitated by his/her work, the owner's desire, and project circumstances?
13. Discuss the purpose and implications of a "minor change." Why should it be documented?
14. Differentiate between a change order and a construction change directive.
15. Is it necessary that the owner sign all changes made in the contract? If yes, explain why.
16. How does a contractor contest rejection of work?
17. Write a specific specification clause that would directly aid you as a contract administrator for the following:

    level concrete floors for resilient tile

    exposed surfaces of concrete foundation walls above grade in an educational building

    hidden (concealed) surfaces of concrete foundation walls

18. Obtain a set or portion of a project specification, and review it for discrepancies and inconsistencies.
19. If the CM has the primary duty to inspect the work, what need is there for the designer to also inspect it?

Chapter 13   Conclusion **353**

20. Discuss dispensing with all inspections by public agencies.
21. Note and describe three kinds of inspections mentioned in contracts. Explain the differences.
22. If inspection by the design professional is only done periodically to ensure that it follows the design, who ensures that the work is done according to contract in other respects? Is this adequate?
23. Examine several project manuals for actual projects. Find and cite examples of unreasonable and inequitable requirements, and explain why you consider them as such.
24. Review some project working drawings or drawings in construction and drafting books and cite unreasonable or unworkable details and other requirements.
25. Write two general requirements regarding periodic and final cleanup on a site with a CM project: one where the CM has some responsibility, and one where only contractors are responsible.
26. Explain and offer effective remedies for some contracts that tend to slow down in the later stages of work.
27. List two specific contract provisions that aid the owner in achieving timely completion.
28. List and show the variations between the duties of the CM supervisor and the prime contractor's supervisor at project close-out.
29. From actual specifications, cite examples of "uncommon" and "particularly significant" (not standard) wording. Explain your choices.
30. For bidders on a CM, what are the two main points of information that need to be communicated to them? How do you achieve this? Cite examples.
31. Select and discuss a particular topic of importance that should be addressed in the contract for a CM project.
32. Write a general condition that you feel is especially required in a contract for a CM project.
33. List 10 skills or abilities that are important and required in a construction superintendent.
34. Why have owners become increasingly interested in their construction projects?
35. How wrong is an owner in accepting a very low bid, simply because the design professional and the local building inspector both have an obligation to see that the work is performed correctly?
36. Why are design professionals usually prohibited from engaging in construction, particularly for the same project?
37. List the duties of the design professional, the owner, and the contractor, based on a standard form of general contract conditions.
38. Should design professionals act as construction managers on projects they design? Explain your position.
39. Briefly explain the obligation of the design professional to (1) the owner and (2) the contractor, based on standard contract forms.
40. Discuss the role of the contract administrator in the process where the design professional is arbiter, per standard General Conditions.
41. Why was construction management introduced and developed?
42. List and briefly explain the functions of the construction manager.
43. In accord with standard General Conditions for the purpose, describe the essential differences between the design professional and the CM.
44. What are the pros and cons of having the CM responsible to the owner for the total project cost? Discuss.

45. From the owner's perspective, discuss CM with other traditional construction modes.
46. List ten college courses, both academic and technical, which would directly aid the contract administrator in her/his work.
47. In giving messages or making inquiries of your office or project architect, is it proper for you to prioritize the question to give a sense of how important a prompt answer is to project progress?
48. Discuss the value of making your site visits or observation on an irregular basis, i.e., not on the same day of the week, at the same time, and in the same sequence (walk the site in the same direction).
49. Often the office of the contract administrator and the construction superintendent will be in the same trailer or shed. In answering the phone, taking messages, or receiving packages directed to the superintendent, what is your responsibility?
50. Discuss the attribute of "firm, but fair."
51. Is it proper for the contract administrator to actively engage in the construction work, i.e., holding a tape, snapping a chalk line, holding an item while it is fastened, etc.?
52. You discover an error on a shop drawing. The drawing was approved by your office; your copy just arrived. Shipment of the product is within a week. How do you proceed with getting this situation corrected?
53. Discuss the value of using humor in a meeting that is usually characterized by stress, anger, and competitiveness among the contractors. Is this a valid way of reestablishing the team concept?
54. How quickly should you attempt to resolve any problem?

APPENDIX A

Field Observation Services and Procedures

I. Duties and Responsibilities

Know the specific duties and responsibilities for your particular position.

    A. Primary

       Important part of the team

       Not an adversary—an ally

       Not just a messenger—solve problems

       Prevent future problems

    B. Observe, report, and assist

       Documentation is critical

       What to do versus how to do

    C. Architect's representative

       Architect's representative duties are identified in AIA B352

       Site inspectors other than architect's representative have similar duties

    D. Be firm, fair, and flexible

II. Review

Be familiar with all the contract-related documents.

    A. Contract documents

       Drawings and specifications

       General conditions

       Supplementary conditions

       Addenda

       Owner-contractor agreement

       Owner-architect agreement

- B. Codes
- C. Responsibilities and limitations
  - Architect's Representative
  - Architect—AIA Handbook, Chapter 8
  - AIA Document B352
- D. Special project conditions

III. Start-Up
- A. Preconstruction conference
- B. Work area
  - Full set of drawings, specs, addenda
  - Prepare job notebook
  - Set up forms
  - Files
    - Correspondence
    - Shop drawings
    - RFIs
    - Clarifications
    - C.O.s
  - Schedule
  - Directory
- C. Basic tools

IV. Site
- A. Foundation drains and waterproofing
- B. Final grade
- C. Plantings
- D. Consultants
- E. Organization

V. Structure
- A. Concrete—poured in place
  - Prepour
  - Slab flatness
  - Slab inserts/openings
  - Water seals/reglets
- B. Concrete—architectural
  - Quality control
  - Columns
- C. Steel
  - Fireproofing
- D. Consultants

VI. Curtain Wall Systems
- A. Definitions
- B. Shop drawings, samples
- C. Mock-ups
- D. Systems tests

- E. Metal
  - Specific system
  - Finish
  - Color vs. sample
  - Joints
  - Weep system
  - Protection
  - Quality control
- F. Glass
  - Specified material
  - Color vs. sample
  - Seals
  - Protection
  - Quality control
- G. Gaskets, sealants
  - Specified material
  - Color vs. sample
- H. Installation
  - Metal-to-metal joints
  - Fasteners
  - Flashings
  - Glass setting techniques
  - Gasket, sealants, temperature restrictions
  - Safing insulation
- I. Storefront
  - Hardware
  - Safety glazing

VII. Exterior Finishes, Sealants
- A. Brick
  - Specified material
  - Sample
  - Protection
    - During installation
    - After installation
- B. Precast concrete
  - Finish and color vs. sample
  - Handling
  - Placement
  - Joints
    - Type
    - Size
- C. Sealants
  - What and where?
  - Approvals

              Type
              Color
          Application
              Weather restrictions
              Manufacturer's specs
              Backer rod
              Experienced mechanics
VIII.   Interior Environments
      A.   Interfacing with building code
        Code analysis
        Fire ratings
        Special requirements
      B.   Walls
        Drywall construction
            Stud height limitations
            Gypsum board types
            Shaft systems
            Rated partitions
            Acoustical
            Installation
            Tracks
            Studs
            Gypsum board
            Trim
            Taping, sanding
        Other inspection considerations
            Cold water pipes in wall insulated
            Pipe flanges clear
            Finished wall flush with built-in items
            Attachment supports
        Masonry
            Joints vs. final finish
            Control joints
      C.   Floors
        Concrete
            Finish
            Sealant
            Control joints
            Expansion joints
      D.   Floor finishes
        Resilient flooring
            Material vs. sample
            Pattern
            Base

Adhesive application
Ceramic, quarry tiles, pavers
  Material vs. sample
  Layout
  Setting bed
  Grout color vs. sample
Wood
Marble/slate
Carpet

E. Ceilings
  Common types
  Ceiling heights
    Per contract drawings
    Conflict
    Lighting clearance
  Suspended ceilings
  Rated assemblies
  Hangers
  Grid system
    Manufacturer
    Layout
    Tolerances
  Ceiling tiles
    Material vs. sample
    Direction
    Edge configuration
    Concealed spline access
  Drywall ceilings, soffits
    Access panels
  Plaster
    Skim coat on concrete, gypsum board
    Lath and plaster
  Combinations

F. Finishes
  Paints/coatings
    Type per location
    Application restrictions
    Lighting
    Substrate condition
  Wallpaper/vinyl wall coverings
    Application restrictions
    Lighting
    Priming, sizing
    Problems to watch for

Wood Paneling
  Quality standards
  Details, shop drawings
  Special consideration
Ceramic tile
  Material vs. sample
  Adhesive
  Substrate
  Special cuts
  Grouting
  Joints with dissimilar materials
Marble/stone
  Material vs. sample
  Installation methods
  Protection
  Critical lighting
G. Lighting Systems
  Inspection considerations
    Support
    Protection
    Use
    Lamping
    Rated assemblies
  Special lighting
    Layouts
    Access
H. Coordinating with mechanical systems
IX. Cabinets
  A. Shop drawings, samples, etc.
  B. Inspection checks:
    Fit
    Laminate
    Tightness of joints
    Door and drawer fit
    Material flaws
    Hardware
      Type
      Operation
    Secured in place
X. Doors, Frames
  A. Frames
    Materials
    Shop drawings, design details
    Masonry installation
    Drywall installation

Fire ratings
  Inspection considerations
 B. Doors
  Materials
  Shop drawings, design details
  Fire ratings
  Inspection considerations
 C. Hardware
  Material vs. schedule
  Finish
  Specialty
  Inspection considerations
 D. Glazed openings
  Metal frames
  Wire glass
 E. Specialty
  Revolving
  Institutional
  Rolling
XI. Special Furnishings
 A. Elevators, escalators
  Primary inspection responsibility
  Temporary use
  Protection
  Finishes
 B. Handrails
 C. Toilet rooms
 D. Other special furnishings
XII. Preparing for Project Close-Out
 A. Punch list inspection
  Assess if ready
  Who
  How
  Tools, aids
  Typical problems to look for
  How to avoid long punch
 B. Coordination
  Operation and maintenance manuals
  Warranties
  Certificate of occupancy
  County/city inspections
  Final payment
  Release of liens
  Certificate of substantial completion

Appendix B

Sample Regulatory Specifications Section

The following sample specification section is included as an example of a single-source section on regulatory requirements. It is not intended to be used as a master or guide specification section. This section must be closely coordinated with the general conditions (00700), supplementary conditions (00800), and other Division 1 sections since it contains information that could conflict with these other documents. Provisions included here should be limited to subjects unique to this section. Consider that Division 1 requirements do affect the cost of the work.

This section is based on the use of Owner-Contractor Agreements, General Conditions, and associated documents published by the AIA and EJCDC that delineate contractual requirements for the work. Refer to CSI *Manual of Practice* (1985 edition with '87, '89, '90 and '91 updates); Part I, Chapter 8, "Division 1, General Requirements."

CSI monograph #01M060 offers a fuller explanation and the legal implications associated with the use of a specification section similar to this sample.

SECTION 01060 - REGULATORY REQUIREMENTS

Instructions: Utilize this Section on each project, regardless of size, to establish the proper regulatory policy for the project.

Part 1 General

Summary

This section specifies procedural and administrative requirements for compliance with governing codes and regulations applicable to the project. These requirements include obtaining permits, licenses, releases, and similar documentation, as well as statements and similar requirements associated with the codes and regulations.

Related Sections

1. Document 00100 - Instructions to Bidders

Instructions: Agreement forms define the obligations and relationships between owner and contractor. In addition, all parties who have fundamental control and responsibility over the project, throughout the course of construction, should be listed on the Title Page (00001).

2. Document 00500 - Agreement Forms
3. Document 00700 - General Conditions
4. Document 00800 - Supplementary Conditions
5. Section 01090 - References
6. Section 01094 - Definitions
7. Section 01200 - Project Meetings
8. Section 01300 - Submittals
9. Section 01400 - Quality Control
10. Section 01600 - Material and Equipment
11. Section 01630 - Product Options and Substitutions

Instructions: The section below is required in every project, but must be modified where the project is located in a jurisdiction requiring that a Certificate of Occupancy be issued prior to use of the project. Ensure that the related section is modified to provide for final regulatory inspections, requisite remedial work, and issuance of certificate or other document permitting occupancy and use.

12. Section 01700 - Contract Closeout

Instructions: This article does not require compliance with the standards, but merely sets the format and context of the standards. This can include, if desired, a list of standards, complete with designations, titles, and date of issue. Coordinate with Section 01090.

References

Reference Standards. Within the Contract Documents, minimum acceptable quality of workmanship and materials has been defined by reference to standards cited in codes; reference to recognized standards, associations, or government standards; and manufacturers' names and catalog numbers.

Instructions: See definitions of "reference standard" and "reference material" in CSI *Manual of Practice.*

Reference standards shall take precedence in the following order:

First: Standards contained, listed, or referenced in applicable codes, their appendices, or similar regulations (note especially the dates of the editions required to be met).

Second: Standards selected by the design professionals, as noted in this Project Manual.

Third: Nonreferenced standards, which are defined as not being applicable to the work except as a general requirement of conformance of the work to recognized construction industry standards.

Fourth: "Reference material," which includes, among others, manufacturers' product specifications, installation instructions, and general recommendations.

Instructions: The date of the standard may have to be changed from that noted, if the regulatory code or other sources remain the same. It is necessary to be aware of changes in standards made during the production of the Contract Documents, as they may influence or change the parameters of the project when actually built.

Standards listed in regulatory codes appear with dates and editions adopted. Compliance with such standards is mandatory.

Unless otherwise noted, comply with reference standards in effect as of the date of the Contract Documents.

Conflicting Requirements. Where compliance with two or more standards is specified, and where these standards establish different or conflicting requirements for minimum quantities or quality levels, the most stringent requirements will be enforced, unless regulatory codes or the Contract Documents specifically indicate otherwise (see above). Before proceeding, refer to the Architect (Engineer) for a decision on which level of quality is more stringent for requirements that are different but apparently equal, and other uncertainties.

Updating Reference Standards. Where standards are revised to new editions or otherwise reissued with new requirements applicable to the project, and the Architect (Engineer) desires to incorporate the new requirements, submit suitable Change Order (Modification) proposals. Such proposals shall contain all changes in work, materials, or equipment compulsory by the new requirements.

Copies of Reference Standards. Where needed for proper performance of work, or requested by Architect (Engineer), or where necessary for enforcement of the requirements, provide copies of such standards directly from publication source, although such copies may be required submittals.

Instructions: Include here definitions of unusual and unique terms, abbreviations, and symbols used in the Contract Documents, which are not normally included in standard references. Also include explanation of acronyms and abbreviations used in the Contract Documents to indicate standards-generating organizations, governing authorities, other entities, and trade associations. Coordinate with Section 01094, if also used.

Definitions

The following definitions are included to explain and clarify the terms, abbreviations, acronyms, etc., used in the text of the Contract Documents. Request explanation of unknown or unclear items, terms, acronyms, and abbreviations from the Architect (Engineer).

LIST OF DEFINITIONS

Instructions: List *all* codes *that are enforced* by the local authority having jurisdiction over the project, or those codes used during the design and working drawing production in the case of prototype projects. In all cases, list the edition of the code used. See the Title Page (00001) for specific information.

Codes and Regulations

The design and construction of this project have been selected and depicted on the Contract Documents in compliance with all applicable codes, which govern the various work, materials, devices, equipment, systems, and procedures. These include but are not limited to the following:

Building Code
Mechanical Code
Plumbing Code
Electrical Code
Energy Conservation Code
Fire Prevention Code

LIST ALL OTHER CODES APPLICABLE

The procedure followed by the Architect (Engineer) has been to contact governing authorities, where necessary, to obtain information needed for the purpose of preparing Contract Documents, recognizing that such information may or may not be of significance in relation to the Contractor's responsibilities for performing Work. Direct contact by the Contractor with such governing authorities can be made for necessary information and decisions that have a bearing on the performance of the Work.

If the Contractor observes or is made aware that a portion of the Contract Documents are at variance with the applicable codes or other regulations, that fact shall be reported, immediately and in writing, to the Design Professional and the Owner. Necessary changes will be accomplished by appropriate Modification.

Further, the Contractor is responsible for compliance with all codes, regulations, and other standards that bear on the performance of the Work and this project, as required in the General Conditions, but not specifically depicted on the Contract Documents. The Contractor shall bear the cost of all necessary corrections arising from any noncompliance on his part, or where appropriate notice of noncompliance is not given.

Where codes or other regulations are revised, updated to new editions, or otherwise reissued with new requirements applicable to the project, and the Architect (Engineer) desires to incorporate the new requirements, submit suitable Change Order (Modification) proposals. Such proposals shall contain all changes in work, materials, or equipment required by the new requirements.

> *Instructions:* List *all* parties that have fundamental control and responsibility over the project throughout the course of construction. Modify and augment the list as required.

Project Responsibility

The following list has been or will be filled with the Code Official having jurisdiction over this project. It is a list of those individuals and/or firms that carry the project responsibilities, as noted:

Overall project

Project design

Structural design

Construction management

HVAC design

Electrical design

Plumbing design

Fire suppression design

While the fundamental responsibility lies with these persons/firms, communications with same shall be established in the preconstruction or initial job meetings. Direct lines of communication and distribution lists shall be established and made available to all parties on the project site.

> *Instructions:* Coordinate with Section 01300, if used, to ensure that there is a request for all relevant data to be furnished by the Contractor, either before, during, or after construction.

Submittals

Acquire and submit required copies of permits, certificates, licenses, inspection reports, releases, jurisdictional settlements, notices, receipts for fee payments and judgments, and similar documents, correspondence, and records pertinent to the work and resulting from compliance with applicable codes and regulations.

> *Instructions:* Although optional, this article can be a great aid in the overall regulatory climate and in coordination of the project as a whole. Coordinate with Section 01600, if used.

Quality Assurance

Qualifications. Statements of qualifications for contractor-employed design personnel, fabricators, installers, and applicators of products and completed work shall be submitted when required elsewhere.

Certifications. Where specified, attach labels of testing authorities, or ensure the attachment by others, to materials or assembly involved. Also, provide statements from manufacturers where required, noting that supplied systems, materials, manufactured units, equipment, components, and accessories comply with specified requirements.

Preinstallation Conference. As specified in Section 01200 or elsewhere, preinstallation conferences may be required for coordinating materials, techniques, and related work.

Part 2 Products (Not Applicable)

Part 3 Execution (Not Applicable)
END OF SECTION

The following are additional regulatory provisions that may be added to the basic sample section, as deemed appropriate:
Instructions: Insert list of pertinent organizations, customized to project and work. See partial listing in Section 01090.

Name Abbreviations

Names of trade associations, standards-generating organizations, governing authorities, and any other entities are frequently referred to in Contract Documents by acronyms and/or abbreviations. Request explanation of unknown terms from Architect (Engineer), where not listed below.

Trade Union Jurisdictions

The manner in which Contract Documents have been organized and subdivided is not intended to be an indication of trade union jurisdiction or preferences. The Owner will rely on the Contractor's experience, judgment, integrity, and local trade contracts and policies to assign work, employ tradespersons and laborers, and in letting all subcontracts in such a manner as to minimize conflict, delays, claims, and losses in performance of the work.

1. The use of certain titles, such as "carpentry" in the specifications text, is not intended to imply that the Work must be performed by accredited or unionized individuals of a corresponding generic name, such as "carpenter." It also is not intended to imply that generic name.
2. In certain circumstances, the specification text requires or implies that specific elements of the Work are to be assigned to specialists who must be engaged to perform that element of Work. Such assignments are special requirements over which the Contractor has no choice or option. They are intended to establish which part or entity involved in a specific element of the Work is considered as being sufficiently experienced in the indicated construction processes or operations to be recognized as "expert" in those processes or operations. Nevertheless, the ultimate responsibility for fulfilling all contract requirements remains with the Contractor.

a. These requirements should not be interpreted to conflict with the enforcement of building codes and similar regulations governing the Work. They are also not intended to interfere with local trade jurisdictional settlements and similar conventions.

Tests and Inspections

Perform or arrange for all tests and inspections as required in Contract Documents, in applicable regulations, or in regulatory procedures. Pay cost of all tests/inspections as required, unless exempt.

Provide 24 hours' notice before all tests and inspections to all interested agencies, organizations, and persons.

Maintain record of each test or inspection. Include time, weather conditions, names of authority performing such work, results of test or inspection, and all other pertinent data.

Quality Assurance

All products and materials supplied for installation on this project shall meet the pertinent requirements of the Contract Documents, the applicable codes and regulations, or testing and listing of recognized authorities (UL, FM, etc.). Submit substantiating data to indicate full compliance with:

1. Loading requirements
2. Fire resistance rating
3. Flame spread/smoke-developed rating
4. Other virtues required by these specifications.

Where specified, attach label of testing authority, or ensure the attachment by others, to material or assembly involved.

Where proprietary material, assemblies, or equipment are specified, obtain all primary material from a single manufacturer.

1. Where a rating or listing indicates proprietary materials used in the test, provide the same materials as used to secure the test results and ratings.

Substitutions

Any proposed substitution must have prior approval of the Owner's representative (agent) and the Architect (Engineer).

Where a substitution requires additional review and approval of the regulatory agency(ies), the submittal to the Architect (Engineer) shall provide sufficient time for such approval, without disrupting construction sequence, nor elongating the time for construction.

Project Finale

Provisions shall be made to include the final inspection of all regulatory agencies and any remedial work required by same, prior to the closeout of the project (as specified elsewhere).

A Certificate of Occupancy/Use or a similar statement of compliance from the regulatory agency(ies) is required. The Contractor shall inspect the Work for compliance, give notice for final inspection, and ensure that everything is in order so the Certificate can be issued without delay.

Submit copy to Architect (Engineer) and Owner prior to request for Final Payment.

APPENDIX C

Detailed Meeting Formats, Procedures, Agenda, and Planning

KEEPING RECORDS OF PRECONSTRUCTION AND PROGRESS MEETINGS

In Section One, this appendix answers the following questions:

Why is the preconstruction conference important?
What information should the owner cover in the preconstruction conference?

In Section Two, this appendix answers the following questions:

How should the contractor prepare and conduct progress meetings?
What should progress meetings cover?

SECTION ONE: PRECONSTRUCTION CONFERENCE

A preconstruction conference takes little time and encourages meaningful communication between owner and contractor. It also provides opportunities for the contractor to initiate effective documentation systems. Here we discuss some specific ways for all parties to get the most from preconstruction meetings.

Purpose

Once the owner awards a contract and before construction begins, the owner needs to hold a preconstruction conference with the contractor. The conference can make the contractor more aware of the owner's policies and procedures and will allow both parties to go through the contract together. Not only will the owner have the chance to answer the contractor's questions, but all parties will become more familiar with how the owner will administer the contract and why.

The contractor can benefit from a preconstruction conference by using it as an opportunity to review and question specifications, drawings, and other contract documents and as a place to discuss potential problems and ways of handling them.

Preparation and Procedures

An engineer who is responsible for arranging the conference should plan the preconstruction conference in the following way (also see Figure C.1):

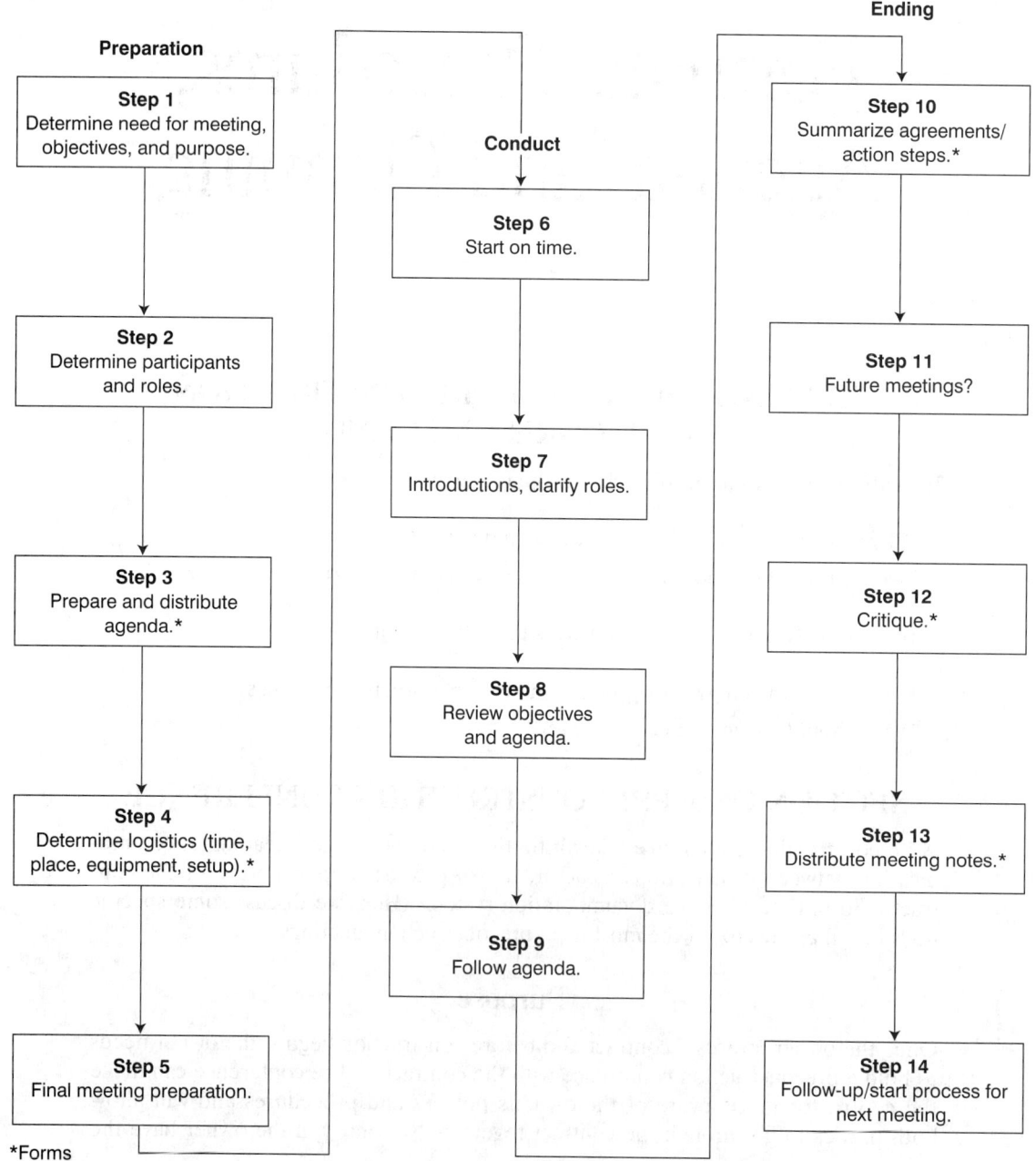

FIGURE C.1 Fourteen steps to effective meetings.

Appendix C Detailed Meeting Formats, Procedures, Agenda, and Planning   **371**

Review project plans and specifications (note unique situations or areas where potential problems may occur).

Meet with staff (or other owner representatives) to discuss the contract.

Develop an agenda for the preconstruction conference (noting areas where different interpretations of the contract exist).

Persons Who Should Attend

The owner should invite the following people to attend the preconstruction conference:

A principal of the prime contractor's firm

The prime contractor's superintendent

A representative of each of the principal subcontractors (if the prime contractor agrees)

The contract administrator

The resident construction engineer

The architect/engineer

Conducting the Preconstruction Conference

The owner should conduct the conference in an informal manner. The following guidelines may be helpful:

Make sure each person has a copy of the agenda (see Figure C.2).

Confine all discussion to the agenda (to issues that involve construction or the contract).

Keep a record of the meeting (including subjects covered, instructions given, agreements reached, and persons who attended).

Send copies of the meeting record to all who attend and ask the prime contractor to return his or her copy by certified mail (return receipt requested).

Opening the Preconstruction Conference

The owner should make sure that everyone understands the goal of the contract—to construct (or renovate) a facility according to the contract plans and specifications for the amount the contract stipulates. The owner should also explain that the goal of the preconstruction conference is to develop ways in which everyone can achieve that goal. The owner should also show the contractor how to identify and refer to the project (by title, project number, contract number, and locations). In addition, the owner should explain how he/she understands the contractor's responsibilities under the contract.

The engineer (in charge of the conference) should introduce all people who attend. Each person should describe the responsibilities designated to him or her by the contract. The preconstruction conference should define the scope of each person's authority.

Construction Progress Charts

The owner should explain contract requirements and the use of a chart or CPM equivalent. The owner should also explain how progress charts should reflect changes and how time extensions are approved.

Cost Breakdown and Progress Payments

The owner should discuss the following:

Meeting Agenda

Meeting Name: _____ Meeting Leader: _____

Meeting Date: _____ Start Time: _____ AM PM

Meeting Location: _____ Finish Time: _____ AM PM

Scope & Purpose: _____

Decision-Making Process: _____

Agenda Item #1	
Desired Outcome:	
Topic Leader:	
Time Required:	

Agenda Item #2	
Desired Outcome:	
Topic Leader:	
Time Required:	

Agenda Item #3	
Desired Outcome:	
Topic Leader:	
Time Required:	

Agenda Item #4	
Desired Outcome:	
Topic Leader:	
Time Required:	

Agenda Item #5	
Desired Outcome:	
Topic Leader:	
Time Required:	

Agenda Item #6	
Desired Outcome:	
Topic Leader:	
Time Required:	

Agenda Item #7	
Desired Outcome:	
Topic Leader:	
Time Required:	

Agenda Item #8	
Desired Outcome:	
Topic Leader:	
Time Required:	

People Participating:
_____
_____
_____
_____
_____
_____

Facilitator: _____ Recorder: _____

FIGURE C.2 Meeting agenda.

How to develop and use a cost breakdown schedule.

How to establish the day of the month that will serve as the cutoff for completed work, which the contractor will include in his/her monthly requests for payment.

How payments will be made and what administrative processes affect the amount of time payments take.

Appendix C Detailed Meeting Formats, Procedures, Agenda, and Planning

How the contractor may be compensated for materials purchased and delivered (both on and off site) but not for those installed. The owner should explain relevant provisions of the "Payments to Contractors" clauses of the contract.

Contractor Submittals

The owner should discuss the following:

The approving authorities.

The number of drawings to be submitted to the contractor, the distribution of approved drawings, and the number of days the owner or A/E will need to check the drawings.

Submittal and approval of catalog cuts.

Identifying and sending samples (as well as the submittal of certificates in lieu of samples).

Owner's policy on the substitution of materials.

The owner should remind the contractor to check carefully all shop drawings he/she receives, to make all necessary coordination drawings, and to note in his or her transmittal letter where his or her work varies from contract requirements.

Subcontractors

The owner should discuss the qualification requirements of any specialty subcontractors that may be called for under the special conditions and the technical specifications.

Contract Modifications

The owner should discuss the following:

Why holding modifications to a minimum is important.

Which procedures the contractor should use to make changes and why time submittals are important. The owner should make sure that the contractor understands *who* has the authority to authorize changes.

When the "Equitable Adjustment" clause is applicable as well as what allowable markups are.

How the contractor requests time extensions (including justification and price proposals) and when time extensions will be considered, what documents will substantiate requests and how they are accepted or rejected.

What differing site conditions are, how changes are processed, and what the contractor's responsibilities are.

Correspondence

The owner should discuss how the contractor should identify and address correspondence, the number of copies he or she should send, the authorized signatures on owner's correspondence, and other relevant issues such as routing and distributing copies.

The owner should also obtain the contractor's address and phone number in the field as well as at the home office. Also, the owner should get telephone numbers for after hours, holidays, and weekend emergencies for all personnel involved in the project.

Miscellaneous

The owner should discuss the following:

Guaranties

Job housekeeping

Substitution of materials and/or methods differing from those specified or previously approved for use. The owner should state the policy clearly—substitution of materials or methods other than those specified (or approved) will not be considered except if the contractor can show that what's specified cannot be obtained. Also, the owner should explain if contractor can receive credit for substitutions.

Delays in completing the contract (refer to relevant contract clauses). The owner should explain that the contractor must document both the cause and extent of each delay.

Salvage

Completion date

Liquidated damages

Utilities for construction, light, heat, elevator use, telephone, etc.

Project documents and survey of outside utility lines

Use of premises and occupancy of owner

General requirements for mechanical and electrical equipment. If a meeting between electrical and mechanical engineers and appropriate subcontractors is necessary, use this conference to schedule it.

The status of specifications, drawings, schedules, large and small scale drawings and details

The furnishing of information and records

Use of owner facilities at the job site, including use of power, lights, telephone, elevators, toilet facilities, and parking, as well as arranging for safety to public, storage and delivery of materials, removal of contractor-owned salvage, and access to the site

Safety and first-aid procedures

Special Contract Provisions

The owner should review any unique administrative or technical provisions with the contractor.

Site Visit

After the preconstruction conference, the owner and contractor should visit the site. This visit can uncover problems that may occur during construction (for example, coordinating agency and contractor activities, delivery of materials, parking, storage, power, heat, lighting, noise).

SECTION TWO: PROJECT MEETINGS

The contractor (working under a fixed-price contract) should prepare for and conduct progress meetings and specially called meetings during the project in the following ways:

Prepare an agenda.

Distribute written notice of each meeting four days in advance of the meeting.

Make physical arrangements for the meeting.

Preside at the meeting.

Keep a record of each meeting. Include significant events and decisions.

Appendix C   Detailed Meeting Formats, Procedures, Agenda, and Planning

Reproduce and distribute copies of each meeting's minutes within three days of the meeting. Send minutes to all participants, all those affected by decisions made at the meeting, and to the A/E.

The representatives for contractor, subcontractors, and suppliers who attend meetings should be qualified and have the authority to act on behalf of their organizations.

The architect/engineer may attend meetings. His/her responsibility is to make sure that the contractor's work conforms to the contract and to construction schedules.

When the CM form of contracting is adopted, all normal construction practices and relationships between owner and contractor change. The nature of these changes must be carefully observed and monitored during progress meetings.

Suggested Agenda for Progress Meetings

Review and approval of minutes of previous meeting

Review of work progress since previous meeting

Field observations and problems

Problems that impede the construction schedule and revisions to the schedule

Review of off-site fabrication and delivery schedules

Coordination of schedules (including progress during succeeding work period)

Review of submittal schedules

Maintenance of quality standards

Pending changes and substitutions

Review of proposed changes, including their effect on the construction schedule and the completion date as well as the effect on other contracts involved

Other business

Appendix D

Complete List of AIA Documents

Documents listed are revised periodically; revision of some documents is due in mid-1997. Contact AIA headquarters or the nearest chapter for the latest editions and prices.

A SERIES/Owner-Contractor Documents

A101   Owner-Contractor Agreement Form—Stipulated Sum (4/87) with instruction sheet wrapped

A101/CM Owner-Contractor Agreement Form—Stipulated Sum—Construction Management Edition (6/80) with instruction sheet wrapped

A107   Abbreviated Owner-Contractor Agreement Form for Small Construction Contracts—Stipulated Sum (4/87) with instruction sheet wrapped

A111   Owner-Contractor Agreement Form—Cost Plus Fee (4/87) with instruction sheet wrapped

A117   Abbreviated Owner-Contractor Agreement Form—Cost Plus Fee (4/87) with instruction sheet wrapped

A171   Owner-Contractor Agreement for Furniture, Furnishings and Equipment (3/79) with instruction sheet wrapped

A177   Abbreviated Owner-Contractor Agreement for Furniture, Furnishings and Equipment (5/80) with instruction sheet wrapped

A191   Standard Form of Agreements Between Owner and Design/Builder (1985)

A201   General Conditions of the Contract for Construction (4/87)

A201/CM General Conditions of the Contract for Construction—Construction Management Edition (6/80)

A201/SC General Conditions of the Contract for Construction and Federal Supplementary Conditions of the Contract for Construction (8/77)

A271	General Conditions of the Contract for Furniture, Furnishings and Equipment (12/77) with instruction sheet wrapped
A305	Contractor's Qualification Statement (12/86)
A310	Bid Bond (2/70)
A311	Performance Bond and Labor and Material Payment Bond (2/70)
A311/CM	Performance Bond and Labor and Material Payment Bond—Construction Management Edition (6/80)
A312	Performance Bond and Payment Bond (12/84)
A401	Contractor-Subcontractor Agreement Form (5/87)
A491	Standard Form of Agreements Between Design/Builder and Contractor (1985)
A501	Recommended Guide for Bidding Procedures and Contract Awards (6/82)
A511	Guide for Supplementary Conditions (incorporates A512 6/87)
A511/CM	Guide for Supplementary Conditions—Construction Management Edition (3/82)
A521	Uniform Location Subject Matter (1981, Reprinted 7/83)
A571	Guide for Interiors Supplementary Conditions (3/82)
A701	Instructions to Bidders (4/87) with instruction sheet wrapped
A771	Instruction to Interiors Bidders (5/80) with instruction sheet wrapped

B SERIES/Owner-Architect Documents

B141	Standard Form of Agreement Between Owner and Architect (4/87) with instruction sheet wrapped
B141/CM	Standard Form of Agreement Between Owner and Architect—Construction Management Edition (6/80) with instruction sheet wrapped
B151	Abbreviated Owner-Architect Agreement Form (4/87) with instruction sheet wrapped
B161	Standard Form of Agreement Between Owner and Architect for Designated Services (11/77) with instruction sheet wrapped
B161/CM	Standard Form of Agreement Between Owner and Architect for Designated Services—Construction Management Edition (12/82) with instruction sheet wrapped
B162	Scope of Designated Services (11/77) with instruction sheet wrapped
B171	Standard Form of Agreement for Interior Design Services (3/79) with instruction sheet wrapped
B177	Abbreviated Interior Design Services Agreement (5/80) with instruction sheet wrapped
B181	Owner-Architect Agreement for Housing Services (6/78) with instruction sheet wrapped
B352	Duties, Responsibilities, and Limitations of Authority of the Architect's Project Representative (5/79)
B431	Architect's Qualification Statement (12/79)
B727	Standard Form of Agreement Between Owner and Architect for Special Services (6/79) with instruction sheet wrapped
B801	Standard Form of Agreement Between Owner and Construction Manager (6/80) with instruction sheet wrapped
B901	Standard Form of Agreements Between Design/Builder and Architect (1985)

Appendix D Complete List of AIA Documents

C SERIES/Architect-Consultant Documents

- C141 Standard Form of Agreement Between Architect and Consultant (4/87) with instructions wrapped
- C142 Abbreviated Form of Agreement Between Architect and Consultant (4/87)
- C161 Standard Form of Agreement Between Architect and Consultant for Designated Services (6/79) with instructions wrapped
- C431 Standard Form of Agreement Between Architect and Consultant (4/82) with instructions wrapped
- C727 Standard Form of Agreement Between Architect and Consultant for Special Services (4/82) with instructions wrapped
- C801 Joint Venture Agreement (6/79) with instructions wrapped

D SERIES/Architect-Industry Documents

- D101 Architectural Area and Volume of Buildings (1/80)
- D200 Project Checklist (8/82)

F SERIES/Architect's Accounting Forms—Manual System

25 Sheets per Unit

- F101 Cash Journal—1949
- F102 Cash Journal—1949
- F103 Cash Journal—1949
- F104 Cash Journal—1953
- F105 Cash Journal—1949
- F106 Cash Journal—1949
- F107 Journal Form—1949
- F201 Payroll Journal—1953
- F202 Payroll Journal—1949
- F203 Payroll Journal—1949
- F301 Ledger Account Form—1949
- F401 Job Expense Record Form—1953
- F402 Employee Record Form—1953
- F403 Fixed Assets Record—1949
- F404 Note and Investment Record—1949

50 Sheets per Unit

- F501 Trial Balance—1953
- F502 Balance Sheet—1953
- F503 Profit and Loss Statement—1949
- F504 Indirect Expense Factor—1953
- F601 Time Record Sheet—1971
- F603 Expense Voucher Nonpersonnel—1953
- F605 Expense Record—1953
- F701 Billing Extract—1972
- F703 Aged Accounts Receivable—1972
- F712 Project Payroll Cost Worksheet—1972

F714	Detail of Expenses—1972
F716	Time Distribution Summary—1972
F721	Project Estimating and Budget Worksheet—1972
F723	Project Progress Report—1972
F725	Project Summary Report—1972

F SERIES/Compensation Guidelines Forms and Worksheets

F800 F810 through F860 (1978) Complete Set of 2 each of 24 Forms Needed

25 Sheets per Unit

F810	Scope of Designated Service Worksheet (unphased)
F820	Phase Compensation Worksheet (unphased)
F850	Project Time and Payment Schedule

F SERIES/Standardized Accounting for Architects Forms

25 Sheets per Unit

F1001	Cash Receipts Journal (1978)
F1002	Cash Disbursements Journal (1978)
F1004	Trial Balance (1978)
F1006	Balance Sheet and Income and Expense Statement (1978)
F2002	Fixed Assets Record (1978)
F2004	Notes and Investment Record (1978)
F3003	Payroll Journal (1978)
F3005	Staff Payroll Record (1978)
F4001	Project Time Distribution (1978)
F4003	Project Expense Record (1978)
F5001	Billing Extract (1978)
F5003	Aged Accounts Receivable (1978)
F6001	Ledger Accounts (1978)
F6002	Journal (1978)
F6003	Accounting Worksheet (1978)
F6004	Accounting Worksheet (1978)

50 Sheets per Unit

F2001	Expense Voucher (1978)
F3001	Time Record (1978)
F3002	Staff Expense Record (1978)
F5002	Invoice for Architectural Services (1978)

G SERIES/Architect's Office and Project Forms

These forms have been greatly expanded and are now sold individually.

G601	Land Survey Requisition (6/79) with instructions
G602	Geotechnical Services Agreement (8/83)

Appendix D   Complete List of AIA Documents

G605/606 Purchase Order and Purchase Order Continuation Sheet
G610   Owner's Instructions for Bonds and Insurance (11/78)
G611   Owner's Instructions Regarding Construction Agreement and Bidding Procedures (9/81) with instruction sheet wrapped

50 Sheets per Unit

G604   Professional Services Supplemental Authorization (3/79) with instructions
G701   Change Order (4/78)
G701/CM Change Order—Construction Management Edition (6/80)
G702   Application and Certificate for Payment (4/78, Rev. 5/83)
G703   Continuation Sheet for G702 (4/78, Rev. 5/83)
G703/CR Continuous Roll Continuation Sheet For G702 (5/83)
G704   Certificate of Substantial Completion (4/78)
G705   Certificate of Insurance (11/78)
G706   Contractor's Affidavit of Payment of Debts and Claims (4/70)
G706A   Contractor's Affidavit of Release of Liens (4/70)
G707   Consent of Surety to Final Payment (4/70)
G707A   Consent of Surety to Reduction in or Partial Release of Retainage (6/71)
G709   Proposal Request (4/70)
G710   Architect's Supplemental Instructions (3/79) with instructions
G711   Architect's Field Report (10/72)
G712   Shop Drawing and Sample Record (10/72)
G713   Construction Change Authorization (3/79) with instructions
G722   Project Application and Project Certificate for Payment (6/80) with instructions
G723   Project Application Summary (6/80) with instructions
G801   Application for Employment (6/75)
G804   Register of Bid Documents (4/70)
G805   List of Subcontractors (4/70)
G807   Project Directory (4/70)
G809   Project Data (4/70)
G810   Transmittal Letter (4/70)
G811   Employment Record (11/73)
G813   Temporary Placement (1/74)

Single copies of the AIA Bookstore Catalog (N711) are available at no cost. Single copies of the Documents Price List (N710) are available at no cost.

How to Purchase Documents from AIA Document Price List

Documents may be obtained from the AIA Order Dept., 9 Jay Gould Court, P.O. Box 740, Waldorf, MD 20601, or from your authorized local distributor for AIA contracts and forms.

All document orders should be directed to your local distributor if you reside in one of the following states: Alabama, Alaska, Arizona, Arkansas, California, Colorado, Connecticut, Delaware, Florida, Georgia, Hawaii, Idaho, Illinois, Indiana, Iowa, Kansas, Kentucky, Louisiana, Maryland, Massachusetts, Michigan, Minnesota, Mississippi, Missouri, Montana, Nebraska, Nevada, New Hampshire, New Jersey, New Mexico,

New York, North Carolina, North Dakota, Ohio, Oklahoma, Oregon, Pennsylvania, Rhode Island, South Carolina, South Dakota, Tennessee, Texas, Utah, Virginia, Washington State, West Virginia, Wisconsin and Bermuda.

Because AIA documents in the A-G series and all chapters from the Architect's Handbook of Professional Practices are revised regularly, some may have been superseded by new editions. Editions listed are the latest available.

In the Washington, D.C. area, documents may be purchased over the counter at the AIA Bookstore, 1735 New York Ave. NW, during business hours, 8:30 A.M.–5:00 P.M., Monday through Friday.

APPENDIX E

Standards-Producing Organizations

SOURCES: TECHNICAL INFORMATION AND REFERENCE STANDARDS

Because of the massive amount of information required for any project, design professionals often utilize information contained in publications produced by various organizations. Trade associations, standards-generating organizations, and governing authorities are invaluable sources of technical information, literature, and audio-visual aids. This information is specific, complete, and in-depth. It includes design, fabrication, processing, production, and installation data, with pertinent details. Usually, there is far more information than required, in that the testing and manufacturing procedures may be noted. Many items are provided gratis, but ask for catalog and applicable price list for available items and complete ordering information.

These documents are generally categorized as "reference standards." The data are promotional in nature, not directed toward sales, but toward understanding and the correct use and implementation of the products involved.

In lieu of repeating all of the necessary information on the drawings or in the specifications, professionals usually use a system of referring to the required materials, often using acronyms or abbreviations to represent the full names of the organizations involved. Following is a list of many such organizations; other organizations exist that are not listed. Names and addresses are subject to change, but are believed to be accurate as of the date of production of this book. To verify or update information readers are advised to consult one of the following:

Encyclopedia of Associations, published by Gale Research Company.

National Trade and Professional Associations of the United States and Canada and Labor Unions, published by Columbia Books, Washington, DC.

ARCAT, published by The Architect's Catalog, Inc., Fairfield, CT.

Architectural Graphics Standards manual.

Most larger public libraries have a directory service (free), which will locate addresses and telephone numbers in various cities.

AA	Aluminum Association 900 Nineteenth St. NW, Suite 300 Washington, DC 2006	(202) 862-5100
AAA	American Arbitration Association 140 West 51st St. New York, NY 10020	(212) 484-4000
AABC	Associated Air Balance Council 1518 K St. NW, Suite 503 Washington, DC 20005	(202) 737-0202
AAMA	American Architectural Manufacturers Association 1827 Walden Office Square, Suite 104 Schaumberg, IL 60195	(847) 303-5664
AAMA	Architectural Aluminum Manufacturers Association 2700 River Rd., Suite 118 Des Plaines, IL 60018	(312) 699-7310
AAN	American Association of Nurserymen 1250 Eye St. NW, Suite 500 Washington, DC 20005	(202) 789-2900
AASHTO	American Association of State Highway and Transportation Officials 444 North Capitol St., Suite 225 Washington, DC 20001	(202) 624-5800
AATCC	American Association of Textile Chemists and Colorists P.O. Box 12215 Research Triangle Park, NC 27709	(919) 549-8141
ACI	American Concrete Institute 38800 Country Club Dr. Farmington Hills, MI 48331	(810) 848-3700
ACIL	American Council of Independent Laboratories 1725 K St. NW Washington, DC 20006	(202) 887-5872
ACPA	American Concrete Pipe Association 8320 Old Courthouse Rd. Vienna, VA 22180	(703) 821-1990
ADC	Air Diffusion Council 230 N. Michigan Ave., Suite 1200 Chicago, IL 60601	(312) 372-9800
AFPA	American Forest & Paper Council American Wood Council (formerly National Forest Products Association) 1111 19th St. NW, Suite 800 Washington, DC 20036	(202) 463-2700
AGA	American Gas Association 1515 Wilson Blvd. Arlington, VA 22209	(703) 841-8400
AGA	American Galvanizers Association 1200 E. Iliff Ave., Suite 204 Aurora, CO 80014	(303) 750-2900
AHA	American Hardboard Association 1210 W. Northwest Highway Palatine, IL 60067	(847) 934-8800

Appendix E Standards-Producing Organizations

AHAM	Association of Home Appliance Manufacturers 20 N. Wacker Dr. Chicago, IL 60606	(312) 984-5800
AHMI	Appalachian Hardwood Manufacturers, Inc. P.O. Box 427 High Point, NC 27261	(910) 885-8315
AI	Asphalt Institute Asphalt Institute Building College Park, MD 20740	(301) 277-4258
AIA	American Institute of Architects 1735 New York Ave. NW Washington, DC 20006	(202) 626-7300
A.I.A.	American Insurance Association 85 John St. New York, NY 10038	(212) 669-0400
AIHA	American Industrial Hygiene Association 475 Wolf Ledges Parkway Akron, OH 44311	(216) 762-7294
AISC	American Institute of Steel Construction One E. Wacker Dr., Suite 3100 Chicago, IL 60601	(312) 670-2400
AISI	American Iron and Steel Institute 1101 Seventeenth St. NW, Suite 1300 Washington, DC 20036	(202) 452-7100
AITC	American Institute of Timber Construction 7012 S. Revere Parkway, Suite 140 Englewood, CO 80110	(303) 792-9559
ALI	American Lighting Institute 435 N. Michigan Ave. Chicago, IL 60611	(312) 644-0828
ALI	Associated Laboratories Eight Brush St. Pontiac, MI 48053	(313) 335-6114
ALSC	American Lumber Standards Committee P.O. Box 210 Germantown, MD 20874	(301) 972-1700
AMCA	Air Movement and Control Association 30 W. University Dr. Arlington Height, IL 60004	(312) 394-0150
ANSI	American National Standards Institute 11 W. 42nd St. New York, NY 10036	(212) 642-4900
APA	American Plywood Association P.O. Box 11700 Tacoma, WA 98411	(206) 565-6600
APFA	American Pipe Fittings Association 7297 Lee Highway, Suite N Falls Church, VA 22042	(703) 533-1321
API	American Petroleum Institute 1220 L St., NW Washington, DC 20005	(202) 682-8000
ARI	Air-Conditioning and Refrigeration Institute 4301 N. Fairfax Dr., Suite 425 Arlington, VA 22203	(703) 524-8800

ARMA	Asphalt Roofing Manufacturers Association 6288 Montrose Rd. Rockville, MD 20852	(301) 231-9050
ASA	Acoustical Society of America 335 E. 45th St. New York, NY 10017	(516) 349-7800
ASC	Adhesive and Sealant Council 1500 Wilson Blvd., Suite 515 Arlington, VA 22209	(703) 841-1112
ASHRAE	American Society of Heating, Refrigerating and Air-Conditioning Engineers 1791 Tullie Circle NE Atlanta, GA 30329	(404) 636-8400
ASME	American Society of Mechanical Engineers 345 E. 47th St. New York, NY 10017	(212) 705-7722
ASSE	American Society of Sanitary Engineering P.O. Box 40362 Bay Village, OH 44140	(216) 835-3040
ASSEI	American Society of Safety Engineers, Inc. 1800 East Oakton St Des Plaines, IL 60016	(708) 692-4121
ASTM	American Society for Testing and Materials 100 BarrHarbor Dr. West Conshohocken, PA 19428	(610) 832-9585
AWA	Fine Hardwoods–American Walnut Association 5603 W. Raymond St., Suite O Indianapolis, IN 46421	(317) 244-3312
AWI	Architectural Woodwork Institute 2310 S. Walter Reed Dr. Arlington, VA 22206	(703) 671-9100
AWPA	American Wood Preservers Association P.O. Box 286 Woodstock, MD 21163	(410) 465-3169
AWPB	American Wood Preservers Bureau P.O. Box 6058 2772 S. Randolph St. Arlington, VA 22206	(703) 931-8180
AWS	American Welding Society 550 NW Le Jeune Rd. Miami, FL 33126	(305) 443-9353
AWWA	American Water Works Association 6666 W. Quincy Ave. Denver, CO 80235	(303) 794-7711
BANC	Brick Association of North Carolina P.O. Box 13290 Greensboro, NC 27415	(919) 273-5566
BHMA	Builders Hardware Manufacturers Association 355 Lexington Ave., 17th Floor New York, NY 10017	(212) 661-4261
BIA	Brick Institute of America 11490 Commerce Park Dr., Suite 300 Reston, VA 22091	(703) 620-0010
BIFMA	Business and Institutional Furniture Manufacturers Association 2335 Burton St. SE Grand Rapids, MI 49506	(616) 243-1681

Appendix E   Standards-Producing Organizations

BSI	Building Stone Institute 420 Lexington Ave. New York, NY 10017	
CAUS	Color Association of the United States 343 Lexington Ave. New York, NY 10016	(212) 683-9531
CAGI	Compressed Air and Gas Institute c/o Thomas Associates, Inc. 1230 Keith Building Cleveland, OH 44115	(216) 241-7333
CBM	Certified Ballast Manufacturers Association Hannah Building, Suite 772 1422 Euclid Ave. Cleveland, OH 44115	(216) 241-0711
CDA	Copper Development Association Box 1840, Greenwich Office Park 2 Greenwich, CT 06836	(203) 625-8210
CGA	Compressed Gas Association 1235 Jefferson Davis Highway Arlington, VA 22202	(703) 979-0900
CISPI	Cast Iron Soil Pipe Institute 5959 Shallowford Rd., Suite 419 Chattanooga, TN 37421	(423) 892-0137
CLFMI	Chain Link Fence Manufacturers Institute 1101 Connecticut Ave. NW Washington, DC 20036	(202) 659-3537
CLPA	California Lathing and Plastering Association 25332 Narbonne, Suite 170 Lomita, CA 90717	(213) 539-6080
CRA	California Redwood Association 405 Enfrente Dr., Suite 200 Novato, CA 94949	(415) 382-0662
CRI	Carpet and Rug Institute Box 2048 Dalton, GA 30720	(404) 278-3176
CRSI	Concrete Reinforcing Steel Institute 933 N. Plum Grove Rd. Schaumburg, IL 60173	(847) 517-1200
CTI	Ceramic Tile Institute of America 700 North Virgil Ave. Los Angeles, CA 90029	(213) 660-1911
C.T.I.	Cooling Tower Institute Box 73383 Houston, TX 77273	(713) 583-4087
DHI	Door and Hardware Institute 14170 Newbrook Dr. Chantilly, VA 20151	(703) 222-2010
DIPRA	Ductile Iron Pipe Research Association 245 Riverchase Prkwy. E, Suite O Birmingham, AL 35244	(205) 988-9870
DLPA	Decorative Laminate Products Association (formerly National Association of Plastic Fabricators) 600 South Federal St., Suite 400 Chicago, IL 60605	(312) 345-1600

DORCMA	Door Operator and Remote Controls Manufacturers Association 1200 Sumner Ave. Cleveland, OH 44115	(216) 214-9333
EIA	Electronics Industries Association 2500 Wilson Blvd., 4th Floor Arlington, VA 22201	(703) 907-7500
EIMA	Exterior Insulation Manufacturers Association P.O. Box 75037 Washington, DC 20013	(202) 783-6582
ETL	ETL Testing Laboratories, Inc. P.O. Box 2040 Route 11, Industrial Park Cortland, NY 13045	(607) 753-6711
FCI	Fluid Controls Institute P.O. Box 9036 Morristown, NJ 07960	(201) 829-0990
FH-AWA	(See AWA)	
FM	Factory Mutual Engineering and Research 1151 Boston-Providence Turnpike Norwood, MA 02062	(617) 762-4300
FPRS	Forest Products Research Society 2801 Marshall Court Madison, WI 53705	(608) 231-1361
FTI	Facing Tile Institute c/o Box 8880 Canton, OH 44711	(216) 488-1211
GA	Gypsum Association 810 First St. NE, Suite 510 Washington, DC 20002	(202) 289-5440
GAMA	Gas Appliance Manufacturers Association, Inc. 1901 North Moore St., Suite 1100 Arlington, VA 22209	(703) 525-9565
GANA	Glass Association of North America 3310 S.W. Harrison St. Topeka, KS 66611	(913) 266-7013
HEI	Heat Exchange Institute c/o Thomas Associates, Inc. 1230 Keith Building Cleveland, OH 44115	(216) 241-7333
HI	Hydronics Institute P.O. Box 218 35 Russo Place Berkeley Heights, NJ 07922	(201) 464-8200
HMA	Hardwood Manufacturers Association 805 Sterick Building Memphis, TN 38103	(901) 525-8221
HPMA	Hardwood Plywood Manufacturers Association P.O. Box 2789 Reston, VA 22090	(703) 435-2900
HVI	Home Ventilating Institute 30 West University Drive Arlington Heights, IL 60004	(708) 394-0150
ICEA	Insulated Cable Engineers Association, Inc. P.O. Box P South Yarmouth, MA 02664	(617) 394-4424

Appendix E   Standards-Producing Organizations

IEC	International Electrotechnical Commission (available from ANSI) 655 Fifteenth Street NW, Suite 300 Washington, DC 20015	(202) 639-4090
IEEE	Institute of Electrical and Electronic Engineers 345 E. 47th St. New York, NY 10017	(212) 705-7900
IESNA	Illuminating Engineering Society of North America 345 E. 47th St. New York, NY 10017	(212) 705-7926
IGCC	Insulating Glass Certification Council RT 11, Industrial Park Cortland, NY 13045	(607) 753-6711
ILI	Indiana Limestone Institute of America Stone City Bank Building Bedford, IN 47421	(812) 275-4426
IMSA	International Municipal Signal Association P.O. Box 8249 Forth Worth, TX 76112	(817) 429-8638
IRI	Industrial Risk Insurers 85 Woodland St. Hartford, CT 06102	(203) 520-7300
ISA	Instrument Society of America P.O. Box 12277 67 Alexander Dr. Research Triangle Park, NC 27709	(919) 549-8411
LIA	Lead Industries Association, Inc. 292 Madison Avenue New York, NY 10017	(212) 578-4750
LPI	Lightning Protection Institute P.O. Box 458 Harvard, IL 60033	(815) 943-7211
LRI	Lighting Research Institute 120 Wall St. New York, NY 10005	(212) 248-5014
MBMA	Metal Building Manufacturers Association 1300 Sumner Ave. Cleveland, OH 44115	(216) 241-7333
MCAA	Mechanical Contractors Association of America 5410 Grosvenor Lane, Suite 120 Bethesda, MD 20814	(301) 897-0770
MFMA	Maple Flooring Manufacturers Association 60 Revere Dr. Northbrook, IL 60062	(708) 480-9138
MHI	Material Handling Institute 8720 Red Oak Blvd., Suite 201 Charlotte, NC 28217	(704) 522-8644
MIA	Marble Institute of America 33505 State St. Farmington, MI 48024	(313) 746-5558
M.I.A.	Masonry Institute of America 2550 Beverly Blvd. Los Angeles, CA 90057	(213) 388-0472
ML/SFA	Metal Lath/Steel Framing Association 600 South Federal St., Suite 400 Chicago, IL 60605	(312) 346-1600

MSS	Manufacturers Standardization Society of the Valve and Fittings Industry 127 Park St. NE Vienna, VA 22180	(703) 281-6613
NAA	National Aggregate Association 900 Spring St. Silver Spring, MD 20910	(301) 587-1400
NAAMM	National Association of Architectural Metal Manufacturers 221 N. LaSalle St. Chicago, IL 60601	(312) 346-1600
NAMM	National Association of Mirror Manufacturers 9005 Congressional Dr. Potomac, MD 20854	(301) 365-4080
NAPA	National Asphalt Pavement Association Calvert Building, Suite 620 6811 Kenilworth Ave. Riverdale, MD 20737	(301) 779-4880
NAPF	National Association of Plastic Fabricators (now DLPA)	
NAWIC	National Association of Women in Construction 327 S. Adams St. Fort Worth, TX 76104	(800) 552-3506
NBGQA	National Building Granite Quarries Association 369 N. State St. Concord, NH 03301	(603) 225-8397
NBHA	National Building Hardware Association (now DHI)	
NCMA	National Concrete Masonry Association 2302 Horse Pen Rd. Herndon, VA 22071	(703) 713-1900
NCRPM	National Council on Radiation Protection and Measurement 7910 Woodmont Ave., Suite 1016 Bethesda, MD 20814	(301) 657-2652
NDMA	National Dimension Manufacturers Association 1000 Johnson Ferry Rd. Marietta, GA 30068	(404) 565-6660
NDPA	National Decorating Products Association 1050 North Lindbergh Blvd. St. Louis, MO 63132	(314) 991-3470
NEC	National Electric Code (by NFiPA)	
NECA	National Electrical Contractors Association 7315 Wisconsin Ave. Bethesda, MD 20814	(301) 657-3110
NEII	National Elevator Industry, Inc. 630 Third Ave. New York, NY 10016	(212) 986-1545
NEMA	National Electrical Manufacturers Association 2101 L. St. NW, Suite 300 Washington, DC 20037	(202) 457-8400
NFiPA	National Fire Protection Association Batterymarch Park Quincy, MA 02269	(617) 770-3000
NFoPA	National Forest Products Association (name change; see AFPA listing)	

Appendix E Standards-Producing Organizations

NOFMA	National Oak Flooring Manufacturers Association 8 N. Third St. 804 Sterick Bldg., Suite 810 Memphis, TN 38103	(901) 526-5016
NHLA	National Hardwood Lumber Association P.O. Box 34518 Memphis, TN 38184	(901) 377-1818
NKCA	National Kitchen Cabinet Association P.O. Box 6830 Falls Church, VA 22046	(703) 237-7580
NPA	National Particleboard Association 18928 Premiere Court Gaithersburg, MD 20879	(301) 670-0604
NPCA	National Paint and Coatings Association 1500 Rhode Island Ave. NW Washington, DC 20005	(202) 462-6272
NRCA	National Roofing Contractors Association 10255 W. Higgins Rd., Suite 600 Rosemont, IL 60018	(847) 299-9070
NRDCA	National Roof Deck Contractors Association 600 S. Federal St. Chicago, IL 60605	(312) 922-6222
NRMCA	National Ready Mixed Concrete Association 900 Spring St. Silver Spring, MD 20910	(301) 587-1400
NSF	National Sanitation Foundation P.O. Box 1468 3475 Plymouth Rd. Ann Arbor, MI 48105	(313) 769-3010
NSPE	National Society of Professional Engineers 1420 King St. Alexandria, VA 22314	(703) 684-2800
NSSEA	National School Supply and Equipment Association 2020 Fourteenth St. N, Suite 400 Arlington, VA 22201	(703) 524-8819
NSPI	National Spa and Pool Institute 2111 Eisenhower Ave. Alexandria, VA 22314	(703) 838-0083
NSWMA	National Solid Wastes Management Association 1730 Rhode Island Ave. NW Washington, DC 20036	(202) 659-4613
NTMA	National Terrazzo and Mosaic Association 3166 Des Plaines Ave., Suite 132 Des Plaines, Il 60018	(312) 635-7744
NWMA	National Woodwork Manufacturers Association (now NWWDA)	
NWWDA	National Wood Window and Door Association (formerly NWMA) 1400 E. Touhy Ave., Suite 470 Des Plaines, IL 60018	(847) 299-5200
PCA	Portland Cement Association 5420 Old Orchard Road Skokie, IL 60077	(847) 966-6200
PCI	Prestressed Concrete Institute 175 W. Jackson Blvd. Chicago, IL 60694	(312) 346-4071

PDCA	Painting and Decoration Contractors of America 3913 Old Lee Highway Fairfax, VA 22030	(703) 359-0826
PDI	Plumbing and Drainage Institute 45 Bristol Dr., Suite 101 Southeastern, MA 02375	(508) 230-3529
PEI	Porcelain Enamel Institute 111 Nineteenth St. Arlington, VA 22209	(703) 527-5257
PI	Perlite Institute 88 New Dorp Plaza Staten Island, NY 10306	(718) 351-5723
PPI	Plastic Pipe Institute 1275 K St. NW, Suite 400 Washington, DC 20005	(202) 371-5200
PWC	Professional Women in Construction 342 Madison Ave. New York, NY 10173	(212) 687-0610
RCSHSB	Red Cedar Shingle and Handsplit Shake Bureau 515 116th Ave. NE, Suite 275 Bellevue, WA 98004	(206) 453-1323
RFCI	Resilient Floor Covering Institute 966 Hungerford Dr., Suite 12-B Rockville, MD 20805	(301) 340-8580
RIS	Redwood Inspection Service 591 Redwood Hwy., Suite 3100 Mill Valley, CA 94941	(415) 381-1304
RMA	Rubber Manufacturers Association 1400 K St. NW Washington, DC 20005	(202) 682-4800
SAMA	Scientific Apparatus Makers Association 1101 Sixteenth St. NW Washington, DC 20036	(202) 223-1360
SDI	Steel Deck Institute P.O. Box 9506 Canton, OH 44711	(216) 493-7886
S.D.I.	Steel Door Institute 30200 Detroit Rd. Cleveland, OH 44145	(216) 899-0010
SFPA	Southern Forest Products Association P.O. Box 64170 Kenner, LA 70064	(504) 443-4464
SGAA	Stained Glass Association of America Box 22642 Kansas City, MO 64113	(816) 333-6690
SGCC	Safety Glazing Certification Council Route 11, Industrial Park Cortland, NY 13045	(607) 753-6711
SHLMA	Southern Hardwood Lumber Manufacturers Association (now HMA)	
SIGMA	Sealed Insulating Glass Manufacturers Association 111 E. Wacker Dr. Chicago, IL 60601	(312) 644-6610
SJI	Steel Joist Institute 1205 48th St. N., Suite A Myrtle Beach, SC 29577	(803) 449-0487

Appendix E   Standards-Producing Organizations

SMACNA	Sheet Metal and Air Conditioning Contractors National Association 4201 LaFayette Center Dr. Chantilly, VA 20151	(703) 803-2989
SPIB	Southern Pine Inspection Bureau 4709 Scenic Highway Pensacola, FL 32504	(904) 434-2611
SPRI	Single-Ply Roofing Institute 175 Highland Ave., 3rd Floor Needham, MA 02194	(617) 444-0242
SSPC	Steel Structures Painting Council 4400 Fifth Ave. Pittsburgh, PA 15213	(412) 578-3327
SWI	Steel Window Institute (c/o Thomas Associates, Inc.) 1230 Keith Bldg. Cleveland, OH 44115	(216) 241-7333
TCA	Tile Council of America P.O. Box 326 Princeton, NJ 08542	(609) 921-7050
TIMA	Thermal Insulation Manufacturers Association 7 Kirby Plaza Mt. Kisco, NY 10549	(914) 241-2284
TPI	Truss Plate Institute 583 D'Onofrio Dr., Suite 200 Madison, WI 53719	(608) 833-5900
UL	Underwriters Laboratories 333 Pfingsten Rd. Northbrook, IL 60062	(312) 272-8800
VA	Vermiculite Association, Inc. 600 S. Federal St., Suite 400 Chicago, IL 60605	(312) 922-6222
WCLIB	West Coast Lumber Inspection Bureau P.O. Box 23145 Portland, OR 97223	(503) 639-0651
WCMA	Wall Covering Manufacturers Association 66 Morris Ave. Springfield, NJ 07081	(201) 379-1100
WIC	Woodwork Institute of California P.O. Box 11428 Fresno, CA 93773	(209) 233-9035
WRCA	Western Red Cedar Association Box 120786 New Brighton, MN 55112	(612) 633-4334
WRCLA	Western Red Cedar Lumber Association 1200-555 Burrand St. Vancouver, BC, Canada	(604) 684-0266
WRI	Wire Reinforcement Institute 8361 A Greensboro Dr. McLean, VA 22102	(703) 790-9790
WSC	Water Systems Council 221 N. LaSalle St. Chicago, Il 60601	(312) 346-1600
WSFI	Wood and Synthetic Flooring Institute 4415 West Harrison St., Suite 242 C Hillside, IL 60162	(312) 449-2933

WWPA	Western Wood Products Association 1500 Yeon Building Portland, OR 97204	(503) 224-3930
W.W.P.A.	Woven Wire Products Association 2515 N. Nordica Ave. Chicago, IL 60635	(312) 637-1359

Federal Government Agencies

Names and titles of federal government standards or specification-producing agencies are frequently abbreviated. The following acronyms or abbreviations as referenced in the contract documents indicate names of standards or specification-producing agencies of the federal government. Names and addresses are subject to change but are believed to be accurate.

CE	Corps of Engineers (U.S. Department of the Army) Chief of Engineers-Referral Washington, DC 20314	(202) 693-6456
CFR	Code of Federal Regulations Government Printing Office Washington, DC 20402 (material is usually first published in the *Federal Register*)	(202) 783-3238
CPSC	Consumer Product Safety Commission 111 Eighteenth St. NW Washington, DC 20207	(202) 634-7700
CS	Commercial Standard (U.S. Department of Commerce) Government Printing Office Washington, DC 20402	(202) 377-2000
DOC	Department of Commerce National Institute of Standards and Technology Gaithersburg, MD 20899	(301) 975-2000
DOT	Department of Transportation 400 Seventh St. SW Washington, DC 20590	(202) 426-4000
EPA	Environmental Protection Agency 401 M St. SW Washington, DC 20460	(202) 260-4700
FAA	Federal Aviation Administration (U.S. Department of Transportation) 800 Independence Ave. SW Washington, DC 20590	(202) 426-4000
FCC	Federal Communications Commission 1919 M St. NW Washington, DC 20554	(202) 632-7000
FHA	Federal Housing Administration (U.S. Department of Housing and Urban Development) 451 Seventh St. SW Washington, DC 20201	(202) 755-5995
FS	Federal Specifications (General Services Administration) Specifications Unit (WFSIS) 7th and D Sts. SW Washington, DC 20406	(202) 472-2205 or 472-2140

Appendix E Standards-Producing Organizations

GSA	General Services Administration F St. and 18th St. NW Washington, DC 20405	(202) 655-4000
MIL	Military Standardization Documents (U.S. Department of Defense) Naval Publications and Forms Center 5801 Tabor Ave. Philadelphia, PA 19120	
NBS	National Bureau of Standards (U.S. Department of Commerce) Gaithersburg, MD 20234	(301) 921-1000
OSHA	Occupational Safety and Health Administration U.S. Department of Labor/OSHA 200 Constitution Ave. NW Washington, DC 20210	(202) 219-7725
PS	Product Standards (National Bureau of Standards) (U.S. Department of Labor) Government Printing Office Washington, DC 20402	(202) 783-3238
REA	Rural Electrification Administration (U.S. Department of Agriculture) 14th St. and Independence Ave. SW Washington, DC 20250	(202) 382-1255
TRB	Transportation Research Board 2101 Constitution Ave. NW Washington, DC 20418	
USDA	U.S. Department of Agriculture Independence Ave. Washington, DC 20250	(202) 447-4929
USPS	U.S. Postal Service 475 L'Enfant Plaza SW Washington, DC 20260	(202) 245-4000

Other Sources

Names and titles of other standards, code, or specification-producing agencies are also frequently abbreviated. The following are subject to change but are believed to be accurate.

CSI	Construction Specifications Institute 601 Madison St. Alexandria, VA 22314
EJCDC/NSPE	Engineers Joint Contract Documents Committee, and National Society of Professional Engineers 1420 King St. Alexandria, VA 22314
ACEC	American Consulting Engineers Council 1015 E. 47th St. New York, NY 10017
BOCA	Building Officials and Code Administrators International, Inc. *National Building Code* 4051 West Flossmoor Rd. Country Club Hills, Il 60477
NFiPA	National Fire Protection Association *Life Safety Code* and *National Electric Code* Batterymarch Park Quincy, MA 02269

CABO	Council of American Building Officials
One and Two Family Dwelling Code
5203 Leesburg Pike
Falls Church, VA 22041
SBBC	Southern Building Code Congress International
Standard Building Code
900 Montclair Road
Birmingham, AL 35213
ICBO	International Conference of Building Officials
Uniform Building Code
5360 S. Workman Mill Rd.
Whittier, CA 90601

Magazine, Digest, and Research Sources

AIA	American Institute of Architects
Architecture/Architectural Technology
Box 2063
Marion, OH 43305
NIBS	National Institute of Building Sciences
Building Science Newsletter
1015 15th St. N.W.
Washington, DC 20005
NRC	National Research Council
Canadian Building Digest
Ottawa, Canada KIA OR6
CSI	Construction Specifications Institute, Inc.
Construction Specifier
601 Madison St.
Alexandria, VA 22314
NIST	National Institute of Standards and Technology
Building Science Series
U.S. Superintendent of Documents
Washington, DC 20402
HHFA	Housing and Home Finance Agency
Housing Research Papers
Washington, DC 20402
SHC	Small Homes Council
Reports
University of Illinois
Champaign, IL 61820
FPL	Forest Products Laboratory
Wood Handbook, USDA Handbook 72
U.S. Government Printing Office
Washington, DC 20402   (202) 783-3328

Index

Abstract of legal paper, 178–181
Administration
    defined, 31
    successful, 76
Alternative dispute resolution, 174–178
American Institute of Architects (AIA)
    contract form B141, 22, 31, 154–164
    document B161/162, 24
    document B352, 84, 108–111
    document G701, 254
    document G710, 137
    document revisions, 16, 169
    General conditions (document A201), 38, 264, 280, 290, 291–301
        outline for owner review, 293
            accurate role descriptions, 293
            legal considerations, 296
        owner-recommended provisions to replace A201, 295
            revised paragraph 3.2.1, 295
            revised paragraphs 3.7.3 and 3.7.4, 295
        owner's changes in, 291–301
    list of documents, 377–382
Arbitration, 175
Architect
    as chief coordinator, 128
    design duties, 42, 181
    as team leader, 101
Architectural immunity, 181–191
    supervision of construction, 182
        to assure substantial conformity, 183
        methods and techniques, 185
            architects' attempt to limit liability, 191
            early cases, 185
            Miller Doctrine, 187
Architecture
    and contract administration, 69
    philosophies, 69
    teaching, 68
    true measure of, 105
Assumptions, 138

Beverly Hills Supper Club disaster, 346–347
Building code
    background, 326
    commitment to comply, 326
    document requirements, 165–166
    enforcement, 350
    errors in construction, 326
    problems with application of, 346–350
        system failure example, 346
Building construction, 108
    imperfections, 347–348
    safety, 340–341
Building inspector
    as authoritative bystander, 333
    certification, 332
    characteristics, 332
    continuing education, 34
    defined, 325, 337
    functions, 335–337
    job parameters, 329–330
    need to know various systems, 331–332
    professionalism, 334
Burke, Edmund, 82

Case law
    effect, 21
    importance, 5, 53
    influence, 169
    summary of recent cases, 54–56
Changes and substitutions, 136
Claims
    generating, areas of work, 126–128
    typical, 171–174
Clerk-of-the-Works, 149
Client expectations, 17
Code compliance, 18
Communications
    as cause of errors, 93
    need, 92, 113
    project system, 89–96
    prompt, 92
Constructability, 259
Construction contract administration. *See also* Contract administration
    defined, 2
    full time, 28
    fundamental, 16
    as hallmark program, 7
    modified, 28
    normal, 27
    as reduced form of project management, 15
    types, 27
    ultimate goal, 12
    use of design professional, 15
Construction management
    agency type, 258
    agency versus GMP type, 263

Construction management, *continued*
  authority, 265
  construction activities, 258
  crux of, 269
  development, 259, 262–263
  duties and relationships, 260–261, 270–273
  guaranteed maximum price (GMP) type, 264
  manager defined, 257
  organization formats, 267–269
  professional (PCM) type, 258
    quality control, 269
    service summary, 258
  role definition, 266
  scope, 80, 149
  third-party type, 258
  university curricula, 211
  work assignments on site, 269
Context (of this book), xxi–xxiii
Continuing education, resources, 196–198
Contract
  elements of, 169
  meeting of the minds, 195
  professional and construction manager, 270
  suggested text inserts, 170
Contract administration. *See also* Construction contract administration
  areas of knowledge, 75
  basic service, 2, 6, 147
  as business policy of professional, 166–169
  as careful business pursuits, 195
  as code administration, 170
  compliance as essence of, 34
  as enforcement of contract, 168
  equal emphasis for, 167
  as feature of control, 200
  good practices, 132–133
  key functions of participants, 36
  lack of, 165
  legal concerns, 113–128
  major functions, 15
  not an added service, 46
  not desirable, 39, 57, 78
  optimal, 167
  participants, 35
  preventing work impediments, 273
  primary benefits, 52
  and professionals, 340
  representative
    personnel selection, 200
    tasks
      full time, 6
      part-time, 6
  standardized format, 221
  use of photography, 208
    digital photos, 210
Contract administrator
  and communications, 90–91
  as conduit for information, 135
  correct enforcement by, 134
  defined, 32
  and documentation system, 143
  duties and AIA document B352, 84, 91
  as educator, 132
  functions, 49, 106
  good practice principles, 132–133
  making assumptions, 138
  precautions for, 150
  responsibilities of, 130
Contract, construction
  best for owner, 104
  conditions, 7
  defined, 30, 110
  normal requirements, 6
  options/types, 102–104
  for renovations, 23
  true measure of, 110
Contract documents
  commitment to comply with codes, 326
  imperfect, 16
  influence, 82
  interpretation, 52
  problems in preparation, 85
  purpose, 7
  relationship and precedence, 51
  standardized forms, 17
Contractor(s)
  abilities and attitudes, 82–83
  administration of subcontracts, 313–314
  consultation with, 81
  as coordinator, 308–309
  defined, 303
  duties of each, 306
  in flow of material, 305
  inspections, 308–309
  involvement in project, 304
  personnel attitudes, 85
  quality control by, 312
  resolving information, 310–312
  responsibilities, 304, 307
  superintendent for, 312–313
    objectives, 315
  tempo of project, 315–316
Cost control, 298

Definitions
  examples, 20
  need, 20
  others for consideration, 38
  sample contract, 71–74
Design concept, acceptance, 106
Design professionals
  basic tasks, 4
  defined, 99
  duty to client, 15
  overall responsibility, 112
  precautions for, 150
  responsibilities for site observation, 24–27, 128–129
  services defined, 17
  staff understanding, 20
Documentation

Index

hints about, 222–227
importance, 200
keys to, 215
sample forms, 228–256
system, 143, 204
written material, 213, 227
    limits, 228

Education, 4
    topics lacking emphasis, 4
Engineers Joint Contract Documents Committee (EJCDC), 16
Errors and omissions insurance, 154
Experience
    and ability, 5
    as teacher, 5

Fast tracking, 79
Faulty work, 112
Feedback, 88, 89, 139–140
    as part of communications, 88, 140
    quality of, 89
Field conditions, adjusting to, 87
Field observation services, 355–361
Fire safety, 341
Ford, Henry, 71
Forms, sample, 229–256
    action request form, 234
    addendum drawing, 242
    architect's supplemental instructions, CSI 13.4A, 241
    bulletin drawing, 243
    change order, AIA G701, 254
    checklist, AIA D200, 201–226
    clarification notice, CSI 13.3A, 248
    color schedule, 253
    daily report, CSI 9.3A, 237
    facsimile cover sheet, 231
    feedback form, CSI 16.0A, 240
    field report
        AIA G711, 235
        CSI 9.1A, 236
    letter of transmittal, 229, 230
    meeting record, 232
    noncompliance notice, CSI 9.8A, 244
    payment application/certification, AIA G702, 255
    phone conversation memo, CSI 7.0A, 239
    project field instruction, 249
    project memorandum, 233
    punch list, CSI 14.1A, 252
    request for information, 246, 247
        log, 247
    request for interpretation, CSI 13.2A, 245
    shop drawing log, CSI 12.1A, 251
    submittal, log, 247
        checklist, CSI 12.1A, 251
        control sheet, 250
    substantial completion certificate, AIA G704, 256
    substitution request, CSI 13.1A, 238

Graves, Michael, 68
Groves, J. Russell, 6, 140, 345

Halls, Peter C., 112–128
Hyatt Regency-Kansas City disaster
    court decision, 59
    sets responsibilities, 72

Inspection, 33, 115
Interpreting contracts
    concepts and rules, 53
    prompt, 144

Kettering, Charles, 79
Kliment, Stephen A., 3, 6

Law, and practice, 5
Legal concerns in construction field administration, 112–128
    areas of work generating claims, 126–128
    communications and recommendations, 113
    documentation issues, 125
    enforcement of test results, 119
    improper supervision, 117
    job safety, 122
    pay requests, 121
    scheduling, 120
    shop drawing and submittal review, 120
    supervision/observation/inspection, 115
Liability. *See also* Professional liability
    assignment, 171
    defined, 153
    third-party, 181–191
List of materials and equipment, 93

Master builder concept, 68
McElroy, Mary J., 291–301
Mediation, 176
Meeting formats, 369–375
Meier, Hans W. (Bill), 42
MGM Grand Hotel fire, 347
Monitor, 33

Nischwitz, Jeffrey L., 181–191

Observation of construction, 25, 35, 115
    defined, 33
    in lieu of supervision/inspection, 29
On-site techniques, 145
    facilities, 148
Owner(s)
    and contracts, 105
    cost, scope, quality quandary, 101
    decision making, 288
    defined, 275
    desires and wants, 100
    education of, 276–279
        development of, 277
        guidelines, 279
        include attorneys, 277

Owner(s), *continued*
  pivotal issues, 279
  primary issues of, 278
  troublesome issues, 278
 first-hired as agent, 265
 inclusion of, 283
 interests protected, 286
 maintenance personnel, 285
 need for legal assistance, 265
 in passive role, 280
 policing contracts, 287
 as project leader, 277
 and project procedures, 283, 285
 and project variables, 87
 representative (staffer), 280
  authority, 284
  potential conflicts, 280
  resident engineer, 282
  site construction manager, 281
 response to use of the A-201 (1987) general conditions, 291–301
 role, 288
 as team member, 283
 understanding of codes, 346–350
 unrealized expectations, 194

Pay requests, 121
Periodic site visits, 90
 as timely observation, 144
Postoccupancy evaluation, 345
Preconstruction conference, 26, 129
Professional actions
 as arbiter, 29
 best qualified administrator, 77
 and code compliance, 334, 349
 commitment to code compliance, 326–329
 consequences, 34
 construction responsibilities, 50, 70
 court decisions, 42
 differences between disciplines, 106–108
 and documentation system, 143
 functions, 35, 49
 goal of, 12
 insight, 15
 knowledge and expertise, 51
 lack of document review, 138
 legal responsibility, 49
 need for quality control coordinator, 167
 negligence, 180
 and the project scheme, 76
 relationship to project, 77
 responsibilities, 61–64
 strict liability of, 42
Professional liability, 14, 154, 192–193
 liability insurance, 154
 philosophies, 14
 "watch-outs," 193
Professional services
 breakdown by phases, 8
 documents only, 39
 on-site duties and responsibilities, 22, 28
 on-site observation, 14
 "plans and specs" only, 14, 59, 78, 154
Professionalism, 32
Project delivery, essential elements, 78
Project meetings, 140–143
Project success, 95

Quality control coordinator, 167

Rand, J. Patrick, 12
Reference standards, 19
Regulatory specification samples, 363–368
Responsibilities, all parties, 72
Risk management, 194
Rule of tens, 8

Safety, 122
Sample forms. *See* Forms, sample
Scheduling, 120
Schuller, Dr. Robert H., 95
Shop drawings
 notations on, 36
 review of, 36
 as submittals, 58, 120
Site visits, periodic, 111
Smith II, Herbert G., 24
Standard of care, 5
Standards-producing organizations, 383–396
Stop work, 25
Strict liability, of professionals, 178–180
Subcontractor(s)
 defined, 317
 inspections by, 323
 legal requirements, 321
 pre-installation meetings, 321
 protection of interests, 319
 relationship to others, 319
 responsibilities of, 318, 320
Substantial completion, 75
Successful project
 defined, 32
 elements of, 81, 95
Superintend, 33
Supervision, 7, 33, 115, 117
Sweet, Justin, 146

Team
 ability, 83
 concept of building, 82
 process, 71
Tone of project, 129, 132

Vavreck, Edward C., 337

Weisbach, Gerald G., 7, 69, 78, 314

Yatt, Barry D., 106